INDUSTRIAL WELDING PROCEDURES

Frank R. Schell
Bill Matlock

Four Books in One
   Oxyacetylene
   Electric Arc
   MIG & TIG
   Design & Special Processes

VAN NOSTRAND REINHOLD COMPANY
NEW YORK CINCINNATI TORONTO LONDON MELBOURNE

Copyright ©1979 by Litton Educational Publishing, Inc.
Library of Congress Catalog Card Number 78-11031
ISBN 0-442-27416-5

Published in 1979 by Van Nostrand Reinhold Company
A division of Litton Educational Publishing, Inc.
135 West 50th Street, New York, NY 10020, U.S.A.

Van Nostrand Reinhold Limited
1410 Birchmount Road
Scarborough, Ontario M1P 2E7, Canada

Van Nostrand Reinhold Australia Pty. Ltd.
17 Queen Street
Mitcham, Victoria 3132, Australia

Van Nostrand Reinhold Company Limited
Molly Millars Lane
Wokingham, Berkshire, England

16   15   14   13   12   11   10 9   8   7   6   5   4   3   2   1

Library of Congress Cataloging in Publication Data

Schell, Frank R.
    Industrial welding procedures.

    1. Welding.   I. Matlock, William, joint author.
II. Title.
TS227.S2793   671.5'2   78-11031
ISBN 0-442-27416-5

WELDING PROCEDURES: OXYACETYLENE

Preface

In 1895, it was discovered that a combination of oxygen and acetylene produced a flame with a higher temperature than any flame yet known. At about the same time methods of producing oxygen and calcium carbide were invented. These two developments opened the field of oxyacetylene welding. Before this time most metal joining was done with rivets.

Since the invention of the oxyacetylene welding process there have been many variations of the process and new methods of producing the heat to fuse metal have been developed. However, the oxyacetylene flame continues to be one of the most important sources of heat for joining and cutting metals. The ability to use oxyacetylene equipment efficiently and safely is considered a requirement for entry into many trades. WELDING PROCEDURES: OXYACETYLENE is designed to help you acquire these abilities.

Each unit is preceded by clearly stated, behavioral objectives which describe what you will be able to do when the unit is completed. All of the subsequent material in the unit is designed to fulfill these objectives.

A concise discussion of any pertinent related information follows the unit objectives. This part of the unit contains such things as new terms, safety information, and a description of problems that might be encountered.

At the heart of each unit is an opportunity for you to learn while performing a job with oxyacetylene equipment. These jobs are designed to enhance one another, each building on the knowledge and skills gained through those preceding it. The jobs include a complete list of all necessary equipment and material; a procedure for performing an oxyacetylene operation; key points, to make the operation more meaningful and to aid you in the successful completion of the job; and, in most cases, a destructive test of the completed joint.

At the conclusion of each unit there are review questions. By answering these questions, you can review some highlights of the unit and be sure all of the material has been learned. In addition to the unit features of the book, each of the five sections is concluded with a comprehensive review.

Frank Schell has been a journeyman welder for twenty-four years. He is currently Curriculum Development Coordinator and Professor of Welding at the College of Southern Idaho. He is a member of the American Vocational Association and the Idaho Vocational Association. He is the author of several pieces of instructional material for welding students. WELDING PROCEDURES: OXYACETYLENE is a product of the author's background in welding, both as an educator and as a journeyman.

Safety Rules

Oxygen Handling

1. Important. Use no oil or grease around oxygen. When mixed with oil or grease, oxygen can cause a violent explosion.
2. Do not use pipe fitting compounds on oxygen connections.
3. Only hoses which are made for welding should be used.
4. Do not force connections which do not fit.
5. If gas cylinders are not clearly marked as to contents, do not use them.
6. Do not use oxygen under high pressure without an oxygen regulator.
7. Be sure that the oxygen cylinder is securely fastened so it cannot fall.
8. Be sure the pressure adjusting screw is turned out (to the left) before the cylinder is opened.
9. Stand to one side when opening a cylinder.

Acetylene Handling

1. Do not use pipe fitting compounds on acetylene equipment.
2. Always use an acetylene pressure reducing regulator.
3. Never force connections which do not fit.
4. Be sure the acetylene cylinder is securely fastened so it cannot fall.
5. Do not release acetylene into the atmosphere when welding is being done in the area.
6. Leave the bottle key on the acetylene cylinder.

General Rules

1. Always use a striker to light a torch.
2. Never lay a lighted torch down.
3. Before lighting the torch be sure it is not pointing at another person.
4. Before lighting the torch be sure that the flame will not come in contact with inflammable material.
5. Make it a habit to hold your hand close to a piece of metal to see if it is hot, before picking it up.
6. Never operate equipment without instruction on its use.
7. Wear safety glasses or grinding shields when using a power grinder.

8. Be aware of the condition of the hoses on the torch. If a leak develops it should be reported immediately, and the hoses taken out of service.

9. Report all burns.

10. Never work with defective equipment.

11. Hammers, chisels and punches wear out. Do not use them if they are defective.

12. Never use any kind of fire around oxygen and acetylene cylinders. Oxygen supports combustion and acetylene will burn or explode.

13. Bronze rod contains zinc. Galvanizing on some steel contains zinc. When zinc is heated it gives off a toxic vapor. Do not breathe these fumes. Be sure the ventilation is adequate.

14. Avoid breathing the fumes when welding painted objects.

15. Always wear goggles when cutting or welding.

16. Always wear protective clothing when cutting or welding.

17. Know the location of fire extinguishers, and how to use them.

18. Some factors which contribute to accidents:

    Poor lighting
    Improper electric outlets
    Poor maintenance of equipment
    Poor ventilation
    Poor housekeeping
    Machines without guards
    Poor arrangement of equipment
    Horseplay
    Slippery, dirty floors
    Damaged equipment
    Inadequate instruction

Contents

SECTION 1: WELDS IN THE FLAT POSITION

Unit

SECTION 2: WELDS IN HORIZONTAL AND VERTICAL POSITIONS

SECTION 3: OVERHEAD WELDS, BRAZE WELDING, AND BACKHAND WELDING

SECTION 4: OXYACETYLENE CUTTING

SECTION 5: WELDING STEEL PLATE AND PIPE

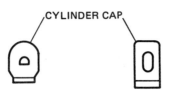

CYLINDER CAP

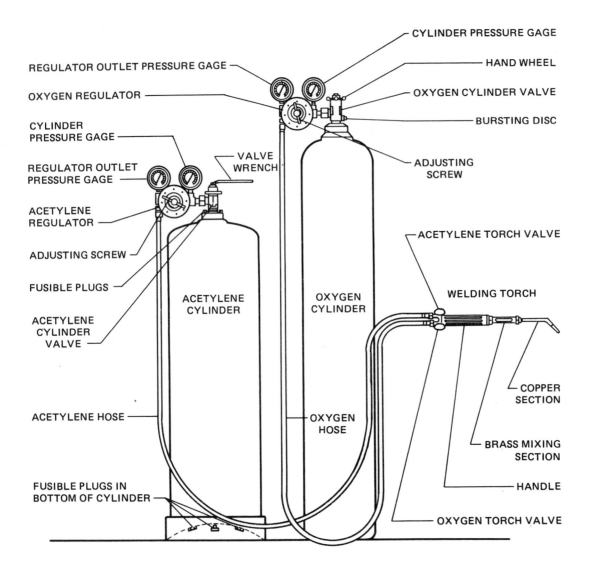

REGULATOR OUTLET PRESSURE GAGE

OXYGEN REGULATOR

CYLINDER PRESSURE GAGE

REGULATOR OUTLET PRESSURE GAGE

ACETYLENE REGULATOR

ADJUSTING SCREW

FUSIBLE PLUGS

ACETYLENE CYLINDER VALVE

ACETYLENE HOSE

FUSIBLE PLUGS IN BOTTOM OF CYLINDER

VALVE WRENCH

ACETYLENE CYLINDER

OXYGEN CYLINDER

CYLINDER PRESSURE GAGE

HAND WHEEL

OXYGEN CYLINDER VALVE

BURSTING DISC

ADJUSTING SCREW

ACETYLENE TORCH VALVE

WELDING TORCH

COPPER SECTION

BRASS MIXING SECTION

HANDLE

OXYGEN TORCH VALVE

OXYGEN HOSE

SECTION 1:
Welds in the Flat Position

Unit 1   Setting Up Oxyacetylene Welding Equipment

OBJECTIVES

After completing this unit the student will:

- Properly set up oxyacetylene welding equipment.
- Observe specific safety precautions when handling oxyacetylene equipment.
- Be able to list the components of an oxyacetylene welding outfit and describe their functions.

The following is a list of the most important equipment for a standard welding outfit:

- Oxygen
- Acetylene
- Oxygen Regulator
- Acetylene Regulator
- Welding Hoses
- Torch (sometimes called a blowpipe)
- Welding tips
- Goggles
- Hammer
- Pliers
- Striker
- Tip cleaners
- Table with steel or firebrick top
- Twelve-inch adjustable wrench (or manufacturer's cylinder wrenches)
- Protective clothing

Fig. 1-1 Oxygen Cylinder

Oxyacetylene Processes

Oxyacetylene welding is based on the principle that, when acetylene gas is burned in the proper proportions, with oxygen gas, a flame is produced,

1

which is hot enough to melt and fuse metals. This proportion is approximately 1 part of acetylene to 2½ parts of oxygen. Oxyacetylene flame cutting uses much of the same equipment, but the principle is different. In oxyacetylene cutting a stream of oxygen is directed against a piece of *ferrous* metal (metal which contains iron is called ferrous), which has been heated to a red heat. This causes the metal to burn.

Oxygen Cylinder

The oxygen cylinder, figure 1-1, is usually green or yellow in color, so that it can be identified. It is made from a single plate of high grade steel, which has been heat treated to develop toughness and strength. When fully charged, the oxygen *bottle*, as it is sometimes called, contains 244 cubic feet of oxygen at a pressure of 2,200 pounds per square inch at 70° Fahrenheit. This oxygen is 99.99% pure and is colorless, odorless and tasteless. Oxygen by itself will not burn, but it does support combustion.

> CAUTION: Because of the extremely high pressure at which oxygen is stored in the cylinder, several precautions must be observed at all times.
>
> - All cylinders must have Interstate Commerce Commission markings, indicating the dates of bottle pressure tests.
> - Cylinders must be stored so they cannot be knocked down.
> - They should not be stored in an area where extreme temperature changes occur.
> - Oxygen cylinders must not be stored near grease, oil or electrical connections. Bringing oxygen into contact with oil or grease may cause a violent explosion.
> - They must never be moved without the cylinder cap in place, on top of the cylinder.
> - Cylinders which are defective in any way should be taken out of service and reported to the supplier.

Valve Protection Cap

The valve protection cap, or *bottle cap*, screws onto the cylinder and completely covers the valve, figure 1-2. It protects the valve from damage when the cylinder is being moved or if it is accidentally knocked over.

Oxygen Cylinder Valve

The oxygen cylinder valve, figure 1-3, is attached to the top of the oxygen cylinder. It is used to turn the flow of oxygen on or off, as needed. These valves are *double seated*. This means that when the valve is completely closed the flow of oxygen from the cylinder is shut off, and when the valve is opened all the way the valve seats and prevents leakage of oxygen around the valve stem. The valve should be opened completely when the cylinder is in use.

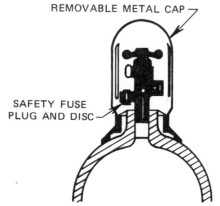

REMOVABLE METAL CAP

SAFETY FUSE PLUG AND DISC

Fig. 1-2 Oxygen Cylinder Cap and Safety Plug

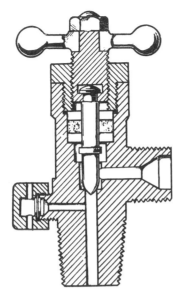

Fig. 1-3 Oxygen Cylinder Valve

A *safety fuse plug* and *disc* are installed in the oxygen cylinder valve. As the temperature of the oxygen in the cylinder increases, the pressure also increases. If the pressure of the gas in the bottle becomes too great, the safety plug and disc will release the pressure.

Acetylene Cylinder

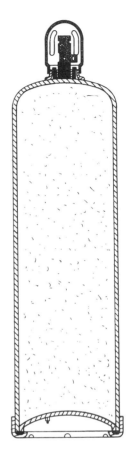

Fig. 1-4 Acetylene Cylinder

The pressure in acetylene cylinders is not as high as that in oxygen cylinders. For this reason, acetylene cylinders are rolled to the size needed and welded at the seams, figure 1-4.

Acetylene cylinders are filled with a porous material, such as fuller's earth or balsa wood. A liquid chemical (acetone) is poured into the bottle and is absorbed by the porous material. Acetone absorbs acetylene gas.

Acetylene gas, which is made by mixing water and calcium carbide (a gray, rock-like substance), has a strong disagreeable odor, resembling garlic. It is highly inflammable and, in combination with oxygen, produces the hottest flame known (5,800° - 6,300° Fahrenheit). When there is not enough oxygen present it burns with a smoky, yellow flame.

CAUTION: Because acetylene gas is highly inflammable and explosive, certain safety precautions must be observed.

- Cylinders must be tested and certified by the Interstate Commerce Commission.
- Cylinders which leak or are defective in any way should be taken out of service and reported to the supplier.
- Free acetylene gas (that which is not absorbed in acetone) must not be stored at pressures above 15 pounds per square inch. Above this pressure acetylene becomes very unstable and may explode.
- If large numbers of acetylene cylinders are stored close to oxygen cylinders a fire resistant wall must be built between the two types of cylinders.

- Cylinders should never be used in any position but the upright position. Liquid acetone can run into the gages and hoses if the cylinder valve is opened while the cylinder is lying on its side.
- Never store acetylene cylinders where excessive heat may contact them.

Acetylene Cylinder Valves

The acetylene cylinder valve, figure 1-5, is attached to the top of the acetylene cylinder. It is used to turn the flow of acetylene on or off as needed. Acetylene cylinder valves are not double seated, because they do not have to withstand the high pressure that oxygen cylinders do.

These valves are of two types. One type has a handwheel, resembling that on the oxygen cylinder valve. The other has a square stem, without the wheel, and is turned on and off with a special wrench, called a *key*.

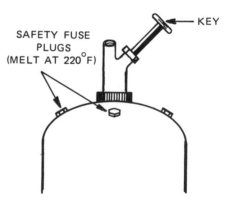

SAFETY FUSE PLUGS (MELT AT 220°F)

KEY

Fig. 1-5 Acetylene Bottle Key and Fuse Plugs

CAUTION: The acetylene cylinder valve should never be opened more than 1½ turns. In this way, it can be turned off quickly in case of fire. For the same reason, the key should always be left on the valve.

Acetylene cylinders have plugs installed in them for safety. These plugs are made of a metal which melts at a low temperature. Any excessive heat, which would cause the gas in the cylinder to reach higher pressure, melts the plugs. This allows the acetylene to escape and prevents an explosion.

Oxygen Regulators

Full oxygen cylinder pressure is 2,200 pounds per square inch. It is impossible to weld with this much pressure, so a regulator is installed on the cylinder, figure 1-6. This regulator allows the welder to set the pressure at reduced amounts. It has a safety device, which vents the pressure if it exceeds safe limits.

Regulators are equipped with two gauges, figure 1-7. One indicates the cylinder pressure, while the other indicates the working or torch pressure. Oxygen gauges are generally built to withstand 3,000 pounds per square inch of pressure. When temperature variations cause the gas in the cylinder to expand, and increase the pressure, the gauge indicates this rise in pressure.

The regulator is equipped with a nut, which screws onto the cylinder valve. The threads are conventional, right-hand threads. To install the regulator, tighten the nut with the wrench supplied by the manufacturer or with an adjustable wrench.

Acetylene Regulators

Acetylene regulators are similar to oxygen regulators, with two exceptions. All acetylene fittings have left-hand threads. This is important to remember, as the fittings may be damaged by attempting to turn them the wrong way. The reason for the left-hand threads on acetylene fittings, is to prevent them from being accidentally installed on oxygen equipment.

The second way in which they are different from oxygen regulators is that acetylene gauges have lower numbers than oxygen gauges. As a general rule, the acetylene cylinder pressure gauge registers to 500 pounds per square inch. The acetylene working pressure gauge registers to a maximum of 15 pounds per square inch. Also, the numbers and graduations on the dial of oxygen gauges are normally green, while on acetylene gauges they are normally red.

Note: Gauges are delicate mechanisms and through mishandling they may not register correctly. However, the regulator will hold the correct pressure, even if the gauge does not indicate it correctly.

Torch

The welding torch (or *blowpipe*) has separate inlets for oxygen and acetylene. It transports the gases to the mixing chamber where they mix in the correct proportion for welding. The mixing is controlled by two valves on the handle, each of which may be opened and closed to regulate the flow of gases for the welding flame.

The most commonly used torch is the equal pressure (or medium pressure) type, figure 1-8. In this type approximately equal amounts of oxygen and acetylene are used for welding.

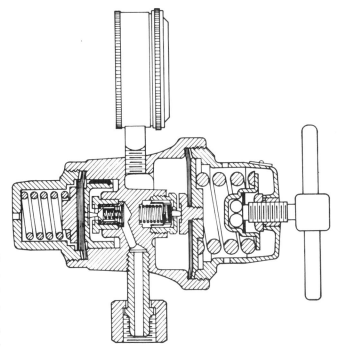

Fig. 1-6 Oxygen Regulator

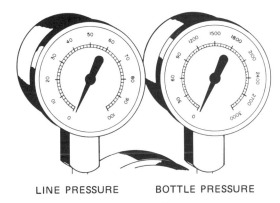

LINE PRESSURE   BOTTLE PRESSURE

Fig. 1-7 Oxygen Gauges

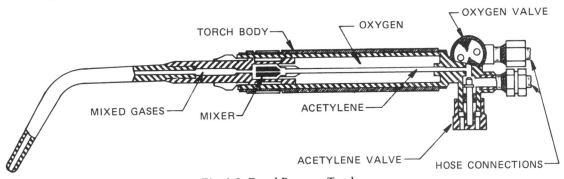

Fig. 1-8 Equal Pressure Torch

As with all other oxyacetylene equipment the torch has left-hand threads on the acetylene connection and right-hand threads on the oxygen connection.

Welding Tip

Different sized tips must be attached to the torch. These tips have a mixing chamber and *orifices* (holes) to supply different flame sizes. The tip concentrates the gases coming from the torch so that the flame can be directed toward the weld to be made.

Tips come with different size orifices, to vary the size of the flame. When a tip with larger holes is used more gas can escape. This provides a larger flame for welding on heavier metal, where more heat is required. Tips should be kept clean and in good working order.

Fig. 1-9 Bottle Cart

Bottle Cart

A bottle cart, figure 1-9, is a two-wheeled cart equipped with a chain for fastening the cylinders securely. When oxyacetylene welding equipment is installed on the bottle cart it makes a portable welding station which can be wheeled to the job.

Hoses

Special hoses are used for oxyacetylene welding equipment. They are made to withstand the high welding pressures. Welding hoses are supplied in 3/16", 1/4", 3/8" and 1/2" sizes. Select the correct size for the equipment to be used. They may be furnished as a double hose, connected by a rubber web and molded together. If single hoses are used, tape them together at 18-inch intervals to keep the unit solid.

Oxygen hoses are green in color and have right-hand threads. Acetylene hoses are red in color and have left-hand threads. Also, acetylene hose fittings have grooves cut into the nuts, to indicate a left-hand threaded connection.

SUMMARY

- Oxygen does not burn, but supports combustion.
- Oxygen in contact with oil or grease may cause a violent explosion.
- Oxygen is colorless, odorless and tasteless.
- Oxygen is bottled under extremely high pressures; the cylinders must be handled carefully.
- Acetylene is highly inflammable.
- Acetylene has a strong, disagreeable odor, resembling garlic.
- Acetylene is very unstable over 15 pounds per square inch.
- The oxyacetylene flame is the hottest flame known.
- Oxygen valves are double seated and should be either fully opened or fully closed.

- Acetylene valves should not be opened more than 1½ turns.
- Oxygen regulators register high pressures.
- Acetylene regulators are made to register lower pressures.
- Oxygen equipment is identified by green markings and has right-hand threaded connections.
- Acetylene equipment is identified by red color and has left-hand threaded connections, with grooves cut in the nuts.

JOB 1: SETTING UP OXYACETYLENE WELDING EQUIPMENT

Equipment:

Oxygen bottle
Acetylene bottle
Oxygen regulator
Acetylene regulator
One set of welding hoses
Torch
Bottle cart
Acetylene bottle key (if required)
12″ adjustable wrench or manufacturer's supplied wrenches

PROCEDURE	KEY POINTS
1. Obtain full oxygen and acetylene cylinders and install them in the bottle cart, or fasten them securely in an upright position.	1. Fasten the cylinders securely.
2. Remove the cap from the oxygen cylinder. Store the cap on the cart.	
3. Open the cylinder valve slightly. Allow a small amount of oxygen to blow through the valve. Close the cylinder valve.	3. Do not stand in front of the valve. Make sure no one else is standing in front of the valve. This is called *cracking* the valve.
4. Install the oxygen regulator, using a suitable wrench.	4. Oxygen cylinders and regulators have right-hand threads. Tighten the nut snugly, but do not apply too much force or the threads may be stripped.
5. Remove the cap from the acetylene cylinder. Store the cap.	
6. Open the valve on the acetylene cylinder slightly, using the cylinder key. Close valve.	6. **CAUTION: This gas is inflammable. No open fire!**

PROCEDURE	KEY POINTS
7. Install the acetylene regulator.	7. Left-hand thread.
8. Install the green hose on the oxygen gauge; install the red hose on the acetylene gauge.	8. Tighten fittings snugly. Over tightening will strip the threads.
9. Open the oxygen cylinder valve slowly until a small amount registers on the gauge, then open it completely. Turn the adjustment screw on the regulator to the right, until a small amount of pressure shows on the low pressure gauge. This will blow the hose clean. Turn the adjustment screw to the left and release the pressure.	9. Sudden release of oxygen pressure may damage the gauges. Open valve very slowly until a slight pressure registers. **CAUTION: Do not stand in front of the gauge face. Pressure may blow face outward.**
10. Open the acetylene valve slowly until a small amount registers on the gauge. Then open it 1 1/2 turns. Turn the adjustment screw on the regulator to the right, until a small amount of pressure shows on the low pressure gauge. This will blow the hose clean. Turn the adjustment screw to the left and release the pressure.	10. Acetylene gas is inflammable. When releasing it into the room, be sure there is no open fire present. Do not open the acetylene valve more than 1 1/2 turns.
11. Install the blowpipe (torch) on the open ends of the hoses.	11. Acetylene fittings have left-hand threads.
12. Be sure the torch valves are closed. Adjust the regulators so that 10 pounds of pressure shows on both the acetylene and the oxygen gauges.	
13. Check all connections with soapsuds. If bubbles appear, a leak is indicated and the connections must be tightened.	13. **CAUTION: Do not use soap with an oil base. Oil and oxygen may cause a violent explosion. Use no oil.**

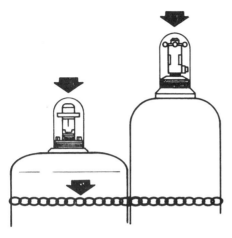

1. TWO BOTTLES SAFETY CHAINED AND CAPS IN PLACE

2. CRACK VALVES — INSTALL REGULATORS

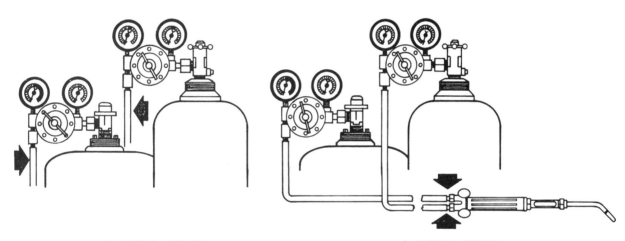

3. INSTALL HOSES.

4. INSTALL TORCH.

5. OPEN BOTTLE VALVES.

6. SET ADJUSTING SCREWS FOR LINE PRESSURE.

Fig. 1-10

Summary: Job 1

- Fasten the bottles securely in an upright position.
- Crack the valves to clean any dirt from them.
- Install the regulators. (Acetylene has left-hand threads)
- Install the hoses.
- Install the torch.
- Set the regulators for 10 pounds per square inch, working pressure.
- Check for leaks with soapsuds.
- Reverse the procedure to disassemble the equipment.

REVIEW QUESTIONS

1. List the steps to be followed in setting up a welding outfit.
2. What are the main differences between the valve on an oxygen cylinder and the valve on an acetylene cylinder.
3. What is acetylene made from?
4. What is the maximum safe pressure of free acetylene?
5. Why are cylinder valves cracked before installing the regulators?
6. Why is "Use No Oil" so strongly stressed?
7. Why should the oxygen cylinder valve be opened very slowly after the regulator is attached?
8. What is the purpose of a regulator?
9. What is the difference between oxygen and acetylene fittings?
10. Is oxygen inflammable?

Unit 2 Lighting the Oxyacetylene Torch

OBJECTIVES

The welding student will:

- Light the oxyacetylene torch following proper safety precautions.
- Be able to identify oxidizing, carburizing, and neutral flames.
- Adjust the torch for each of the three types of flame.
- Discuss the effect of each of the three types of flame on the metal.

Flames

Three basic flames can be made by adjusting the valves on the welding torch.

- The *carburizing flame.*
- The *neutral flame.*
- The *oxidizing flame.*

Carburizing Flame

A *carburizing flame*, figure 2-1, is the result of too much acetylene gas in the flame. This flame may be recognized by a long streamer of green colored gas which burns around the inner cone of the flame. This is called an acetylene feather. A carburizing flame is used to make the outside of metal hard, but is not good for a weld. The addition of the extra acetylene to the melted weld adds carbon to the metal and makes a hard, brittle weld. When this flame is used on melted parent metal, it causes the puddle to turn dark red, and gives it a boiling action.

Neutral Flame

The *neutral flame*, figure 2-2, is the welding flame. This flame can be recognized by a sharp inner cone and the absence of an acetylene feather. It is made up of 2 1/2 parts of oxygen and 1 part of acetylene. One part of the oxygen in the flame and one part of the acetylene come from the bottles. The other 1 1/2 parts of oxygen are picked up from the air around

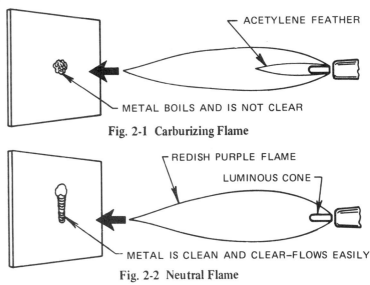

Fig. 2-1 Carburizing Flame

Fig. 2-2 Neutral Flame

the welding tip. A neutral flame does not add anything to or subtract anything from the *parent metal* (the metal being welded). The acetylene torch is adjusted for a neutral flame, for most welding jobs that require the metal to be melted and mixed together.

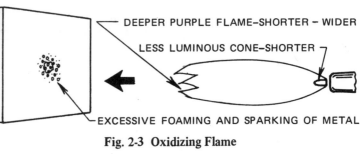

DEEPER PURPLE FLAME–SHORTER – WIDER

LESS LUMINOUS CONE–SHORTER

EXCESSIVE FOAMING AND SPARKING OF METAL

Fig. 2-3 Oxidizing Flame

Oxidizing Flame

An *oxidizing flame,* figure 2-3, is the result of having too much oxygen in the gas mixture. This flame can be

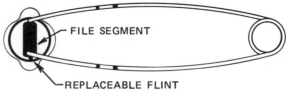

FILE SEGMENT

REPLACEABLE FLINT

Fig. 2-4 Striker

identified by a shorter inner cone and a whistling sound. It causes the molten metal to boil and spark. The additional oxygen in the flame causes the metal to burn, resulting in a brittle weld. A slightly oxidizing flame may be used for brazing, but it is not used for fusion welding.

Striker

The striker, figure 2-4, produces a spark by dragging a piece of flint across a file. A striker must always be used to light the oxyacetylene torch. The use of matches creates a hazard and may result in personal injury. The flints in most strikers may be replaced when the original one is worn out.

Goggles

Goggles are to be worn when welding. They are made in many shapes and sizes, to suit the individual welder. Welding goggles have dark lenses which filter out ultraviolent and infrared rays. For oxyacetylene welding they should be equipped with #4 or #5 filter lenses, either blue or brown. A clear glass cover lens fits over the filter lens, to protect the more expensive filter lenses from hot weld spatter.

Ultraviolet rays are given off by the welding flame. These are the same invisible rays which make it dangerous for the human eye to look directly at the sun. These rays can cause severe burns and possible blindness.

Infrared rays can cause a burn which looks like sunburn. The skin must be covered by clothing and the eyes protected by dark lenses to avoid the dangers of burns from infrared rays while welding.

JOB 2: LIGHTING THE OXYACETYLENE TORCH

Equipment:

Oxyacetylene outfit as assembled in Unit 1
Welding tips
Goggles
Gloves
Striker

PROCEDURE	KEY POINTS
1. Set up the welding equipment, following the procedures outlined in JOB 1.	1. Review KEY POINTS of JOB 1.
2. Install the tip on the torch body.	2. Tighten the tip retaining nut hand tight only.
3. Open the oxygen cylinder valve slightly until pressure registers on the high pressure gauge, figure 2-5, then open the valve fully.	3. Do not stand in front of the gauges when opening the cylinder valve.

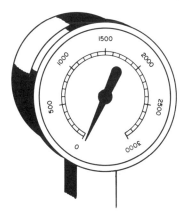

Fig. 2-5 High Pressure Gauge (Oxygen)

| 4. Turn the adjusting screw on the regulator to the right until pressure registers on the low pressure gauge, figure 2-6. | 4. Set the oxygen pressure at about 5 psi. |

Fig. 2-6 Low Pressure Gauge (Oxygen).

| 5. Open the oxygen needle valve on the torch handle and readjust the regulator until about 5 pounds registers, with the needle valve open. Close the needle valve. | 5. Always adjust the pressure with torch valve open. When it is closed, pressure may register higher on the gauge, but the welding pressure will be correct with the valves open. |

6. Open the acetylene cylinder valve slightly until pressure registers on the high pressure gauge, figure 2-7, then open it 1/2 turn.

6. Never open the acetylene cylinder valve over 1/2 turn. Leave the key on the valve.

Fig. 2-7 High Pressure Gauge (Acetylene)

7. Turn the adjusting screw to the right until pressure registers on the low pressure gauge, figure 2-8.

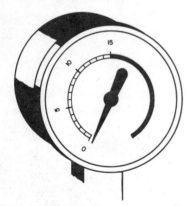

Fig. 2-8 Low Pressure Gauge (Acetylene)
(Note Calibration Stops at 15 lbs.)

8. Open the acetylene needle valve on the torch handle and readjust the screw on the regulator until about 5 pounds registers, with the needle valve open. Close the needle valve on the torch.

8. Set at about 5 psi.

9. Open the acetylene needle valve on the torch about 1/2 turn. Hold the striker in the left hand (if right-handed), the torch in right hand, and strike a spark in front of the escaping gas.

9. Be sure the torch is not pointed toward any inflammable material or people. Wear gloves and goggles.

10. Open the needle valve on the torch until the flame jumps away from tip about 1/8″.

10. The flame will appear turbulent, but will not smoke.

11. Open the oxygen needle valve on the torch slowly, adding oxygen to the burning acetylene.

12. As oxygen is added to the acetylene, observe the luminous cone at the tip, and the long greenish-color envelope around it. The green envelope is the excess acetylene of the carburizing flame.

12. A carburizing flame is used for hardening steel. It is generally used with the acetylene feather about 3 times as long as the inner cone.

13. Continue to add oxygen by opening the oxygen needle valve until the feather of acetylene just disappears. The inner cone will now appear soft and luminous. This is a neutral flame.

13. The torch should make a soft, even blowing sound.

14. If more oxygen is added now, the flame becomes pointed and white in color. In addition, it makes a sharp whistling sound. This is an oxidizing flame.

14. An oxidizing flame is harmful to the weld. It is never used for welding.

15. Practice adjusting the torch to carburizing, neutral, and oxidizing flames.

16. Shut off the acetylene needle valve on the torch; shut off the oxygen needle valve on the torch; shut off both cylinder valves completely; open the needle valves on the torch handle and drain the hoses. (Watch the gauges until they register 0.)

16. Shut off the acetylene first, so the escaping oxygen will blow any soot or impurities from tip.

17. Close the needle valves on the torch, release the pressure adjusting screws on the regulators by turning the handles to the left; coil the hoses and hang them up on the hose holder.

17. Do not hang the hoses on the gauges. The weight may damage the gauges. Make sure no fires are burning around the work area.

Summary: Job 2

- Set up the welding equipment.
- Install the tip on the torch.
- Open the oxygen cylinder valve slowly.

- Turn the adjusting screw on the regulator until the gauge indicates 5 psi.
- Open the oxygen needle valve on the torch.
- Readjust the gauge pressure to 5 psi with the valve open.
- Open the acetylene cylinder valve.
- Turn the adjusting screw on the regulator until the gauge indicates 5 psi.
- Open the acetylene needle valve on the torch.
- Readjust the pressure to 5 psi with the valve open.
- Close the valve on the torch.
- Open the acetylene valve on the torch and light the flame.
- Continue to open the acetylene needle valve until the flame jumps slightly away from the tip.
- Open the oxygen needle valve on the torch and adjust the flame to neutral.
- Reverse the procedures for shutting off the equipment.
- Check for fires.

REVIEW QUESTIONS

1. What is a neutral flame?
2. What is a carburizing flame?
3. What is an oxidizing flame?
4. Why is the pressure on the line gauges adjusted with the needle valves open?
5. When shutting off the torch, which needle valve should be closed first? Why?
6. Why should the operator stand behind, or to one side of, the oxygen cylinder when opening the cylinder valve?
7. What is the maximum working pressure of acetylene gas?

Unit 3   Running a Bead with Filler Rod, Flat Position

OBJECTIVES

The student will:

- Be able to define penetration and its importance.
- Be able to describe filler rod and its use.
- Be able to describe the flat position for welding.
- Run a bead in the flat position using filler rod.

Penetration

An oxacetylene weld should always have 100% penetration. This means that the weld appears on the bottom of the parent metal as well as on the top. Poor penetration causes the metal to break in the weld. However, too much penetration causes the molten metal to drip through and hang down below the parent metal. This condition is known as "icicles". Excessive drop-through of molten metal causes oxidation, as the molten metal takes oxygen out of the atmosphere into the weld. This oxidation causes a brittle weld which is easily broken.

Mild Steel

Steel which contains 0.30% carbon, or less, is called mild steel. It is the most commonly used steel for construction purposes. It is also called low-carbon steel and black iron. Most of the steel used for construction purposes has a black appearance, caused by hot rolling at the steel mills, resulting in the formation of oxide on the metal.

Forehand Welding

Right-handed welders usually hold the torch in the right hand and weld from right to left, adding the filler rod at the front of the flame. This procedure is called *forehand* welding. Left-handed operators weld from left to right, but the rod is still added at the front of the flame, since the torch is held in the left hand and the filler rod in the right.

Flat Weld Position

A weld made on the topside of the parent metal and within 30 degrees of horizontal is called a *flat weld*. Flat welding is the most desirable position for welding, since the welder can control penetration and bead appearance easily. In many welding shops, machines called *positioners* are used to hold the work so that it can easily be turned into the flat position.

Filler Rod

Filler rod is commonly called welding rod. It is added to the molten puddle to build up the cross section of weld where the penetration has forced the molten metal below the surface of the parent metal. To insure good penetration, filler rod should be added only after the puddle has been formed. The weld should have a cross section thicker than the original parent metal, so that strength is added at the point of the weld. The word *convex* means a rounded-up bead. Filler rods most commonly used are 1/16", 3/32" and 1/8" diameters.

Sheet Metal

Metal which has been rolled in the steel mill to a thickness of 1/8" or less is commonly called sheet metal. This is the metal most commonly used with the oxyacetylene welding process, since any thickness over 1/8" is more easily joined by electric arc welding. Sheet metal is designated by gauges. A few of the most common gauges are listed below:

12-gauge	. . . .1120"	. . . .	approx. 1/8" thickness
14-gauge	. . . .0821"	. . . .	approx. 5/64" thickness
16-gauge	. . . .0635"	. . . .	approx. 1/16" thickness
18-gauge	. . . .0508"	. . . .	approx. 3/64" thickness

Neutral Flame

A neutral flame is used for welding. On an equal pressure torch, the gauges should be set at approximately the same pressure, if the welding is not to be done in a confined area. Welding in a restricted area, such as a corner of the metal, uses up the atmospheric oxygen rapidly and changes the character of the flame from neutral to carburizing. More oxygen pressure may be needed for such a welding condition. In this case the oxygen pressure must be higher than the acetylene pressure, until the welding in the restricted area is completed. If the oxygen pressure has been increased, it must be lowered after the welding in the corner is finished.

JOB 3: RUNNING A BEAD WITH FILLER ROD

Equipment and Material:

> Standard Oxyacetylene Welding Equipment.
> 16-gauge mild steel sheet metal
> 1/16" mild steel filler rod.

PROCEDURE	KEY POINTS
1. Place the sheet metal sample on 2 firebricks.	
2. Wear goggles to protect the eyes. Light the torch and adjust to a neutral flame.	
3. Hold the torch in the right hand, if right-handed, and a piece of filler rod in the left hand.	3. Grasp rod as you would a pencil, with 6" to 8" extending below the left hand.

PROCEDURE

4. Hold the flame 1/16″ to 1/8″ above the parent metal at the right edge of the sample until a molten puddle is formed. Dip the end of the rod into the puddle, allowing a small amount of the rod to melt off and fuse with the base metal, figure 3-1.

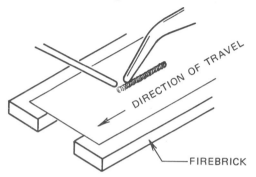

Fig. 3-1 Welding a Bead on Sheet Metal

5. Weld across the sample, from right to left, adding filler rod to the molten parent metal.

6. Melt enough rod into the puddle to build up the bead evenly in height and width. Alternately raise the rod as the base metal melts and dip it into the front of the puddle when penetration is completed.

7. When good penetration is obtained the underside of the metal will look as though a bead has been run on it. Addition of the rod to the molten puddle is the key to good welding. Practice this job until the technique is mastered.

KEY POINTS

4. Do not add rod until a molten puddle is formed. Keep the end of the rod in the outer end of flame envelope to protect it from oxidation. The tip should be at an angle of 45° to 60° with the parent metal, figure 3-2.

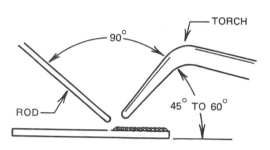

Fig. 3-2 Torch Angle for Welding

5. This is called forehand welding.

6. Get penetration before adding the filler rod. Filler metal melted onto the sheet metal is not a weld, unless it is fused into the parent metal. A slight circular motion of the flame will make a better weld, figure 3-3.

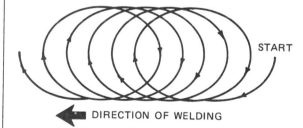

DIRECTION OF WELDING

Fig. 3-3 Circular Torch Movement

7. The edges of the weld should be feathered smoothly into the parent metal. The rod should not be piled up. The bead should look like fish scales overlapping each other.

Summary: Job 3

- Weld with a neutral flame.

- Weld from right to left (left to right if left-handed).

- The torch should be at an angle of 45° to 60° with the parent metal.

- Circular motion of the flame is used to control the heat and bead appearance.

- If the metal becomes too hot the torch may be flashed off the weld. Momentarily direct the flame away from the puddle, but keep the puddle fluid.

- When good penetration has been developed, the underside of the sheet metal will look like a bead has been run on it.

REVIEW QUESTIONS

1. What is parent metal?

2. Why is it necessary to get penetration before the rod is added to the weld?

3. Why is the addition of filler rod necessary for a strong weld?

4. What type of flame is used for welding?

5. What percentage of penetration is necessary for the best weld?

Unit 4 Butt Weld on Mild Steel Sheet Metal, Flat Position

OBJECTIVES

The student will be able to:

- List and define several terms applying to butt welding.
- Discuss the effects of expansion and contraction.
- Compensate for expansion and contraction in butt welding.
- Make a butt weld which withstands a bend test.
- Perform a bend test on a butt weld.

Gapping

One of the greatest concerns the welder has in butt welding steel sheet metal is how expansion and contraction are dealt with. When steel is heated, as in welding, it *expands,* or increases in size. As it cools it *contracts,* or decreases in size.

When butt welding sheet metal it is a good practice to gap the metal. At the side where the weld is to begin leave about a 1/16-inch gap between the two pieces of parent metal. At the side where the weld is to end leave about a 1/8-inch gap. (These figures apply to a piece of 16-gage sheet metal 6-inches long.) The extra gap at the finishing end of the weld allows for the additional heat which is absorbed by the metal as the weld moves along.

Tack the pieces securely, so the gap will remain for welding. Tack welds are small completely fused welds made at several places along the line of the weld to hold the pieces in place.

Attempting to weld sheet metal without gapping the butt weld will produce a scissors effect, where one piece of sheet metal will be warped over the top of the other.

Keyhole

At the start of the weld, as the puddle is formed, the edges ahead of the weld should be melted away a small amount. This creates a place for the penetration to work through the gap. This keyhole should remain small. As it is carried along in front of the puddle, it insures 100% penetration.

Testing

A good weld will bend 180° over the bead, without breaking and will stand as much pull as the parent metal when tested for tensile strength. Most of the tests used in this text are *destructive tests.* They will bend the welded jobs out of shape, so the metal cannot be used again without considerable preparation.

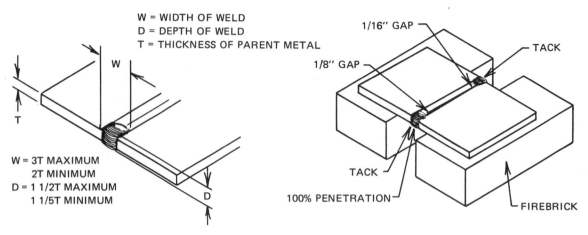

W = WIDTH OF WELD
D = DEPTH OF WELD
T = THICKNESS OF PARENT METAL

W

T

W = 3T MAXIMUM
    2T MINIMUM
D = 1 1/2T MAXIMUM
    1 1/5T MINIMUM

D

1/16" GAP

TACK

1/8" GAP

TACK

100% PENETRATION

FIREBRICK

Fig. 4-1 Setup for Welding the Butt Weld

JOB 4: BUTT WELD IN MILD STEEL SHEET METAL, FLAT POSITION

Equipment and Material:

Standard Oxyacetylene Welding Equipment
16-gauge mild steel samples, 2" wide by 6" long
1/16" or 3/32" mild steel filler rod

PROCEDURE	KEY POINTS
1. Tack weld two pieces of 16-gauge metal as shown in figure 4-1.	1. Leave a gap between the pieces.
2. Beginning at right side, form a puddle and add a little rod. Progress across the joint, alternately melting the puddle and adding rod at the leading edge of the puddle.	2. Keep a keyhole ahead of the puddle at all times so that penetration is achieved before the rod is added to the puddle. 100% fusion and penetration are necessary for a strong weld.
3. Get complete fusion of the metal to the bottom of the gap, and add filler rod so that the weld will be above the top of the parent metal.	3. The weld should always be thicker than the parent metal.
4. Practice this exercise until the butt weld is mastered.	
5. Test the weld. Place the welded sample in a vise with the weld at the top of the vise jaws. Bend the metal against the bead a full 180° No cracks or tears should appear.	5. Both good appearance and strength are necessary. Be sure the sample is cold before testing. Hot metal bends easier, but has less strength.

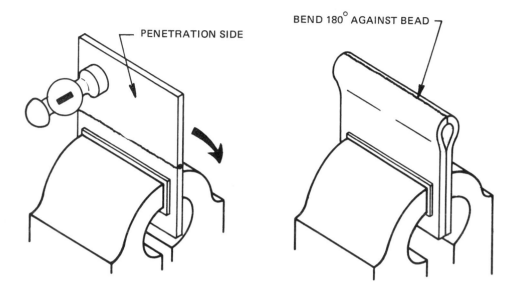

Fig. 4-2 Testing Butt Weld in Mild Steel Sheet Metal

REVIEW QUESTIONS

1. Why is it necessary to gap a butt weld?

2. How high should the bead be built up on this type of weld?

3. What percentage of penetration should be achieved?

4. What is the keyhole?

5. What type of flame is used for this weld?

Unit 5 Fillet Weld on Mild Steel Sheet Metal, Flat Position

OBJECTIVES

The student will be able to:

- Explain undercutting and how to prevent it.
- Discuss the consequences of a lack of oxygen in a confined area and tell how to compensate for this condition.
- Make a fillet weld capable of withstanding a bend test.
- Perform a bend test on a fillet weld.

Fillet Weld

A *fillet weld* is a weld made on two pieces of metal which are joined in any way other than in a flat plane. A fillet is a reinforcement and a weld made in an inside corner is called a fillet weld. Fillet welds are sometimes called T-welds, when one piece forms a 90° angle with the other.

The type of welded joint most often used is the fillet weld. Generally, the fillet is the most difficult type of joint to weld successfully. Penetration must be made completely through the corner to insure that the joint will have full strength. Sometimes welds are made on both sides of the upright piece, but in many cases it is not possible to reach both sides, so the weld made from one side must be strong enough to hold the pieces together.

Undercutting

Gravity exerts a force on fluid metal, so there is a tendency for the metal to drop away from the vertical piece on a T-weld. The metal will be made thinner by the amount that drops away. This thinned section is called an *undercut*.

All fillet welds are subject to undercutting. Great care must be taken to add enough filler metal to completely build up the bead and the torch must be directed into the weld in such a way that excessive melting does not take place in the upper part of the fillet joint. Undercutting, figure 5-2, reduces the thickness of the metal, making it subject to breakage under stress.

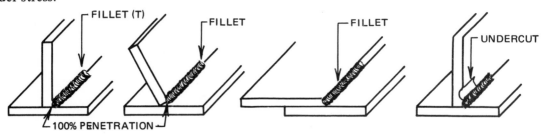

FILLET (T)   FILLET   FILLET   UNDERCUT

100% PENETRATION

Fig. 5-1   **Fig. 5-2**

Popping the Torch

Because welding in confined areas, such as corners, exhausts the available oxygen in the atmosphere the torch tip may become overheated. Frequently the flame goes out momentarily from lack of oxygen. As it reignites, from the hot metal, a popping action results which blows the molten metal out of the weld. This condition may be corrected by (1) using a little more oxygen pressure on the torch, or (2) by changing to the next size larger tip. If more oxygen pressure is used, care must be taken so the flame does not become oxidizing and burn the weld.

JOB 5: FILLET WELD IN MILD STEEL SHEET METAL (T-WELD)

Equipment and Material

Standard oxyacetylene welding equipment
16-gauge mild steel sheet metal samples, 2" x 6"
1/16" or 3/32" mild steel filler rod

PROCEDURE	KEY POINTS
1. Tack weld two pieces of metal as shown in figure 5-3.	

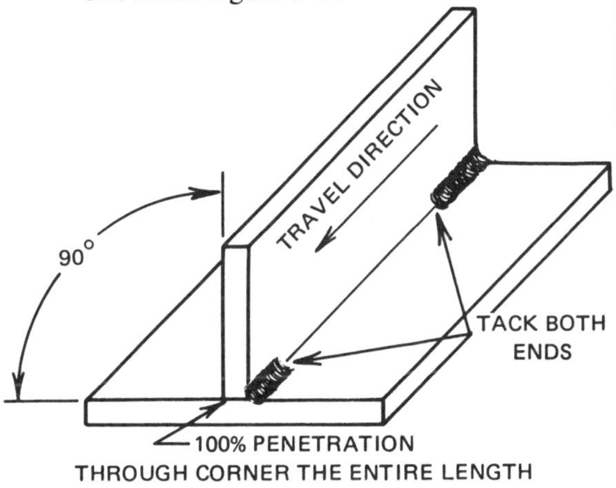
TRAVEL DIRECTION
90°
TACK BOTH ENDS
100% PENETRATION THROUGH CORNER THE ENTIRE LENGTH

Fig. 5-3

PROCEDURE	KEY POINTS
2. Beginning at the right side, play the flame on the flat piece until it is red; then with a slight circular motion, gradually heat both pieces until fusion is accomplished.	2. Since the weld is being made in a corner, more heat will be required. Use one size larger tip if necessary Do not direct the flame completely into corner. The burning gases will overheat the tip.
3. Move at a regular rate of speed across the joint, melting the two pieces together, then adding rod.	3. The rod should be added slightly above the center of the puddle to avoid undercutting.

PROCEDURE	KEY POINTS
4. Practice this weld until penetration and bead appearance are mastered.	4. Be sure to fill the puddle completely. Fuse the edges of weld into both pieces of parent metal. Get complete fusion of parent metals when welding the corner before the rod is added. If the rod is piled up there is no way to get back into the corner to melt the parent metal.
5. Test the weld. Place the welded sample in a vise and bend both pieces of metal against the weld, figure 5-4. No cracks, tears or bead separation should appear.	

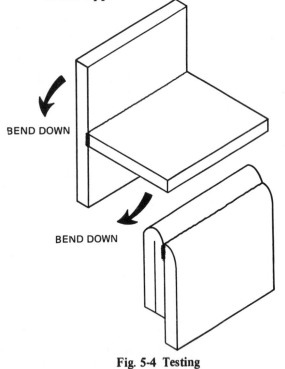

BEND DOWN

BEND DOWN

Fig. 5-4 Testing

REVIEW QUESTIONS

1. Why does this type of weld require more heat than a butt weld?
2. Why is there danger of overheating the tip in this exercise?
3. What is an undercut?
4. Why should undercutting be avoided?
5. What is a fillet weld?
6. What type of flame is used for this weld?

Unit 6 Lap Weld on Mild Steel Sheet Metal, Flat Position

OBJECTIVES

The student will be able to:

- Discuss burn-away and how it is prevented.

- Tell why lap welds are avoided where possible.

- Make a lap weld capable of withstanding a bend test.

- Perform a bend test on a lap welded joint.

A *lap weld* is a weld made with one plate of metal overlapping the other.

The lap joint is not considered to be a desirable method of construction. Even when the two pieces are clamped together tightly, there is often danger of moisture collecting in the space between them. This moisture promotes oxidation (rusting). Corrosion can also occur between the two pieces.

Burn-Away and Torch Angle

Because lap welds are made on the edge of one piece of metal and on the flat side of the other, the top piece absorbs heat faster than the bottom piece. This causes faster melting and burn-away of the edge of the top piece. Burn-away can be corrected by (1) pointing the flame onto the bottom piece, (2) keeping the filler rod between the flame and the edge of the top piece. The filler rod should be added from the top into the molten puddle formed on the bottom piece.

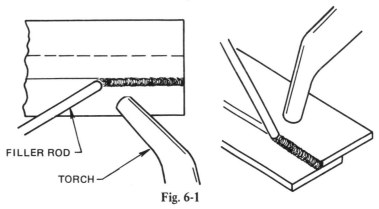

FILLER ROD

TORCH

Fig. 6-1

JOB 6: LAP WELD IN MILD STEEL SHEET METAL, FLAT POSITION

Equipment and Material

Standard oxyacetylene welding equipment
16-gauge mild steel samples, 2″ x 6″
1/16″ or 3/32″ mild steel filler rod

PROCEDURE	KEY POINTS
1. Lay one piece of sheet metal over the other, so that it overlaps slightly, figure 6-2.	1. The two pieces of metal must fit tightly. Gaps will cause unsightly welds and poor fusion.

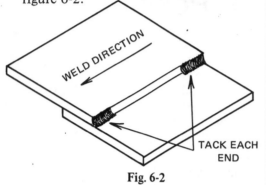

Fig. 6-2

2. Tack weld the two pieces together at each end.	
3. Start the weld on right side of the plate, melting the upper piece into the lower, and adding filler rod as necessary.	3. Play the flame on the lower piece. The upper edge will melt more rapidly, so the welding rod should be added at the upper edge. Build a full bead. A shallow bead will cause the weld to fail under stress.
4. Test the weld by bending both pieces of metal 180° over the face of the weld.	4. The weld will fail if it has not been built up enough, or if penetration is shallow in the lower plate.

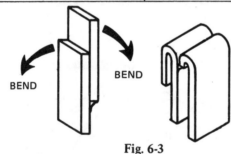

Fig. 6-3

REVIEW QUESTIONS

1. Why is the lap weld considered to be a poor joint design?

2. Why should two pieces of metal in a lap weld be clamped tightly together?

3. Where should the flame be directed in this type of weld?

4. Why is a full bead important on this type of joint?

5. What type of flame is used for this weld?

Unit 7 60° Fillet Weld on Mild Steel Sheet Metal, Flat Position

OBJECTIVES

The student will be able to:

- Discuss bridging and how it is prevented.
- Make a 60° fillet weld with 100% penetration and capable of withstanding a bend test.
- Bend test a 60° fillet welded sample.

A 60° weld is a fillet weld made in a corner where two pieces of metal join at a 60° angle.

This is a difficult weld to make. The lack of space between the two pieces restricts the torch motion and reduces combustion as the oxygen is burned away in the confined area. Extreme care must be taken to melt the parent metal in the corner completely before the filler rod is added.

Bridging

If the filler rod is added before penetration is achieved in the corner, it will bridge across the gap, rather than melting into it. A bridged weld will not stand a bend test, since the angled parent metal is not fused into the flat plate. It is important when welding in confined areas to have the molten puddle achieve complete fusion before the filler rod is added. Filler rod should be added in small quantities so the puddle is not cooled off by the filler rod.

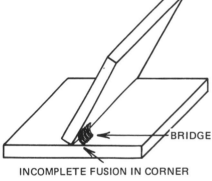

INCOMPLETE FUSION IN CORNER

Fig. 7-1

JOB 7: 60° WELD IN MILD STEEL SHEET METAL, FLAT POSITION

Equipment and Materials:

Standard oxyacetylene welding equipment
2 samples 16-gauge mild steel, 2″ x 6″
1/16″ or 3/32″ mild steel filler rod

PROCEDURE

1. Tack weld one piece of sheet metal in the center of another, at an angle of 60°, figure 7-2.

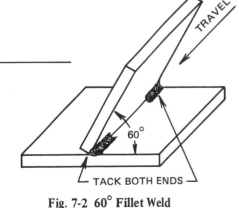

TACK BOTH ENDS

Fig. 7-2 60° Fillet Weld

PROCEDURE	KEY POINTS
2. Beginning at the right edge, make a smooth, continuous weld from one end to the other.	2. Add filler rod only after the puddle has been established. Use a tip one size larger than would ordinarily be required. The confined area requires more heat.
3. Keep the bead the same width.	3. Slow, careful concentration is required to make a good weld at this angle.
4. Test the weld, as in figure 7-3.	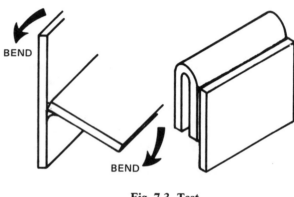

Fig. 7-3 Test

REVIEW QUESTIONS

1. Draw a sketch of a 60° fillet weld between two pieces of sheet metal. Show the angle between the two pieces.

2. What does the term bridging mean in the welding field?

3. What should be done if the metal does not melt in the corner of a 60° fillet weld?

SECTION 1: WELDS IN THE FLAT POSITION, COMPREHENSIVE REVIEW

A. RUNNING A BEAD IN THE FLAT POSITION

Equipment and Materials:

> Standard oxyacetylene welding equipment
> 1 piece 16-gauge sheet metal, 2" x 6"
> 1/16" or 3/32" mild steel filler rod

PROCEDURE

1. Place a piece of 16-gage metal in the flat position.

2. Using a neutral flame and 1/16" filler rod, make a single pass bead across the 6" length of the sheet metal.

3. Cool the sample and brush the weld with a wire brush.

4. Have the sample inspected by the instructor.

B. BUTT WELDING IN THE FLAT POSITION

Equipment and Materials:

> Standard oxyacetylene welding equipment
> 2 pieces 16-gauge sheet metal, 2" x 6"
> 1/16" or 3/32" mild steel filler rod

PROCEDURE

1. Tack weld 2 pieces of 16-gage sheet metal in position for a butt joint.

2. Using a neutral flame and 1/16" filler rod, weld the two pieces together with a butt weld in the flat position.

3. Cool the sample and brush with a wire brush.

4. Have the sample inspected by the instructor for bead appearance and penetration.

5. Test the weld as shown in the drawing.

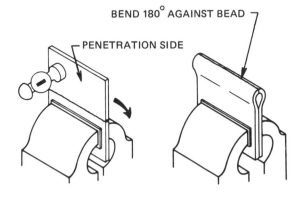

BEND 180° AGAINST BEAD

PENETRATION SIDE

SECTION 2:
Welds in Horizontal and Vertical Positions

Unit 8   Bead on Mild Steel Sheet Metal, Horizontal Position

OBJECTIVES

The student will be able to:

- Describe cold lap and its cause.
- Discuss the use of torch angle and filler rod manipulation in controlling cold lap.
- Run a horizontal bead with good penetration and free of cold lap.

A *horizonal weld* is a weld which is done in a horizontal line and against an approximately vertical surface, figure 8-1.

Welding in the horizontal position is more difficult than welding in the flat position, because the force of gravity tends to cause the molten metal to run down onto the plate. This run-down can be controlled by using the correct torch angle and feeding the rod correctly.

Torch Angle

For horizontal welding, the tip and flame should point slightly upward, with the tip positioned at an angle of 45° to 60° with the work. The filler rod should be fed from the top of the puddle.

Cold Lap

When the molten metal runs down onto the cold metal and it is not mixed with the parent metal, a cold lap results, figure 8-2. This is a problem in horizontal welding, because

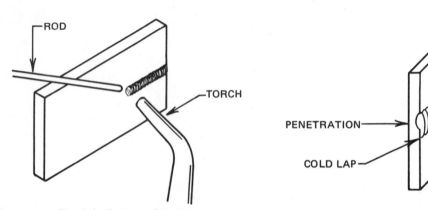

Fig. 8-1 Horizontal Bead   Fig. 8-2 Cold Lap

gravity tends to pull the hot metal downward. The edge of the bead rolls onto the parent metal, without fusion, resulting in a poor weld, with little or no strength.

JOB 8: RUNNING A HORIZONTAL BEAD IN MILD STEEL SHEET METAL

Equipment and Materials:

Standard oxyacetylene welding equipment
2 pieces, 16-gauge mild steel, 2" x 6"
1/16" or 3/32" mild steel filler rod

PROCEDURE	KEY POINTS
1. Position a piece of steel as shown in figure 8-3.	1. The metal may be tacked in a T-position on another plate, to hold it upright.

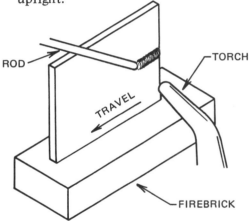

Fig. 8-3 Position of Work for Running a Horizontal Bead.

PROCEDURE	KEY POINTS
2. Beginning at the right side, weld a bead across the 6" length, being careful to get a smooth appearance and 100% penetration.	2. Special attention must be paid to the direction and angle of the torch. The filler rod must be added above the center of the puddle.
3. Examine the back of the metal to check the penetration of the bead.	

REVIEW QUESTIONS

1. What is cold lap?

2. What percent of penetration is required on a horizontal bead?

3. What type of flame is used to weld the horizontal bead?

Unit 9 Butt Weld on Mild Steel Sheet Metal, Horizontal Position

OBJECTIVES

The student will be able to:

- Manipulate the torch and filler rod to control cold lap.

- Make a butt weld in the horizontal position with 100% penetration.

A horizontal butt weld is similar to a butt weld in the flat position. The only difference is that the pieces to be welded are in a vertical position, with the joint running in a horizontal line.

Practice is necessary to master the art of horizontal butt welding. The welder must use the correct tips and pressures and handle the torch and filler rod skillfully. The horizontal butt weld is a common application of oxyacetylene welding. The welding student should concentrate on perfecting this type of weld.

Torch Angle and Cold Lap

Horizontal butt welding involves running a horizontal bead, so cold lap is controlled in the same manner. The torch is aimed upward, at an angle of 45° to 60° with the work, and the filler rod is fed from the top.

Crown

Weld beads should be built up above the surface of the parent metal to insure strength. This buildup is called the *crown* of the weld.

JOB 9: HORIZONTAL BUTT WELD IN MILD STEEL SHEET METAL

Equipment and Materials:

Standard oxyacetylene welding equipment
2 pieces, 16-gauge mild steel, 2″ x 6″
1/16″ or 3/32″ mild steel filler rod

PROCEDURE	KEY POINTS
1. Tack weld two pieces of sheet metal for a butt weld.	1. The metal should be gapped approximately 1/16″ at the right side and 1/8″ at the left.

2. Place the material in a horizontal position, figure 9-1.

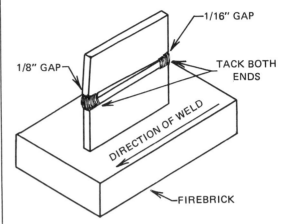

Fig. 9-1 Horizontal Butt Weld

3. Make a butt weld across the 6″ length of the two pieces.

3. Be careful that the molten metal does not roll to the bottom of the bead. Weld slowly and carefully and add the filler rod from the top of the puddle.

4. The weld should appear uniform and smooth, with 100% penetration.

5. Test the weld according to figure 9-2.

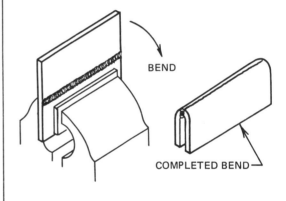

Fig. 9-2 Test

REVIEW QUESTIONS

1. What is the crown of a weld?

2. When making a horizontal butt weld, how much gap should be left between the two pieces of sheet metal?

   a. Right-hand side _____ ″.

   b. Left-hand side _____ ″.

3. When testing a butt weld, the sheet metal should be bent over the _____ side of the welded sample.

4. What is the proper torch angle for horizontal butt welding?

Unit 10 Bead on Mild Steel Sheet Metal, Vertical Position

OBJECTIVES

The student will be able to:

- Describe the vertical position for welding.
- Discuss methods to overcome the force of gravity in vertical welding.
- Run a bead in the vertical position with 100% penetration.

A vertical weld is any weld done in a vertical line. Vertical beads are usually run from the bottom to the top of the work.

Welding in the vertical position presents the problem of overcoming the force of gravity. Torch position and tip angle are used to overcome the tendency of the molten puddle to flow downward. The flame, fueled by gas under pressure, provides some support for the molten puddle. Feeding the filler rod from the top of the weld also helps, by providing a ledge of slightly hardened, cooled metal under the flame. As this ledge is formed the next layer of the weld is built upon it, figure 10-1.

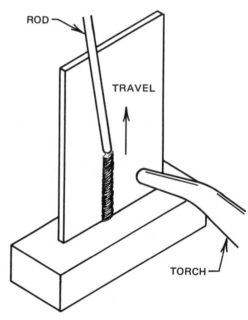

Fig. 10-1 Vertical Bead

JOB 10: VERTICAL BEAD ON MILD STEEL SHEET METAL

Equipment and Materials:

Standard oxyacetylene welding equipment
16-gauge sheet metal sample, 2″ x 6″
1/16″ or 3/32″ mild steel filler rod

PROCEDURE	KEY POINTS
1. Place a piece of 16-gage sheet metal in a vertical position, so that the 6″ length is perpendicular to the bench.	

PROCEDURE	KEY POINTS
2. Beginning at the bottom of the sheet, weld upward, adding rod as necessary.	2. To overcome the effects of gravity, which pulls the molten puddle down the sheet, add the rod from the side or top of the puddle. Keep the upward progress slow and even and do not add too much rod. The bead should be uniform in width, smoothly feathered in at the sides, and should have 100% penetration.

REVIEW QUESTIONS

1. Describe the position of the work in vertical welding.

2. How much penetration is needed on a vertical weld?

3. What is the purpose of the ledge formed in vertical welding?

4. Describe the action of the torch which helps to hold the puddle of molten metal from flowing downward.

5. What direction should the filler rod be added from when welding a vertical bead?

6. Is it possible for vertical beads to be welded so they are as strong as flat beads?

Unit 11   Butt Weld on Mild Steel Sheet Metal, Vertical Position

OBJECTIVES

The student will be able to:

- Define flash off and use it effectively.
- Restart a bead in the vertical position insuring uniform appearance of the finished bead.
- Make a vertical butt weld capable of withstanding a bend test.

A vertical butt weld is similar to a flat or horizontal butt weld, except that the weld is made in a vertical line.

A major difficulty when butt welding in the vertical position is the tendency of the edges to melt away and leave holes ahead of the torch. Careful application of the heat and the angle of the torch contribute to making a successful weld. Vertical butt welding is a common practice in sheet metal welding and must be mastered by the welding student.

Sheet metal should always be gapped and tack welded to prevent warpage. Welding in the vertical position requires concentration of the flame into a small area, so that the heat travelling up the joint does not become too intense and

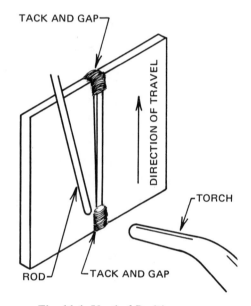

Fig. 11-1 Vertical Position

melt holes ahead of the puddle. The metal should be held rigidly in the vertical position by a clamp or a holding device called a *jig*.

The filler rod should be melted into the puddle from above the center of the weld. This helps maintain the ledge, on which the weld progresses. Also, the torch tip must be directed upward into the weld, in the same manner as for running a vertical bead. This will help to prevent the molten metal from flowing down.

Flash Off

Frequent flash off from the weld prevents hole burning. However, when the flame is flashed off, care should be taken to insure that the puddle does not completely solidify. If the puddle does solidify, the flame should be concentrated on the bead slightly below the puddle. The bead should be remelted at this point and the welding started only after a new puddle has been formed. The weld should then progress upward into the place where the previous welding stopped. Using this method insures complete penetration where the welding stopped and eliminates cold laps, which might occur at starting and stopping points.

JOB 11: BUTT WELD IN MILD STEEL SHEET METAL, VERTICAL POSITION

Equipment and Materials:

Standard oxyacetylene welding equipment
2 pieces, 16-gauge sheet metal, 2" x 6"
1/16" or 3/32" mild steel filler rod

PROCEDURE	KEY POINTS
1. Tack weld two pieces of material together, leaving a gap, as required for butt welding.	1. The gap should be 1/16" at the bottom and 1/8" at the top.
2. Place the material in a jig, so that the joint is perpendicular to the ground.	2. If no jig is available tack the work to a piece of scrap metal, so that it is held in a vertical position.
3. Make a continuous weld from bottom to top, being careful to maintain an even bead and 100% penetration.	3. Work in a comfortable position. The hoses may be draped over the welder's shoulders, to relieve the wrists of the weight of the hoses. Weld at the same rate of speed as for a flat butt weld. The heat travelling ahead may have a tendency to melt the metal faster.
4. Flash off the weld if there is a tendency to burn holes.	
5. Test the weld the same as a flat butt weld.	

REVIEW QUESTIONS

1. What is done to prevent warpage of sheet metal during butt welding?

2. What is the greatest difficulty encountered in butt welding in the vertical position?

3. From what direction should the filler rod be melted into the puddle of a vertical weld?

4. Describe how the torch flame is used to help support the molten puddle.

5. What is a jig?

6. How is a vertical weld restarted to prevent cold lap?

7. What should the percentage of penetration be on a vertical butt weld?

8. What may be done to prevent hole burning?

Unit 12 Lap Weld on Mild Steel Sheet Metal, Vertical Position

OBJECTIVES

The student will be able to:

- Manipulate the flame correctly to prevent burn away in vertical lap welding.

- Manipulate the filler rod to help control heat in vertical lap welding.

- Lap weld a joint in the vertical position which can withstand a bend test.

A vertical lap weld is a weld made between two pieces of metal, one overlapping the other with the joint running in a vertical line.

This is a difficult weld to make, but it can be mastered with practice. A finished vertical lap weld should have the same appearance as one made in the flat position. The bead should be smooth and the rippled, fish scale appearance uniform.

The raw edge of the overlapping plate has a tendency to burn away rapidly while the flat surface of the other plate is heating. The torch flame must be directed at the flat surface of the second plate. When the molten puddle is started it is carried along in the direction of travel.

The filler rod should be added from above the puddle and between the flame and the raw edge of the overlapping piece, figure 12-1. In this manner the filler rod will absorb some of the heat which might burn away the raw edge. The gas pressure will help support the molten puddle. Progress up the weld should be fairly rapid.

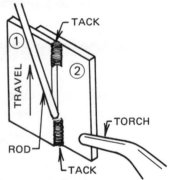

Fig. 12-1 Vertical Lap Weld Bend Test

JOB 12: LAP WELD IN MILD STEEL SHEET METAL, VERTICAL POSITION

Equipment and Materials:

Standard oxyacetylene welding equipment
2 pieces, 16-gauge sheet metal, 2" x 6"
1/16" or 3/32" mild steel filler rod

PROCEDURE	KEY POINTS
1. Position the metal samples with one piece lapped slightly over the other and tack weld.	

PROCEDURE	KEY POINTS
2. Weld the joint in the vertical position, from bottom to top.	2. Direct the torch at the surface of the second plate, melting the edge off the overlapping plate. Do not add too much rod; just enough to fill the bead completely.
3. Test the weld as in figure 12-2.	

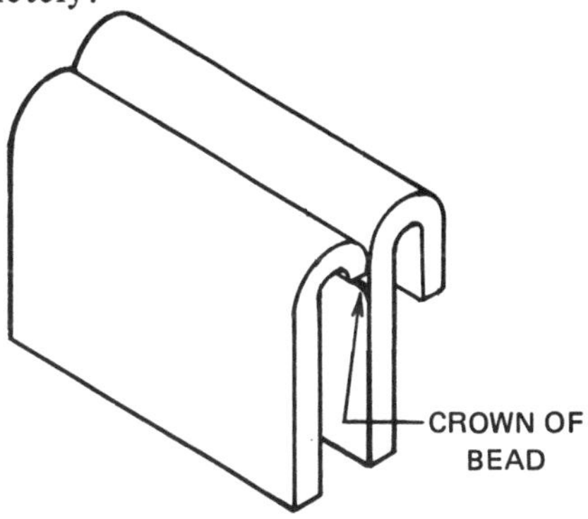

CROWN OF BEAD

Fig. 12-2 Vertical Lap Weld Bend Test

REVIEW QUESTIONS

1. How should the appearance of a finished vertical lap weld compare with the appearance of one made in the flat position?

2. What percentage of penetration is required for full strength in a lap weld?

3. To make a good vertical lap weld, which plate should the flame be directed toward?

4. What direction should the filler rod be added from when making a vertical lap weld?

5. What are the reasons that proper torch angle is important when making a vertical lap weld?

Unit 13 T-Weld on Mild Steel Sheet Metal, Vertical Position

OBJECTIVES

The student will be able to:

- Make a vertical T-weld with 100% penetration.
- Make a vertical T-weld without melting holes in either plate.
- Make a vertical T-weld capable of withstanding a bend test.
- Discuss the importance of preheating in making a vertical T-weld.

A *vertical T-weld* is a fillet weld made between two pieces which are at a 90° angle with each other, with the joint running in a vertical line, figure 13-1.

A vertical T-weld requires careful concentration. The corner, where the two pieces of metal join, must be completely penetrated. Areas in which the penetration is poor will cause the weld to fail when it is tested. To insure complete penetration, the heat should be directed into the corner, so that complete fusion of the two plates is accomplished before the filler rod is added.

Pointing the flame of the torch at one plate more than the other tends to burn holes in the metal. Flash off of the torch is necessary when there is danger of burning through the sheet metal, but 100% penetration is important for full-strength joints.

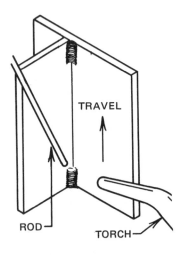

Fig. 13-1 Vertical T-Weld

As with other vertical welds, the torch should be tipped upward slightly and the filler rod added from above the puddle. This helps to support the molten puddle.

JOB 13: T-WELD IN MILD STEEL SHEET METAL, VERTICAL POSITION

Equipment and Materials

Standard oxyacetylene welding equipment
2 pieces. 16-gauge sheet metal, 2" x 6"
1/16" or 3/32" mild steel filler rod

PROCEDURE

1. Tack weld one piece of sheet metal to the other piece at a 90° angle. Position the first piece approximately in the center of the second, with the joint running perpendicular to the welding table, figure 13-2.

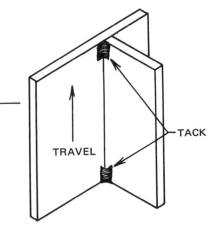

Fig. 13-2 Setup of T-Weld

PROCEDURE	KEY POINTS
2. Weld from the bottom to the top.	2. Hold the torch so the heat travels up the joint. Preheating helps penetration.
3. Concentrate on obtaining 100% penetration and smooth bead appearance.	
4. Test the weld as shown in figure 13-3.	

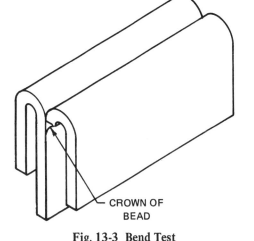

CROWN OF BEAD

Fig. 13-3 Bend Test

REVIEW QUESTIONS

1. How much penetration of a T-weld is necessary for full strength?

2. A T-weld is also called a _____ weld.

3. What will happen if the flame is pointed at one plate of a T-weld more than the other?

4. In order to preheat the metal, the flame should be pointed _____ slightly when making a vertical T-weld.

5. Why should the metal be preheated when making a vertical T-weld?

6. What is the direction of travel on a vertical T-weld?

SECTION II: WELDS IN HORIZONTAL AND VERTICAL POSITIONS, COMPREHENSIVE REVIEW

A. HORIZONTAL BUTT WELD

Equipment and Materials

Standard oxyacetylene welding equipment
2 pieces, 16-gauge mild steel, 2″ x 6″
1/16″ or 3/32″ mild steel filler rod

PROCEDURE

1. Tack weld two pieces of sheet metal for a butt weld.

PROCEDURE

2. Position the material for a horizontal weld. See drawing of horizontal butt weld.

3. Make a butt weld across the 6″ length of the two pieces.

4. Cool the weld and have it inspected by the instructor.

5. Test the weld according to the procedure given for testing a butt weld.

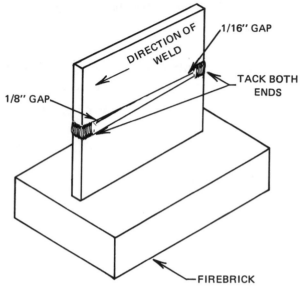

Setup for Horizontal Butt Weld

B.   VERTICAL LAP WELD

Equipment and Materials

Standard oxyacetylene welding equipment
2 samples, 16-gauge mild steel, 2″ x 6″
1/16″ or 3/32″ mild steel filler rod

PROCEDURE

1. Tack weld the samples at each end with one piece overlapping the other slightly as shown in the drawing of a vertical lap weld.

2. Position the material for a vertical weld.

3. Weld the joint from bottom to top.

4. Cool the weld and have it inspected by the instructor.

5. Test the weld according to the procedure given for testing a lap weld.

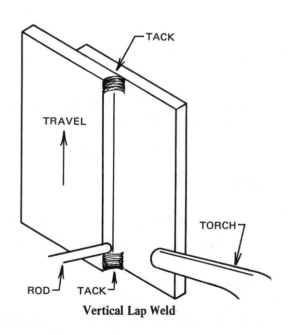

Vertical Lap Weld

SECTION 3:
Overhead Welds, Braze Welding, and Backhand Welding

Unit 14   Bead on Mild Steel Sheet Metal, Overhead Position

OBJECTIVES

The student will be able to:

- List the additional safety precautions necessary for overhead welding.
- Describe the overhead welding position.
- Run a bead in the overhead position with 100% penetration.

An *overhead weld* is one which is made on the underside of the joint and running in a horizontal line. Welds which are inclined 45° or less are considered to be in the overhead position, figure 14-1.

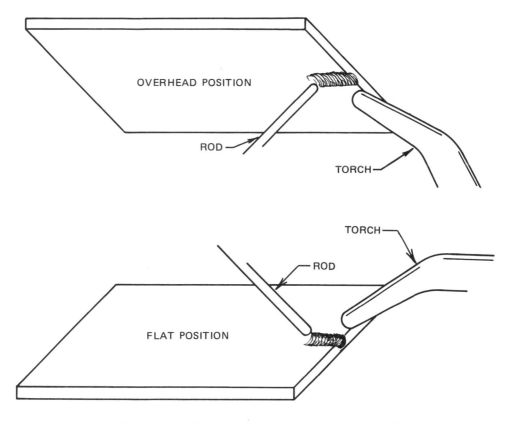

Fig. 14-1 Overhead Welding (Transition from Flat to Overhead Position)

CAUTION: There is a great danger of burns when welding in the overhead position. Sparks and hot metal fall on the welder frequently. Flame retardant clothing must always be worn.

Overhead welding is done with as much ease as flat welding. The awkward position of working with the arms above the head, with no support, makes it seem more difficult.

The weld should be made from right to left (for right-handed welders). The flame should be close to the metal and pointed slightly in the direction of travel. The filler rod should be added in small quantities.

The weight of the hoses may tire the welder. Draping the hoses over the shoulder will relieve this weight and the torch will handle more easily. A jig may be used to hold the material in position for overhead welding.

JOB 14: BEAD ON MILD STEEL SHEET METAL, OVERHEAD POSITION

Equipment and Materials
Standard oxyacetylene welding equipment
16-gauge mild steel sample, 2" x 6"
1/16" or 3/32" mild steel filler rod

PROCEDURE	KEY POINTS
1. Use a jig to clamp a 16-gage sheet metal sample in the overhead position.	
2. Weld completely across the 6" length with one bead, adding filler rod as necessary.	2. CAUTION: Overhead welding may be dangerous. Sparks and molten metal fall from the weld. Wear protective clothing and stay alert. The weight of the hoses may be carried on the welders shoulders. Use only enough rod to build up a slight crown.
3. Practice this until overhead welding is mastered.	

REVIEW QUESTIONS
1. Describe the burn hazards which may be encountered when welding in the overhead position.
2. Why does overhead welding seem more difficult than flat welding?
3. When overhead welding, the flame should be:
   a. Close and pointed in the direction of travel.
   b. Far away and pointed directly into the puddle.
   c. Close and pointed directly into the puddle.
   d. Far away and pointed in the direction of travel.

4. What percent of penetration is required when welding in the overhead position?

5. How much crown should an overhead weld have?

Unit 15 Butt Weld on Mild Steel Sheet Metal, Overhead Position

OBJECTIVES

The student will be able to:

- Handle the hoses in a manner that prevents the wrists from tiring.
- Manipulate the torch and filler rod in a manner that prevents holes from being melted through the parent metal.
- Make an overhead butt weld capable of withstanding a bend test.

An overhead butt weld, figure 15-1, is the same as a butt weld in the flat position, except that the weld is made on the underside of the parent metal. Overhead welds are made at an angle of not more than 45° from horizontal.

CAUTION: Sparks and hot metal may drop from the weld. Wear protective clothing.

The flame should be directed into the gap and the torch should be angled so the flame points in the direction of the weld. The filler rod should be added from the front of the puddle in small quantities. The bead should be small with about a 1/16" crown.

The following suggestions may help prevent holes from melting through the parent metal:

- Use the unmelted filler rod to cool the puddle.
- Flash the torch off the puddle momentarily.
- Use a smaller torch tip.

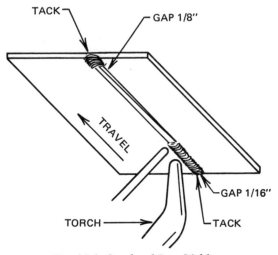

Fig. 15-1 Overhead Butt Weld

Using a smaller torch tip will also help control the amount of molten metal dripping from the weld. However, care must be used to insure that the tip is large enough to allow for 100% penetration.

JOB 15: BUTT WELD ON MILD STEEL SHEET METAL, OVERHEAD POSITION

Equipment and Materials

Standard oxyacetylene welding equipment
2 pieces, 16-gauge mild steel, 2" x 6"
1/16" or 3/32" mild steel filler rod

PROCEDURE	KEY POINTS
1. Prepare the material for a standard butt weld.	1. Gap the pieces the same as for a butt weld in the flat position. The tack welds should be solid.
2. Using any necessary jigs, clamp the material in the overhead position. The wider part of the gap should be on the left side (for right-handed welders).	
3. Weld the joint in one pass, adding filler rod as necessary.	3. Use only enough rod to build up a slight crown. The weight of the hoses may be carried on the welder's shoulders. Be sure penetration is 100%.
4. Test the weld as shown in figure 15-2.	

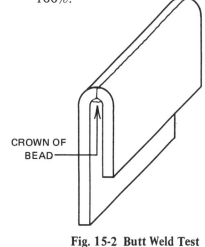

CROWN OF
BEAD

Fig. 15-2 Butt Weld Test

REVIEW QUESTIONS

1. Why is welding in the overhead position dangerous?

2. Where should the flame be directed when butt welding in the overhead position?

3. From what direction should the filler rod be added to the puddle?

4. How much crown should be built up on an overhead butt weld?

5. What can be done to control burn through when making an overhead butt weld?

6. How many passes should be used to complete an overhead butt weld on 16-gage sheet metal?

7. How can the weight of the hoses be relieved from the wrists when welding in the overhead position?

Unit 16   Lap Weld on Mild Steel Sheet Metal,Overhead Position

OBJECTIVES

The student will be able to:

- Use oxyacetylene welding equipment with a smooth rhythm.
- Manipulate the flame and filler rod correctly to make an overhead lap weld.
- Make an overhead lap weld capable of withstanding a bend test.

An overhead lap weld is made on the underside of the metal and runs in a horizontal direction, figure 16-1. This is a difficult weld to make. It requires control of the flame and the puddle, and the filler rod must be added in a smooth rhythm.

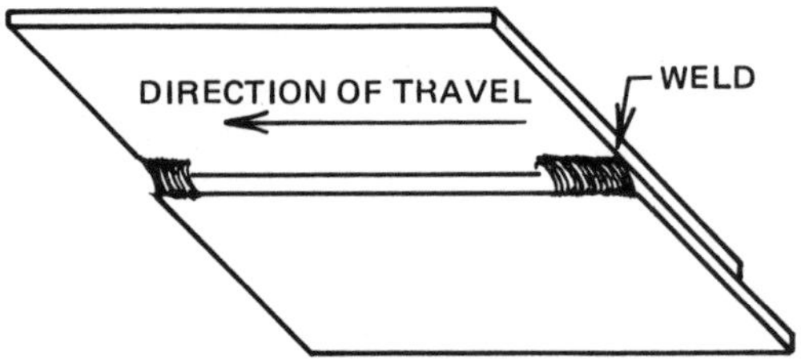

Fig. 16-1 Overhead Lap Weld
(Tack Both Ends)

Welding Rhythm

Welding rhythm is developed through practice of oxyacetylene welding. It is the ability to progress across a weld smoothly and uniformly. The puddle is melted and the filler rod is added in a rhythm which produces a smooth, rippled bead. Frequent pauses or the addition of rod in a ragged manner will spoil the appearance of a weld and also prevent uniform penetration of the joint.

The following is a list of hints for making a good overhead lap weld:

- Be careful to avoid burning away the edge of the lapped metal.
- Do not add filler rod too fast. Adding the filler rod too fast will cool the puddle and result in a bead with poor appearance. However, in order to withstand destructive testing, the bead must be full.
- The heat must be directed onto the flat surface of the second piece and the torch should always be pointed in the direction of travel.
- Keep the flame close to the metal.
- Drape the hoses over the shoulder to remove their weight from the wrist.

JOB 16: LAP WELD IN MILD STEEL SHEET METAL, OVERHEAD POSITION

Equipment and Materials

Standard oxyacetylene welding equipment
2 pieces, 16-gauge mild steel, 2″ x 6″
1/16″ or 3/32″ mild steel filler rod

PROCEDURE	KEY POINTS
1. Prepare the material for a standard lap weld.	1. Tack the samples at both ends.
2. Using any necessary jigs, clamp the material in the overhead position.	
3. Beginning at the right edge, weld the joint in one pass.	3. **CAUTION: Sparks and hot metal may drop from the weld. Wear protective clothing** Keep the flame pointed at the flat surface of the second plate, so the raw edge of the other plate melts slowly. Add only enough rod to fill the bead without cooling the puddle too much.
4. Test the weld as shown in figure 16-2.	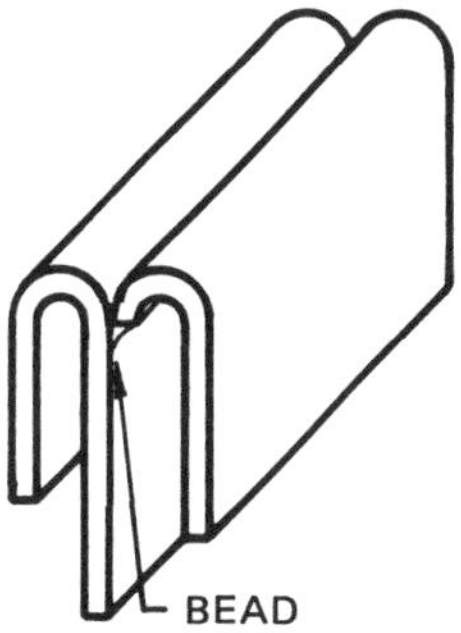

Fig. 16-2 Overhead Lap Weld Test

REVIEW QUESTIONS

1. What is meant by welding rhythm?
2. What might cause the appearance of an overhead lap weld to be poor?
3. Should an overhead lap weld have a full or shallow bead?
4. Where should the heat be directed when making an overhead lap weld?
5. How many passes should be used to complete an overload lap weld on 16-gage mild steel?
6. Describe the destructive test for an overhead lap weld.

Unit 17   Corner Welds on Mild Steel Sheet Metal

OBJECTIVES

The student will be able to:

- Weld a corner joint using no filler rod and obtaining 100% penetration.
- Down weld a joint with 100% penetration.
- Close small holes which may appear in a corner weld.
- Explain why down welding is not done on butt and lap joints.

Melt Welds

One of the most often used applications of oxyacetylene welding is the corner weld. Outside corners are frequently joined by *melt welding.* This refers to fusing two pieces without the use of filler rod. Smooth, strong beads, with excellent appearance can be made on sheet metal corners with little or no filler rod added.

When the pieces have been tacked, the flame is used to start a puddle. Then, with the flame pointing slightly in the direction of travel, the torch is moved across the joint, continuously melting the two edges, figure 17-1.

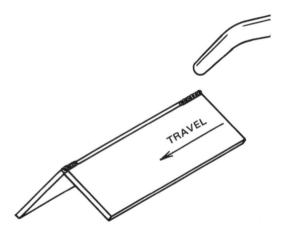

Fig. 17-1 Corner Weld-Flat Position

Care should be taken to keep the motion of the torch slow enough so that the melted metal can soak through the corner. This will insure 100% penetration.

If a hole develops in a weld, a drop of metal, melted from 1/16″ filler rod, may be added. This will close the hole, so the bead can be carried on across the joint.

Tip Size

Tip size is very important for corner welding. Usually, a small tip will give excellent results. Remember, however, that a puddle must be formed and the penetration must be 100%.

Position

Corner welds can be made in any of the four positions with little difficulty. In addition, corner welds are sometimes made in the vertical position from the top down. This is called *down welding.* Down welding is never done on any other type joint, since penetration is more difficult on the lap or butt weld, using this technique.

JOB 17: MELT WELDING CORNERS IN MILD STEEL SHEET METAL

Equipment and Materials

Standard oxyacetylene welding equipment
4 pieces, 16-gauge sheet metal, 2" x 6"
1/16" mild steel filler rod

PROCEDURE	KEY POINTS
1. Tack weld the sheet metal samples to make two assemblies, as shown in figure 17-2.	1. Tack the pieces together securely, using no filler rod. Melt the corners of the samples to form the tack welds.

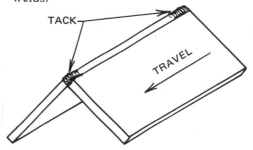

Fig. 17-2 Tack Welded Corner Joint

2. Weld one of the assemblies in the position shown in figure 17-2.	2. Using a neutral flame, start a puddle and work it slowly across the joint. Penetration should be 100%.
3. Down weld the other assembly, as shown in figure 17-3.	3. Weld from the top to the bottom of this joint.

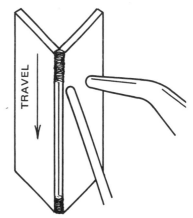

Fig. 17-3 Down Weld, Corner Joint

4. Do not use filler rod unless a hole appears which cannot be closed with parent metal.

PROCEDURE	KEY POINTS
5. Test both welds according to figure 17-4.	5. The welds should be smooth and uniform.

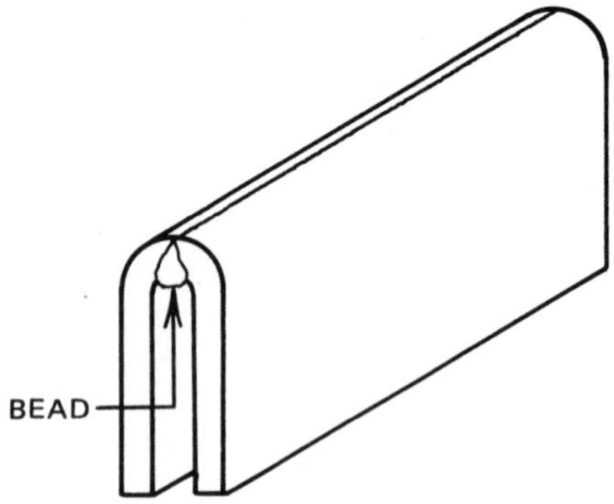

Fig. 17-4 Corner Joint Weld Test

REVIEW QUESTIONS

1. What is melt welding?

2. What is down welding?

3. For what reason is filler rod used in melt welding?

4. What type of flame is used to weld corner joints?

5. Why is down welding not recommended for welding butt and lap joints?

Unit 18   Braze Welding Sheet Metal

OBJECTIVES

The student will be able to:

- Describe the difference between brazing, braze welding, and fusion welding.
- Tell how and why flux is used in braze welding and brazing.
- Define capillary attraction.
- Make a bead with bronze filler metal.

Brazing

Brazing is a process in which a bond is produced between two pieces, which are in very close contact with each other, using a nonferrous filler metal. This filler metal has a lower melting point than that of the parent metal. As the filler metal melts it is drawn into the joint by *capillary attraction.*

Capillary Attraction

Capillary attraction is the property that allows a liquid to be drawn into a close fitting joint. This is different from the *cohesive* action that causes filler metal to adhere to a surface in braze welding. In order for this capillary attraction to take place it is necessary for pieces to be brazed to be in close contact with one another.

Braze Welding

In *braze welding* capillary attraction does not occur and the joint is designed the same as for fusion welding. The metal parts are joined by a molecular union, known as cohesion. Braze welding also uses a nonferrous filler metal, which melts at a lower temperature than the parent metal.

In both brazing and braze welding the parent metal is heated to a dull red heat. Then the filler rod is applied along with a suitable flux. Because the melting point of bronze is approximately 1600°F, it melts as it comes in contact with the hot parent metal.

Braze welding does not produce as strong a joint as fusion welding, because the bronze has less strength.   However it is a widely used process for many applications, such as joining cast iron.

Flux

Brazing flux, which has the appearance of melted glass, is usually made of a mixture of borax and boric acid. It is a cleaning agent which removes *mill scale* (an oxide coating, which forms on molten iron and steel) and other impurities from the surface of the metal. Flux can easily be cleaned from the weldment after it cools.

CAUTION: Bronze contains zinc, which gives off posionous fumes when it is heated. Inhaling these fumes causes stomach cramps, vomiting and shortness of breath. Do not breathe brazing fumes.

JOB 18: BRAZE WELDING SHEET METAL

Equipment and Materials

Standard oxyacetylene welding equipment
1 piece, 16-gauge mild steel 2″ x 6″
1/16″ or 3/32″ Bronze filler rod
Brazing flux

PROCEDURE	KEY POINTS
1. Grind the mill scale from a metal sample, so that the entire surface is shiny.	1. Wear goggles while grinding.
2. Place the sample on firebrick.	
3. Adjust the torch to a slightly oxidizing flame.	3. A slightly oxidizing flame is normally used for brazing.
4. Heat about 2″ of the end of a piece of bronze rod with the flame. Dip the heated rod into a can of flux, so that some of it sticks to the end of the rod.	4. Plenty of flux is the key to good brazing. CAUTION: Brazing fumes are poisonous. Avoid breathing them. Work in a well ventilated area and warn others, so that they can taken precautions.
5. Heat the right edge of the metal to a dull red, then add filler rod, melting it off in the heat of the torch, figure 18-1.	5. Keep plenty of flux on the filler rod, heating it and redipping it frequently.

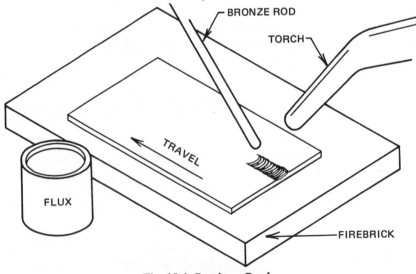

Fig. 18-1 Brazing a Bead
(Use Slightly Oxidizing Flame)

PROCEDURE	KEY POINTS
6. Keep the sample red hot and add bronze until the bead runs the full length of the plate.	6. Keep the flame small and point it in the direction of the weld. Do not melt the parent metal. Bronze will sputter and fume if overheated. Flash the torch off the bead to prevent overheating.
7. Practice until a bead can be applied with a smooth uniform appearance and with no evidence of burning.	7. The finished bead should have a shiny appearance and the metal should not look burned.

REVIEW QUESTIONS

1. Briefly describe what brazing is.
2. Why is flux used in brazing?
3. What is capillary attraction?
4. What is the difference between brazing and fusion welding?
5. How does the strength of a braze welded joint compare with that of a fusion welded joint?
6. Does bronze melt above or below the melting point of steel?
7. List the safety precautions which should be taken when braze welding.

Unit 19   Braze Welding Butt Joint on Mild Steel Sheet Metal

OBJECTIVES

The student will be able to:

- Gauge the proper heat for braze welding by using the parent metal color as an indicator.

- Properly clean and prepare metal for braze welding.

- Braze weld a butt joint capable of withstanding a bend test.

A braze welded joint is made in the same general fashion as a fusion welded joint. Braze welding requires special preparation which is not ordinarily used in the fusion process. The braze welded butt joint is common in many applications because it is faster than fusion welding and requires less training to complete.   However, it is not as strong as a fusion welded joint.

Bronze rod will not flow properly if the metal is not thoroughly cleaned before brazing. Grinding the edges and both sides of the metal near the edge where the braze is to be made cleans the parent metal so that the bronze will flow through the joint and onto the bottom side.

Do not overheat the joint.   Keep the color dull red and flash the torch away from the braze if it appears to be overheating. If the parent metal becomes molten, the bronze will burn into it and make the parent metal weak. Overheating the metal will also cause the bronze to drip through the joint and build up too much underbead.

> **CAUTION: The fumes given off by the brazing process are poisonous. No braze welding should be done unless proper ventilation is provided.**

JOB 19: BRAZE WELDED BUTT JOINT ON MILD STEEL SHEET METAL

Equipment and Materials

Standard oxyacetylene welding equipment
2 pieces, 16-gauge mild steel, 2″ x 6″
1/16″ or 3/32″ Bronze filler rod
Brazing flux

PROCEDURE	KEY POINTS
1. Clean the surfaces to be joined, by grinding.	1. Grind the edges to be butted and both the top and bottom surfaces.

PROCEDURE	KEY POINTS
2. Tack the two pieces in position for a butt joint, gapping the pieces uniformly about 1/16″, figure 19-1.	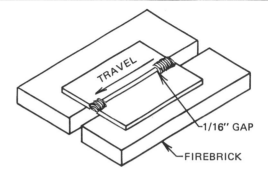 **Fig. 19-1 Braze Welded Butt Joint**
3. Apply flux to the rod and bring the parent metal to a red heat. Add rod to the metal, so that the bronze flows through the gap. Work the braze across the entire joint. 4. Test the braze in a vise, figure 19-2.	3. Do not overheat the metal. This is not a fusion weld; do not melt the parent metal. Build the bead up slightly above flush. 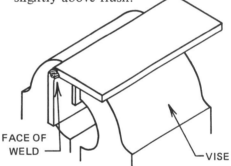 **Fig. 19-2 Braze Welded Butt Joint Test** **(Bend Against Face of Weld, Full 90°)**

REVIEW QUESTIONS

1. Why is braze welding faster than fusion welding?

2. What special preparation of the metal is required before braze welding a butt joint?

3. Why is a braze welded joint not as strong as a fusion welded joint?

4. How is the temperature of the parent metal determined for braze welding?

5. Why is good ventilation necessary when brazing?

Unit 20   Braze Welding Cast Iron

OBJECTIVES

The student will be able to:

- List the properties of cast iron that make it more difficult to weld than other ferrous metals.

- Prepare the surfaces of cast iron for braze welding.

- Braze weld cast iron assemblies capable of withstanding destructive testing.

Cast iron is an alloy, containing from 1.7% to 4.5% carbon and small amounts of silicon.   The addition of large amounts of carbon to iron make it brittle and hard. Cast iron is easily formed into shapes, but being brittle, it will not withstand bending and twisting.   Because brazing is done at much lower temperatures than fusion welding, it causes less expansion and contraction. For this reason cast iron is often braze welded.

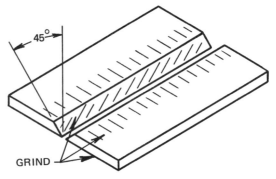

Fig. 20-1 V-Grooving and Grinding

V-Grinding

To present a wider surface for bonding and to assist in getting 100% penetration, the pieces are often *V-ground*. V-grinding refers to beveling the edges of the pieces to be joined, so that when they are put together the joint forms a V, figure 20-1.

The following tips will contribute to good, full-strength braze welded joints:

- Preheat the casting, so cracks do not appear as a result of expansion and contraction.

- Braze welding is not fusion welding. Do not melt the parent metal.

- 100% penetration of the bronze through the joint is required for full strength.

JOB 20: BRAZE WELDING CAST IRON

Equipment and Materials

Standard oxyacetylene welding equipment
2 pieces broken cast iron
3/32″ or 1/8″ Bronze filler rod
Brazing flux

PROCEDURE	KEY POINTS
1. Grind the edges of the two pieces of cast iron to a V of about 45°.	1. Grind the surfaces back from the edges of the V, so that the bead will be wider than the V.
2. Use a large tip, depending on the thickness of the cast iron.	2. Use a slightly oxidizing flame. If the puddle bubbles, there is too much acetylene in the flame.
3. Tack each end of the cast iron pieces so a small gap appears at the bottom of the V.	
4. Heat the area near the joint to a dull red.	4. Play the flame over the pieces. Be sure the entire casting is preheated.
5. Heat the bronze rod and dip it into the flux, then transfer the flux to the heated V.	5. If the joint is long, scatter flux along the bottom of the V by hand. The metal should be hot enough so the flux melts and is not blown off the joint by the torch.
6. While the parent metal is red-hot, melt a small amount of bronze into the V and allow it to flow over the surface of the clean metal.	6. If the bronze does not flow smoothly, the metal may not be hot enough. Applying a thin coat of metal in this manner is called *tinning*.
7. Continue to tin the area ahead of the bead, then back over the area, filling the joint with bronze.	
8. Do not work too rapidly. Cast iron must be at the right temperature to make the bronze bond with the cast iron.	8. Overheating is indicated by a white powder deposit and smoking. **CAUTION: Do not breathe zinc fumes.**
9. Cool the work slowly to avoid cracks.	
10. The bronze must completely penetrate the joint, forming a small bead on the bottom.	
11. Test the cooled assembly by breaking over the bead.	11. A good braze weld should break the parent metal outside the joint.

REVIEW QUESTIONS

1. What property of cast iron makes it more difficult to weld than other ferrous metals?

2. What characteristic of the braze welding process makes it better than fusion welding for joining cast iron?

3. Give two reasons why cast iron joints are often V-ground.

4. What is indicated if a white powder appears or if the braze smokes?

5. What is indicated if the bronze does not flow smoothly?

Unit 21   Backhand Welding on Mild Steel Sheet Metal

OBJECTIVES

The student will be able to:

- Explain how stresses develop in metal during welding.

- Describe backhand welding and torch angle.

- Discuss the use of backhand welding for stress relief.

- Weld a bead using the backhand method.

Backhand welding is frequently used for welding heavy plate and pipe joints. In backhand welding the flame is directed back against the completed weld. The filler rod is added into the flame at the head of the weld and the bead forms behind the filler rod. The direction of the weld is from left to right for right-handed welders (right to left for left-handed welders), figure 21-1.

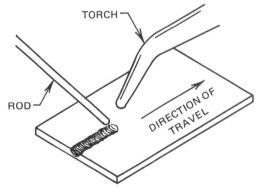

Fig. 21-1 Backhand Welding

Stress Relief

Heating and cooling can cause stresses to develop within the metal. Because the metal next to the weld cools at a different rate than the metal away from the weld, the expansion and contraction is not uniform. The resulting stresses can cause cracks to appear as the hotter metal is pulled away by the cooling process.

These stresses may be relieved by reheating the joint to a red heat and allowing it to cool slowly. In backhand welding the flame is directed onto the newly formed weld. This allows the weld to cool more slowly and stress relieves the joint.

JOB 21: BACKHAND WELDING ON MILD STEEL SHEET METAL

Equipment and Materials

Standard oxyacetylene welding equipment
16-gauge sheet metal sample, 2" x 6"
1/16" or 3/32" Mild steel filler rod

PROCEDURE	KEY POINTS
1. Start a molten puddle at the left end of the sample. (Left-handed welders start at the right end of the sample.)	

PROCEDURE	KEY POINTS
2. Add filler rod and continue moving the bead to the right.	2. Hold the torch at an angle of 30° to 45°, so the flame is directed back over the work which has been welded.
3. Run the bead across the entire 6″ length of the sample.	3. Backhand welding requires practice.

REVIEW QUESTIONS

1. Describe the difference between forehand welding and backhand welding.

2. What causes stresses to develop in welding?

3. How does backhand welding help relieve stress?

4. Name two types of welding that are commonly done by the backhand method.

Unit 22   Backhand Butt Weld on Sheet Metal Mild Steel

OBJECTIVES

The student will be able to:

- List the advantages of backhand welding over forehand welding.

- Control penetration of a backhand weld by torch and filler rod manipulation.

- Make a backhand butt weld capable of withstanding bend testing.

Material for backhand butt welding is prepared in the same fashion as that for forehand welding. Two sheet metal samples are tacked together at each end, with approximately a 1/16" gap at one end and 1/8" gap at the other end. However, welding is done from left to right, so the sample should be set up for welding with the 1/16" gap at the left-hand side and welding should progress toward the 1/8" gap. The torch is pointed so that the flame points over the weld as it is being completed, figure 22-1.

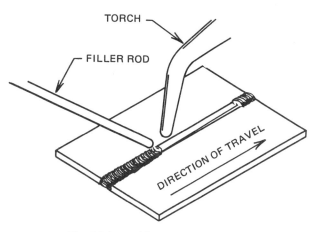

Fig. 22-1 Backhand Butt Weld

The filler rod should be added in front of the flame, and flashing off the weld may be necessary to control the amount of penetration. However, penetration of this weld, as of all oxyacetylene welds, must be 100%.

JOB 22: BACKHAND BUTT WELD ON MILD STEEL SHEET METAL

Equipment and Materials

Standard oxyacetylene welding equipment
2 pieces, 16-gauge mild steel, 2" x 6"
1/16" or 3/32" Mild steel filler rod

PROCEDURE	KEY POINTS
1. Tack weld the two samples of mild steel together for a butt weld.	1. Gap the pieces 1/16" at the left side and 1/8" at the right.

PROCEDURE	KEY POINTS
2. Beginning on the left side of the plates, weld them together using the backhand method.	2. Concentrate on smooth bead appearance and 100% penetration.
3. Test the weld by bending the plates 180° over the crown of the weld.	3. Penetration should be complete, with no cracks or holes.

REVIEW QUESTIONS

1. From what direction is the filler rod added in backhand welding?
2. Compare the appearance of a backhand bead with that of a forehand bead.
3. How much penetration is required in backhand welds?
4. What are the advantages of backhand welding over forehand welding?

SECTION 3: OVERHEAD WELDS, BRAZING AND BACKHAND WELDS, COMPREHENSIVE REVIEW

A. OVERHEAD BUTT WELD

Equipment and Materials

Standard oxyacetylene welding equipment
2 pieces, 16-gauge mild steel, 2" x 6"
1/16" or 3/32" Mild steel filler rod

PROCEDURE

1. Prepare the material for a standard butt weld. Leave a gap for contraction.
2. Clamp the material in an overhead position.
3. Weld the joint with one pass.
4. Test the weld by bending 180° over the bead.
5. Have the instructor inspect the weld.

B. BRAZE WELDED BUTT JOINT IN MILD STEEL

Equipment and Materials

Standard oxyacetylene welding equipment
2 pieces, 16-gauge mild steel, 2" x 6"
1/16" or 3/32" Bronze filler rod
Brazing flux

PROCEDURE

1. Grind the edges to be joined and both faces near the edge.
2. Tack the two pieces in position for a butt joint.

3. Heat the filler rod and apply the flux to it.

4. Heat the parent metal to a red heat, then braze the joint.

5. Test the joint in a vise.

6. Have the instructor inspect the joint.

C. BACKHAND BUTT WELD

Equipment and Materials

Standard oxyacetylene welding equipment
2 pieces, 16-gauge mild steel, 2″ x 6″
1/16″ or 3/32″ Mild steel filler rod

PROCEDURE

1. Prepare the material for a butt weld.

2. Position the material for backhand welding in the flat position.

3. Weld the joint backhand with one pass.

4. Test the weld by bending 180° over the bead.

5. Have the instructor inspect the weld.

SECTION: 4
Oxyacetylene Cutting

Unit 23   Straight Cutting with the Oxyacetylene Cutting Torch

OBJECTIVES

The student will be able to:

- Explain how an oxyacetylene flame cuts ferrous metal.
- Assemble an oxyacetylene cutting outfit.
- Make clean, smooth cuts in mild steel plate.

Oxyacetylene cutting is done by directing a stream of oxygen onto the ferrous metal, which has been preheated. The oxygen burns the metal. By controlling the amount of preheat and the size of the stream of oxygen, a cut may be made with clean, smooth sides. The cut is called a *kerf*

Oxyacetylene cutting is one of the most used oxyacetylene processes. The cutting torch can be used to cut intricate shapes or to make straight, clean cuts. The cutting, or burning, process does not change the chemical composition of the metal. Therefore, any ferrous metal can be welded immediately after it has been cut. However, *slag* (oxidized metal) is sometimes left at the bottom edge of the cut. This must be removed by grinding or chiseling. If oxidized metal is included in the puddle it will contaminate the weld. Metal should always be clean before welding.

Cutting Torch

The cutting torch, figure 23-1, is designed only for cutting ferrous metals. To install the cutting torch the welding torch handle must be removed from the hoses. The cutting torch is then installed in its place. The cutting torch is designed for heavy cutting and performs better over long periods of time than the cutting head.

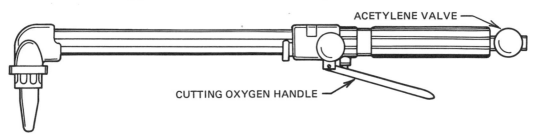

Fig. 23-1 Cutting Torch

Fig. 23-2 Cutting Head   **Fig. 23-3 Cutting Tip**   **Fig. 23-4 Four-Hole Cutting Tip, Installed for Straight-Line Cut. Make Cut Right to Left.**

Cutting Head

The cutting head, figure 23-2, is an attachment to the welding torch. By removing the tip from the welding torch the cutting head may be screwed onto the torch handle. In this manner a welding torch may be used for cutting.

Oxygen Pressure

Since large amounts of oxygen are required to burn the metal, more oxygen pressure is needed for cutting than for welding. When using a cutting torch, or cutting head, the gas pressures should be regulated according to the manufacturer's specfications for the torch being used.

Cutting Tip

The cutting tip, figure 23-3, is designed especially for cutting and cannot be used for welding. Cutting tips are made with a hole in the center, through which the stream of oxygen is directed at the cut. A group of holes around the center hole give off a neutral flame which preheats the metal, figure 23-4. Depending on the size of the tip, there are 4, 6, 8, or 12 preheat holes. Each of these is like a miniature welding tip and when the torch is lighted, it should be adjusted so that each of the preheat holes makes a neutral flame.

When the metal to be cut has been preheated to red-hot, the cutting oxygen valve is pressed. The stream of oxygen will burn (cut) the metal as long as the preheat is maintained. Oxyacetylene cutting must be done at a slow, even rate of speed. If the cut is made too rapidly, the metal may cool down and the cutting action will stop. If this happens the torch should be moved back into the kerf and the metal preheated again. The cut may then be started again.

JOB 23: OXYACETYLENE CUTTING

Equipment and Materials

Oxyacetylene outfit with cutting head
Straightedge
Hammer
Center punch
Soapstone or chalk
Scrap pieces of 1/2-inch mild steel plate

PROCEDURE	KEY POINTS
1. Install the cutting head on the torch handle.	1. The procedure for using a cutting torch is the same as for a cutting head.
2. Install the cutting tip in the position shown in figure 23-4.	
3. Use the acetylene and oxygen pressures recommended by the manufacturer of the equipment.	3. The high pressure of the cutting oxygen helps blow the kerf clear.
4. Light the cutting torch in the same manner as the welding torch and adjust it for a neutral flame.	4. Wear gloves and goggles. All preheat flames should be neutral.
5. Depress the cutting oxygen valve and with the valve wide open, adjust the preheat flames to neutral.	5. With the cutting oxygen valve wide open the working pressure should be adjusted to the manufacturer's specifications.
6. Shut off the torch.	
7. Mark a line 1/2 inch from the end of the steel plate, using the straightedge and soapstone.	7. Center punch the line. Soapstone or chalk will burn off, but center punch marks can be followed after the soapstone is gone.
8. Position the plate on a suitable table so that the marked end overhangs the edge of the table.	
9. Relight the cutting torch.	
10. Hold the flame about 1/8" above the edge of the plate.	10. Position the torch so the tip is at the beginning of the line, but do not let it touch the plate.
11. Preheat the metal to a red heat.	11. Brace the torch by using the bench behind the plate for support.
12. Depress the oxygen lever and, as the metal burns and is blown through the kerf, move across the plate. Follow the marked line until the cut is completed.	12. **CAUTION: Stand clear of the material; when the cut is completed the material will fall off. Falling sparks can ignite clothing. Do not cut or direct the stream of oxidized material toward inflammable objects or toward oxygen and acetylene containers.**
13. Practice cutting pieces from the plate until a smooth, even, straight cut is achieved.	

Summary: Job 23

- The preheat flames must be kept neutral.

- The torch must be held at an angle of 90° with the cut so that a straight edge is made.

- Keep the oxygen trigger fully depressed so that the kerf blows clean.

- The tip must be kept clean. Use tip cleaners if the holes become plugged.

- Be sure the tip is installed for straight cutting.

- The metal must be kept red-hot, or the cutting action will stop.

REVIEW QUESTIONS

1. How does the oxyacetylene flame cut metal?

2. How is a cutting head installed on an oxyacetylene welding outfit?

3. What is the result when a cut is made too fast and the preheat is lost?

4. What is a kerf?

5. What is slag?

6. Make a sketch of the end of the cutting nozzle, showing how it is installed for a straight cut.

7. Is the acetylene pressure increased when the cutting torch is used?

8. Why should a line be center-punched for cutting with the cutting torch?

9. List the safety measures which should be taken when cutting with the torch?

Unit 24   Beveling Plate with the Oxyacetylene Cutting Torch

OBJECTIVES

The student will be able to:

- Describe the position of the cutting torch tip for cutting a bevel.

- Define a land and tell why it is used.

- Cut a straight, smooth bevel with the oxyacetylene torch.

Ordinarily, 100% penetration is required of an oxyacetylene weld in order to insure strength in the welded joint. Metal which is over 1/8 inch thick is very difficult to melt through, so some method has to be provided to insure complete penetration. Metal 1/8 inch to 3/16 inch thick is frequently gapped for welding, but the edge of metal over 3/16 inch thick should be *beveled.* This is done by cutting the edge of the metal on an angle.

Cutting straight through a piece of steel leaves a cross section the same width as the thickness of the original metal. However, when the edge is beveled, the cross section is increased, figure 24-1.

Beveling leaves the bottom of the plate with a very thin edge, which has a tendency to melt off during welding. This edge is generally ground square to a thickness of 1/16 inch, to prevent the edge from melting off. This ground shoulder is called a *land.*

Sometimes when bevels are made, a small increase of oxygen pressure is necessary to cut the larger cross section of the bevel. To cut a bevel, the cutting tip should be turned so the holes line up as shown in figure 24-2.

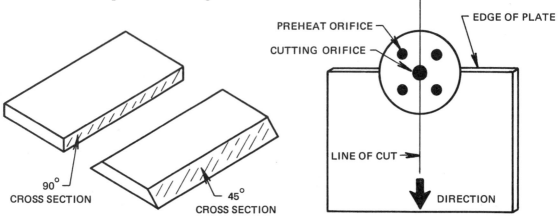

Fig. 24-1 Cross Section of Straight and Beveled Edges

Fig. 24-2 Cutting Tip Installed for Bevel Cut

JOB 24: BEVELING PLATE WITH THE OXYACETYLENE CUTTING TORCH

Equipment and Materials:

Standard oxyacetylene outfit with cutting head
Straightedge
Hammer
Center punch
Soapstone
Scrap pieces of 1/2″ thick mild steel plate

PROCEDURE	KEY POINTS
1. Mark a straight line with soapstone and a center punch.	
2. Preheat the edge of the plate and cut along the mark, with the torch held at a 45° angle with the plate.	2. Be sure the cutting tip is installed on the torch correctly for a bevel cut. Keep the torch the same distance from the plate at all times. Move slowly and steadily, so the preheat is not lost. Visibility will be best if the torch is drawn toward the operator.
3. Practice cutting a bevel until a smooth, regular bevel is achieved.	3. An angle iron guide may be used to maintain a straight, even bevel, figure 24-3.
	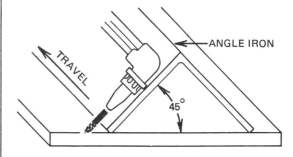
	Fig. 24-3 Method of Using Angle Iron to Cut Bevel

REVIEW QUESTIONS

1. Why is metal over 3/16 inch thick beveled to prepare it for a butt weld?
2. How does beveling affect the cross section of a piece of metal?
3. What is the land of a beveled plate?
4. If a piece of angle iron is used as a guide for cutting a bevel what will the angle of the bevel be?
5. Draw a sketch showing the placement of the preheat holes in a cutting tip as it is used for beveling metal.

Unit 25 Cutting Holes with the Oxyacetylene Cutting Torch

OBJECTIVES

The student will be able to:

- Pierce steel plate with the oxyacetylene cutting torch.
- Cut round holes within 1/16 inch of the right diameter.
- Holes cut will have clean, straight sides and be free of slag.

The oxyacetylene cutting torch is a good tool for cutting holes in steel, where a precision fit is not necessary. Round, square, rectangular, and odd-shaped holes can be cut equally well.

As the cutting oxygen valve is opened, after the metal has been preheated to a red heat, the torch tip must be

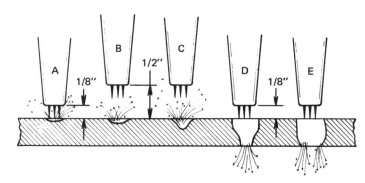

Fig. 25-1 Torch Handling Sequence for Hole Piercing

moved upward away from the cut. This is to keep the slag from blowing up into the tip. When the hole is burned completely through the plate, lower the torch until the preheat flames are about 1/8 inch from the surface of the metal, figure 25-1. With the hole pierced and the torch in position to keep the metal preheated, cut away from the hole to the mark and around the circle. Clean the slag from the underside when the cut is completed.

JOB 25: CUTTING HOLES WITH THE OXYACETYLENE CUTTING TORCH

Equipment and Materials

    Standard oxyacetylene cutting outfit
    Soapstone
    Center punch
    Hammer
    Scraps of 1/4-inch or 1/2-inch mild steel plate

PROCEDURE	KEY POINTS
1. Mark 1-inch and 2-inch circles on the plate, using soapstone.	1. Center punch the soapstone lines.
2. With a neutral flame, preheat the inside of the circle. When the metal is red-hot, depress the oxygen lever.	

PROCEDURE	KEY POINTS
3. When the cut starts, raise the tip up about 1/2″.	3. This prevents the oxide from blowing back into the tip.
4. When the hole has been burned through, lower the torch until the preheat flame is about 1/8″ from the surface of the metal.	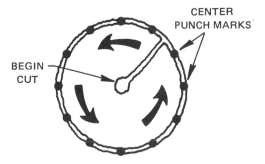
5. Cut outward to the punched line and follow the mark around the circle, figure 25-2.	
6. Practice this until a smooth, slag-free cut is achieved.	**Fig. 25-2 Cut from Center Outward to Rim**

REVIEW QUESTIONS

1. What type of preheat flame is used for oxyacetylene cutting?

2. What is done to prevent the slag from blowing up into the tip as a hole is started?

3. When cutting a hole in the center of a plate, where is the cut started?

4. When making a cut, how far should the preheat flames be from the surface of the metal?

5. How can the operator tell when the metal is preheated enough to begin a cut?

SECTION 4: OXYACETYLENE CUTTING, COMPREHENSIVE REVIEW

A. OXYACETYLENE FLAME CUTTING

Equipment and Materials

Standard oxyacetylene cutting outfit
Soapstone
Center punch
Hammer
1/4-inch mild steel plate

PROCEDURE

1. Mark all cuts with soapstone and center punch the marks. See oxyacetylene cutting evaluation drawing.

2. Complete all cuts slowly and carefully.

3. Remove all slag.

4. If the cuts are not acceptable, practice cutting on scrap until enough skill is developed to complete this evaluation.

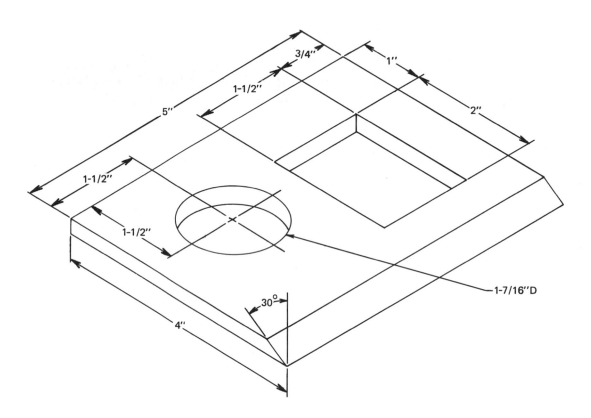

SECTION 5:
Welding Steel Plate and Pipe

Unit 26   Welding Mild Steel Plate

OBJECTIVES

The student will be able to:

- Explain the disadvantage of using oxyacetylene for welding heavy plate.

- Prepare steel plate for a butt weld.

- Make a butt weld having good appearance and penetration in steel plate.

Oxyacetylene welding is not generally used to join plates over 3/16 inch thick. It is, in fact, most often used for metals less than 1/8 inch thick. However, the method described in this unit may be used for any thickness of metal.

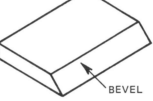

Fig. 26-1 Bevel 30° to 45°

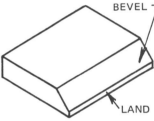

Fig. 26-2 Land Ground
on Bevel

Plate (metal over 1/8 inch thick) welding requires concentration of the heat. For this reason, most plate welding is done by the electric arc process.

Preparation

It is impossible to achieve 100% penetration of thick metal without leaving a gap between the pieces, or grinding the edges. For this reason, plate preparation is as important as the weld itself. Steel plate is beveled, to insure 100% penetration. This is done by grinding the edges at an angle, figure 26-1.

Beveling leaves a very thin edge, which can melt off during welding and allow excessive penetration. To prevent this edge from melting off, a land, or square shoulder, is ground on the bevel, figure 26-2.

JOB 26A: WELDING MILD STEEL PLATE

Equipment and Materials

Standard oxyacetylene welding equipment
2 Mild steel plates, 3/16" x 4" x 6"
1/8-inch mild steel filler rod

PROCEDURE	KEY POINTS
1. Grind a 45° bevel on one edge of each plate.	1. A bevel is necessary for 100% penetration.
2. Place the material on a firebrick so that the sides are separated about 1/8″ at the left end and 1/16″ at the right end, figure 26-3. (Left-handed welders reverse this position.)	2. Gapping the plates allows for expansion and contraction caused by the welding heat.

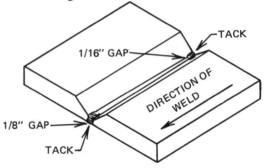

Fig. 26-3 Setup for Butt Weld

PROCEDURE	KEY POINTS
3. Tack the plates securely at each end.	3. Tack welds should have 100% penetration and good fusion.
4. Place the heat over the right end of the joint until the metal is molten. Hold the end of the filler rod in the cone of the flame to heat it for welding.	
5. When the molten metal penetrates the gap entirely, add the filler rod, moving the torch in a slow arc.	5. Insure 100% penetration at all times, but do not overheat the metal enough to cause excessive drop-through. It is important that fusion is good at the bottom edge of the plates.
6. Continue in this manner until the weld is completed.	6. Stir the molten metal by working the torch in a circular motion, keeping the metal rolling. Thoroughly mix the metal from both plates to insure good fusion.
7. Allow the sample to cool slowly.	

JOB 26B: TESTING A BUTT WELD IN MILD STEEL PLATE

Equipment and Material

Oxyacetylene cutting equipment or metal saw
Straightedge
Soapstone
Center punch
Hammer

PROCEDURE	KEY POINTS
1. With the soapstone and straightedge, mark three one-inch strips the length of the plate, figure 26-4.	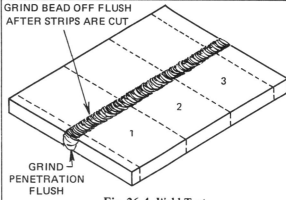 GRIND BEAD OFF FLUSH AFTER STRIPS ARE CUT GRIND PENETRATION FLUSH **Fig. 26-4 Weld Test**
2. Carefully cut the strips from the plate with the cutting torch or the saw.	2. If the cutting torch is used, be careful not to leave deep gouges in the edge of the strips. Discard the outside strips.
3. Grind the slag and crown of the bead off from the plate, so that it is the original thickness for its entire length.	3. All grinding marks should be parallel to the length of the plate. Gouges or grinding marks across the plate will cause it to fail at that point.
4. Place the center strip in a tensile tester and pull until the sample breaks.	4. The weld should hold more than the parent metal.
5. Bend one of the outside strips against the penetration and the other with the penetration.	5. Both samples should bend 180° without failure.

REVIEW QUESTIONS

1. What is the disadvantage of oxyacetylene for welding heavy steel plate?

2. Why must the edges of plate be specially prepared for welding?

3. What is the land of a beveled plate?

4. What is plate?

5. What is a bevel?

6. What type of flame is used for plate welding?

7. Why should a circular motion of the torch be used to weld plate?

8. How much penetration is required for a good plate weld?

Unit 27 Butt Welding Pipe in the Vertical Fixed Position

OBJECTIVES

The student will be able to:

- Describe the vertical fixed position for pipe welding.
- Properly prepare pipe for welding in the vertical fixed position.
- Butt weld pipe in the vertical fixed position, making a joint capable of withstanding a 180° bend.

The oxyacetylene welder should be able to weld mild steel pipe of 2 inch or less diameter. Most welding on pipe which is over 2 inch diameter is done with electric arc welding, because of speed of production. However, a great amount of 1-, 1-1/2 and 2-inch pipe is welded with the oxyacetylene torch.

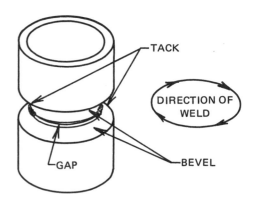

It is sometimes possible to make pipe butt joints where the pipe can be rolled to keep the weld on the topside and easily available. However, many times the pipe cannot be moved, and the welder must work around the pipe, changing the angle of the torch to fit the pipe circumference. This is called *fixed position welding*. When a pipe is within 10 degrees of vertical and in the fixed position, it is in the vertical fixed position, figure 27-1.

Fig. 27-1 Butt Weld, Pipe in the Vertical Fixed Position

The pipe ends should be grooved to insure 100% penetration of the butt joint. Also, the pieces must be solidly tacked together, to hold them in a straight line. With the pipe in the vertical position, the weld will be horizontal.

JOB 27: BUTT WELDING PIPE IN THE VERTICAL FIXED POSITION

Equipment and Materials

Standard oxyacetylene welding equipment
Jig for holding pipe sample
Oxyacetylene cutting torch or metal saw.
1-inch pipe samples, 8″ long
1/8-inch mild steel filler rod

PROCEDURE	KEY POINTS
1. Grind the ends of the two pipe samples to a 30° bevel.	

PROCEDURE	KEY POINTS
2. Tack weld the beveled ends together.	2. Keep pipe samples in a straight line when tacking. Tacks should have 100% penetration and good fusion.
3. Position the pipe in the vertical fixed position.	3. Welding pipe in the vertical fixed position is the same as making a horizontal butt weld, except that the weld travels around the pipe, instead of across the plate.
4. Weld around the pipe, walking around it.	4. Keep a keyhole at all times. Overlap the beginning and end of the weld.
5. Cut three 1/2-inch strips from the welded pipe and bend test them in a vise, figure 27-2.	

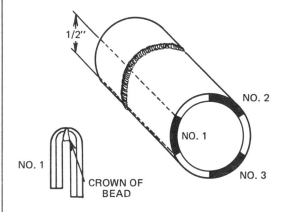

Fig. 27-2 Pipe Weld Test

6. If the strips break in the bend test, examine the weld and determine the reason for failure. **CAUTION: Be careful with the lighted torch when walking around the joint. Be especially careful of other people in the area when flashing off the weld.**	6. Incomplete penetration or too much heat may cause failure. Pipe joints with incomplete penetration may leak under pressure.

REVIEW QUESTIONS

1. Why are the ends of pipe grooved in preparation for welding?

2. In the vertical fixed position does the weld run in a vertical or horizontal direction?

3. Why are lengths of pipe tacked together before they are welded?

4. Other than making a weak joint, what might be the result of poor penetration in a pipe weld?

5. What sizes of pipe are most often welded with the oxyacetylene torch?

Unit 28   Butt Welding Pipe in the Horizontal Fixed Position

OBJECTIVES

The student will be able to:

- Describe the horizontal fixed position for pipe welding.
- Overlap beads on a pipe weld with complete fusion.
- Weld pipe in the horizontal fixed position, making a joint capable of withstanding a 180° bend.

When pipe is welded in the horizontal position it is often impossible to roll the pipe for welding. The welder must make an acceptable joint with the pipe in the horizontal fixed position. This is the position when the pipe is within 30° of horizontal and cannot be rolled.

To achieve complete penetration, begin the weld at the bottom of the joint and weld vertically around the side to the center of the top. Then begin at the bottom and weld up the other side to complete the weld, figure 28-1. Keep a keyhole in front of the weld for penetration.

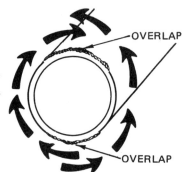

Fig. 28-1 Welding Pipe in the Horizontal Fixed Position

Whenever a weld overlaps another, great care must be taken to thoroughly melt the beads together for good fusion. Incomplete fusion (cold laps) will cause weld failure or leakage.

JOB 28: BUTT WELDING PIPE IN THE HORIZONTAL FIXED POSITION

Equipment and Materials

Standard oxyacetylene welding equipment
Jig for holding pipe sample
Oxyacetylene cutting torch or metal saw
1-inch pipe samples, 8″ long
1/8-inch mild steel filler rod

PROCEDURE	KEY POINTS
1. Bevel one end of the two pipes.	
2. Tack the two beveled sections together.	2. Make sure samples are in a straight line.
3. Position the sample in the horizontal fixed position, figure 28-2.	

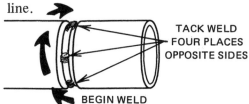

Fig. 28-2 Horizontal Fixed Pipe Weld

PROCEDURE	KEY POINTS
4. Start the weld at the bottom of the groove, on the under side.	4. Most of the weld is vertical. Make sure the overlaps are completely fused. Keep a keyhole for 100% penetration.
5. Cut three 1/2 inch wide strips from the pipe and bend test them in the vise.	
6. If the strips break in the bend test, examine them and determine the cause of failure.	

REVIEW QUESTIONS

1. What is meant by beveling?

2. What is meant by fixed position?

3. Is a horizontal pipe butt weld made from top to bottom, or bottom to top?

4. What precaution must be taken when welding an overlap on a bead?

5. How much penetration is required for pipe butt joints?

6. Why is a keyhole kept open on a butt weld?

SECTION 5: WELDING MILD STEEL PLATE AND PIPE, COMPREHENSIVE REVIEW

A.   WELDING MILD STEEL PLATE

Equipment and Materials

Standard oxyacetylene welding equipment
Two samples, mild steel plate 1/4″ x 4″ x 6″
1/8-inch mild steel filler rod

PROCEDURE

1. Bevel one edge of each plate.

2. Tack weld the plates for a butt joint.

3. Play the heat on the end of the joint until a puddle is established.

4. Add filler rod and move the torch in a slow arc, being sure to achieve 100% penetration.

5. Continue welding in this manner until a slight crown is built up.

6. When the weld has cooled, grind both sides flush with the surface of the parent metal.

7. Cut three test strips from the weldment.

8. Perform a tensile test on one strip and bend the other two in opposite directions.

B. WELDING 2-INCH PIPE

Equipment and Materials

Standard oxyacetylene welding equipment
Jig for holding samples
Oxyacetylene cutting torch or metal saw
Four samples of 2-inch pipe, 8 inches long
1/8-inch mild steel filler rod

PROCEDURE

1. Bevel one end of each piece of pipe, and tack weld for two assemblies.

2. Position one sample for a vertical fixed pipe weld and the other for a horizontal fixed pipe weld.

3. Weld both samples.

4. Cut three strips from each sample.

5. Bend test the strips and have them checked by the instructor.

Glossary

Backhand Welding: A method of welding in which the torch is directed in the opposite direction from that of the welding progress.

Bead: The line of metal deposited by welding.

Bevel: To grind the edges of a plate at an angle for welding.

Brazing: A process using a filler metal which melts at a lower temperature than the parent metal, but above 800 degrees Fahrenheit. Brazing relies on capillary attraction instead of fusion.

Bridging: A condition resulting when the parent metal is not completely fused into the corner of a fillet weld, and the bead has been built across the gap.

Butt Weld: A weld joining the edges of two pieces of metal which are in line with one another.

Calcium Carbide: A gray, rock-like substance obtained by smelting coke and lime in an electric furnace. It produces acetylene gas when mixed with water.

Carburizing Flame: An oxyacetylene flame containing an excess of acetylene. A carburizing flame is characterized by an acetylene feather.

Cold Lap: Incomplete fusion between the filler rod and the parent metal.

Crown: The buildup of the bead above the thickness of the parent metal.

Cutting Head: An attachment to be used on a welding torch. The cutting head has a cutting oxygen valve and holds cutting tips, so that it can be used as a cutting torch.

Cutting Tip: A special tip to be used on a cutting torch or a cutting head, for cutting ferrous metals. The cutting tip has a center, cutting oxygen hole, and several preheat holes.

Cutting Torch: A specially designed torch to be used for flame cutting ferrous metals.

Cylinder Valve: A valve installed in the top of acetylene and oxygen cylinders. The cylinder valve is used to turn on and off the flow of gas from the cylinder.

Ductile Strength: The ability of a metal to withstand breakage.

Filler Rod: Also called welding rod. It is generally of the same metallic composition as the parent metal.

Fillet Weld: An inside corner weld made where two pieces of metal join at any angle. When one plate forms a 90 degree angle with the other, it is often called a T-weld.

Flash Off: Momentarily removing the flame from a molten puddle when the melting process becomes too rapid.

Flat Position: The position of a weld made on the topside of the parent metal and within 30 degrees of horizontal.

Flux: A cleaning agent which removes impurities from the surface of metal for brazing.

Forehand Welding: A method of welding in which the torch is directed in the direction that the weld progresses.

Fusion Welding: The complete mixing of two or more pieces of metal, melted to a liquid state by the oxyacetylene torch, and allowed to cool.

Goggles: Eye protection devices made from spark-resistant material, equipped with shaded filter lenses.

Horizontal Position: A weld made within 45 degrees of horizontal and against a vertical surface.

Jig: Any device used for holding material while work is being done. (A holding fixture).

Kerf: The cut made by any tool, such as a saw or cutting torch.

Keyhole: A small hole remaining open ahead of a butt weld to insure penetration.

Land: A small square edge ground at 90 degrees with the bottom edge of a bevel. It is also referred to as the "nose" of a bevel.

Lap Weld: A weld made on two pieces of metal with the edge of one overlapping the other.

Melt Weld: A weld made by fusing the parent metal without adding filler rod.

Mild Steel: Steel which contains 0.30% carbon, or less.

Mill Scale: Refers to the surface coat of oxide, which forms on molten iron or steel. Mill scale turns black when cool, giving metal the black coating which leads to the term "black iron".

Neutral Flame: A flame resulting from a balance of oxygen and acetylene. A neutral flame is used for all oxyacetylene fusion welding.

Non-Ferrous Metal: Any metal which does not contain iron. (Aluminum, copper, brass, etc.)

Overhead Position: The position of a weld made on the underside of the parent metal and within 45 degrees of horizontal.

Oxidizing Flame: An oxyacetylene flame containing an excess of oxygen. An oxidizing flame is characterized by a short inner cone and a whistling sound.

Parent Metal: The metal being welded.

Penetration: The depth the welding puddle melts into the parent metal.

Plate Metal: Metal which is more than 1/8 inch in thickness.

Regulator: A device which reduces the pressure of gas coming from a cylinder to the desired working pressure. Separate regulators are required for acetylene and oxygen.

Sheet Metal: Metal which has been rolled in the steel mills to a thickness of 1/8 inch, or less.

Slag: Oxides which are usually deposited on the surface of the metal after flame cutting or welding.

Striker: A tool used to light the torch. It makes a spark by moving a piece of flint across a file.

Tack Welding: To hold two or more pieces of metal in position for welding, small, completely fused welds (tacks) are made at various places along the line of the weld.

Tensile Strength: The ability to withstand a pull.

Torch: A welding tool with inlets for oxygen and acetylene, valves to control the flow of the gases, and a means of attaching the tip.

Torch Angle: The angle at which the torch tip and flame are pointed into the weld.

Undercut: The thinned, metal section resulting when fluid metal drops from the vertical surface of a weld.

V-Grinding: Grinding the edges of pieces to be welded so that a V is formed when they are put together.

Vertical Position: The position of a weld which is made on a vertical surface with the bead perpendicular to the ground or floor.

Welding Tip: A torch attachment equipped with a mixing chamber and orifice to regulate the size of the flame. It concentrates the gases coming from the torch so the flame can be directed toward the weld.

Weldment: An assembly of two or more parts which have been welded together.

OXYGEN SAFETY PRECAUTIONS

Oxygen is a colorless, odorless, and tasteless gas. It makes up about 21 per cent of our atmosphere.

WARNING:

OXYGEN SUPPORTS AND CAN GREATLY ACCELE-RATE COMBUSTION.

OXYGEN AS A LIQUID OR COLD GAS MAY CAUSE SEVERE FROSTBITE TO THE EYES OR SKIN. DO NOT TOUCH FROSTED PIPES OR VALVES. IF ACCIDENTAL EXPOSURE TO LIQUID OXYGEN OCCURS, CONSULT A PHYSICIAN AT ONCE. IF A PHYSICIAN IS NOT READILY AVAILABLE, WARM THE AREAS AFFECTED BY FROSTBITE WITH WATER THAT IS NEAR NORMAL BODY TEMPERATURE.

USE A PRESSURE-REDUCING REGULATOR WHEN WITHDRAWING GASEOUS OXYGEN FROM A CYLIN-DER OR OTHER HIGH-PRESSURE SOURCE.

KEEP COMBUSTIBLES AWAY FROM OXYGEN AND ELIMINATE IGNITION SOURCES.

Many substances which do not normally burn in air and other substances which are combustible in air may burn violently when a high percentage of oxygen is present. DO NOT permit smoking or open flame in any area where oxygen is stored, handled, or used. Keep all organic materials and other flammable substances away from possible contact with oxygen, particularly oil, grease, kerosene, cloth, wood, paint, tar, coal dust, and dirt which may contain oil or grease.

KEEP ALL SURFACES WHICH MAY COME IN CONTACT WITH OXYGEN CLEAN TO PRE-VENT IGNITION.

Even normal industrial soot and dirt can constitute a combustion hazard. Do not place liquid oxygen equipment on asphalt, or on any surface which may have oil or grease deposits. Use cleaning agents which will not leave organic deposits on the cleaned surfaces. In handling equipment which may come in contact with oxygen, use only clean gloves or hands washed clean of oil. Do not lubricate oxygen equipment with oil, grease, or unapproved lubricants.

MAINTAIN ADEQUATE VENTILATION.

To prevent accumulation of oxygen in areas containing oxygen equipment and to minimize combustion hazards adequate ventilation must be provided.

LIQUID OXYGEN IS EXTREMELY COLD.

(297 deg. F. below zero).
COVER EYES AND SKIN.
Accidental contact of liquid oxygen or cold oxygen gas with the eyes or skin may cause severe frostbite. Handle liquid so that it will not splash or spill. Protect your eyes with safety goggles or face shield, and cover the skin to prevent contact with the liquid or cold gas, or with cold pipes and equipment. Clean, protective gloves that can be quickly removed, and long sleeves are recommended. Cuffless trousers should be worn outside boots or over high-top shoes to shed spilled liquid. If clothing should be splashed with liquid oxygen or otherwise saturated with oxygen gas, air out clothing immediately. Such clothing should not be considered safe to wear for at least 30 minutes, since it will be highly flammable and easily ignited while the concentrated oxygen remains.

CONTAINERS, EQUIPMENT, AND REPLACE-MENT PARTS MUST BE SUITABLE FOR OXYGEN SERVICE.

Use only equipment, cylinders, containers and apparatus designed for use with oxygen. Many materials, especially some non-metallic gaskets and seals, constitute a combustion hazard when in oxygen service, although they may be acceptable for use with other gases. Make no substitutions for recommended equipment, and be sure all replacement parts are compatible with oxygen and cleaned for oxygen service. Keep repair parts in sealed clean plastic bags until ready for use.

REGULATORS

Before attaching regulator to cylinder, inspect the regulator very carefully. Make visually certain that the regulator and the inlet filter are free of oil, grease or other hydrocarbon type contaminants. These contaminants may be ignited when the cylinder valve is opened and would burn violently in an enriched oxygen atmosphere. Replace the inlet filter if broken, missing or found contaminated. When filter is missing or damaged, the regulator should also be reconditioned and the high pressure gauge replaced. Before attaching the regulator to the cylinder valve, crack the cylinder valve momentarily to blow out any dust or dirt that might have accumulated in the cylinder valve outlet. Connect the regulator to the valve, back out the pressure adjusting screw until it turns freely then open the cylinder valve very slightly and very slowly so the inlet pressure gauge moves slowly to the cylinder pressure — then open the cylinder valve all the way. To minimize chance of injury stand to one side of the regulator when opening the cylinder valve.

In the rare instances when oxygen cylinder regulators catch fire, experience has shown that those with aluminum bodies burn far more violently than those with brass bodies. Brass-bodied regulators are preferred to reduce injury potential.

OBSERVE ALL APPLICABLE SAFETY CODES WHEN INSTALLING OXYGEN EQUIPMENT.

Before installing, become thoroughly familiar with NFPA (National Fire Protection Association) Standard No. 50, "Bulk Oxygen Systems at Consumer Sites," NFPA No. 56-F, "Non-Flammable Medical Gas Systems," Standard No. 51, "Oxygen-Fuel Gas Systems for Cutting and Welding," American National Standards Institute Pamphlet No. Z49.1, and with all local safety codes. For further information, refer to Linde Form 9888, "Precautions and Safe Practices — Liquefied Atmospheric Gases," and Form 2035 "Precautions and Safety Practices in Welding and Cutting with Oxygen-Fuel Gas Equipment."

IF NECESSARY TO DISPOSE OF WASTE GAS OR LIQUID, EXERCISE CAUTION.

Gaseous oxygen should be released only in an open outdoor area. Liquid oxygen should be dumped into an outdoor pit filled with clean, grease-free and oil-free gravel, where it will evaporate safely.

ACETYLENE AND LINDE FG-2* SAFETY PRECAUTIONS

Acetylene is a colorless gas with a distinctive garlic-like odor. LINDE FG-2 is a colorless gas with a sweet, ether-like odor.

WARNING:

ACETYLENE AND LINDE FG-2 (PROPYLENE) ARE FLAMMABLE GASES. A MIXTURE OF ACETYLENE OR LINDE FG-2 WITH OXYGEN OR AIR IN A CONFINED AREA WILL EXPLODE WHEN BROUGHT IN CONTACT WITH A FLAME OR OTHER SOURCE OF IGNITION.

USE A PRESSURE-REDUCING REGULATOR WHEN WITHDRAWING ACETYLENE OR LINDE FG-2 FROM A CYLINDER OR PIPELINE AS RECOMMENDED BY THE REGULATOR MANUFACTURER. NEVER ADJUST THE ACETYLENE REGULATOR TO OBTAIN A DELIVERY PRESSURE GREATER THAN 15 PSIG. IF ACETYLENE IS USED IN AREAS WITH ELEVATED AMBIENT PRESSURES, BE SURE THAT GAUGE PLUS AMBIENT PRESSURE DOES NOT EXCEED 30 PSIA.

UNDER CERTAIN CONDITIONS, ACETYLENE FORMS READILY EXPLOSIVE COMPOUNDS WITH COPPER, SILVER, AND MERCURY. CONTACT SHOULD BE AVOIDED BETWEEN ACETYLENE AND THESE METALS, THEIR SALTS, COMPOUNDS, AND HIGH-CONCENTRATION ALLOYS.

KEEP ACETYLENE AND LINDE FG-2 AWAY FROM SOURCES OF IGNITION, AND DO NOT PERMIT ANY ACCUMULATION OF GAS.

Concentrations of acetylene between 2.5 per cent and 81 per cent, and of LINDE FG-2 between 2.0 per cent and 11.1 per cent by volume in air are relatively easy to ignite by a low-energy spark and may cause an explosion. Smoking, open flames, unapproved electrical equipment, and other ignition sources must not be permitted in acetylene or LINDE FG-2 storage areas. Store cylinders outdoors or in other well ventilated areas and away from heat sources, such as furnaces, ovens, hot-metal ladles, and radiators and away from flammable materials, such as gasoline, kerosene, oil and combustible solids.

NEVER USE EQUIPMENT OR A CYLINDER THAT IS LEAKING ACETYLENE OR LINDE FG-2.

Be certain that the regulator-to-cylinder valve, hose-to-regulator, and torch-to-hose connections are leaktight before starting work. Regulators, hoses and torches must be properly maintained to work correctly and safely. If an acetylene valve should leak around the cylinder-valve spindle when the valve is opened, close the valve and tighten the gland nut. If this does not stop the leak, close the valve and return the cylinder to the supplier. If a LINDE FG-2 valve should leak around the cylinder-valve spindle when the valve is opened, close the valve and return the cylinder to the supplier.

DO NOT TAMPER WITH FUSIBLE PLUGS OR VALVES ON CYLINDERS.

Acetylene cylinders are equipped with fusible-metal safety plugs which melt at about 212°F, the boiling point of water. These plugs, usually of hexagonal shape, are threaded into the cylinder head and bottom on most cylinders. Fusible-metal channels may also be provided in the valve body on smaller cylinders. Do not tamper with these fusible plugs or permit a torch flame to come into contact with them. Keep cylinders away from overhead welding and cutting operations which could permit hot slag to drop onto the cylinder head and melt the plugs. Locate cylinders away from overhead and ground-level welding and cutting operations to prevent flying sparks and slag from accumulating on or around the cylinder and causing release of the fusible plugs. LINDE FG-2 cylinders are equipped with pressure-relief valves. Protect all cylinders from falling objects and avoid rough handling, dropping or knocking of cylinders to prevent damage to the cylinder, fusible plugs, relief valves or cylinder valves. Store, transport and use acetylene and LINDE FG-2 cylinders in a vertical position.

KEEP EQUIPMENT AREA WELL VENTILATED.

Although acetylene and LINDE FG-2 are non-toxic, they are anesthetics and can cause asphyxiation in a confined area that does not have adequate ventilation. Any atmosphere which does not contain enough oxygen for breathing (at least 18 per cent) can cause dizziness, unconsciousness, or even death. If adequate ventilation is not provided, acetylene and LINDE FG-2 may displace normal air. Acetylene gas can be detected by its distinctive garlic-like odor, and LINDE FG-2 by its sweet, ether-like odor. If the odor of either gas is noticed immediately attempt to locate the source of leakage and correct it. If a leak in a cylinder or connected apparatus cannot be stopped, move the cylinder and apparatus outdoors to a safe location away from ignition sources and immediately contact the gas supplier. Do not store or transport acetylene or LINDE FG-2 cylinders in confined or unventilated spaces, such as cabinets, closets, tool boxes, or automobile trunks.

NEVER USE CONTAINERS, EQUIPMENT, OR REPLACEMENT PARTS OTHER THAN THOSE SPECIFICALLY DESIGNATED FOR USE IN ACETYLENE OR LINDE FG-2 SERVICE.

Never attempt to transfer gas from one cylinder to another nor to mix any other gas with it in a cylinder. Use the proper T-wrench or key for opening acetylene cylinder valves, and leave the wrench or key on the valve for emergency shutoff.

OBSERVE ALL APPLICABLE SAFETY CODES WHEN USING ACETYLENE AND LINDE FG-2.

Before installing or using equipment and cylinders, become thoroughly familiar with ANSI (American National Standards Institute) standard Z49.1, "Safety in Welding and Cutting," NFPA (National Fire Protection Association) standard No. 51, "Oxygen-Fuel Gas Systems for Welding and Cutting" and standard No. 58, "Storage and Handling of Liquefied Petroleum Gases." For further safety information, refer to Linde Form 2035, "Precautions and Safe Practices in Welding and Cutting with Oxygen-Fuel Gas Equipment," and follow apparatus manufacturers' instructions.

* Linde FG-2 is Union Carbide's designation for its liquefied hydrocarbon industrial fuel gas.

WELDING PROCEDURES: ELECTRIC ARC

by Frank R. Schell

Preface

Electric arc welding includes all of the processes which use the heat generated by electric current jumping through an air gap. Metal arc welding is a form of electric arc welding, using a metal electrode as part of the welding circuit and as filler metal. In terms of materials consumed, people employed, and equipment in use, metal arc welding is the most widely used welding process. WELDING PROCEDURES: ELECTRIC ARC is designed to help the student acquire the ability to weld in a variety of situations, using the metal arc process. This ability is considered a prerequisite to employment in many trades.

The text is divided into four sections, each of which deals with a particular position or situation. These sections are subdivided into units which cover a particular joint design or welding condition. Each section also contains a comprehensive review of the material covered in the section. These comprehensive reviews are comprised of two or more procedures to be followed by the student. In addition, the Instructor's Guide contains comprehensive review questions about the material covered throughout the section. These procedures and review questions provide a logical means by which the instructor can test the student's knowledge and ability.

Each unit of the text is preceded by clearly stated behavioral objectives that describe what the student will be able to do when the unit is completed. All of the subsequent material in the unit is designed to fulfill these objectives. A concise discussion of any pertinent related information follows the unit objectives. This part of the unit covers such things as new terms, safety information, and problems that might be encountered. The units also include procedures to be followed by the student. These procedures are supplemented with key points which correspond with the procedural steps. By performing these jobs the student has the opportunity to learn through experience. Each job includes a list of the necessary equipment and material to complete the job. Most of the jobs include a destructive test of the welded specimen, providing the student with a means of self-evaluation. Following the job is a series of review questions. By answering these questions the student can review some of the highlights of the unit and be sure all of the material has been learned, without requiring special testing by the instructor.

After the last section there are two final review projects which require knowledge and ability in most of the areas covered in the textbook. These projects should be valuable for review at the end of a course, evaluation, or for additional practice.

All new or technical terms are explained as they are introduced, but in addition the textbook includes a glossary. This glossary serves as a ready reference so that the student need not refer back to material previously covered or ahead to that which has not yet been covered in order to locate the definition of a term.

Frank Schell has been a journeyman welder for twenty-four years. He is currently Curriculum Development Coordinator and Professor of Welding at the College of Southern Idaho. He is a member of the American Vocational Association and the Idaho Vocational Association. He is author of several other pieces of instructional material, including WELDING PROCEDURES: OXYACETYLENE.

Contents

SECTION 1:
Welds in the Flat Position

Unit 1   Setting up
Arc Welding Equipment

OBJECTIVES

> After completing this unit the student will be able to:

- set up arc welding equipment according to the standards of the welding industry.
- explain the fundamental theory of arc welding.
- list six basic safety rules for electric arc welding.

Arc Welding Process

Arc welding uses the heat produced as electricity jumps the gap from one conductor of electricity to another. As the electricity passes through this gap intense and concentrated heat (6,500 to 7,000°F) is produced. The two basic types of shielded-arc (stick-arc) welding are those that use AC (alternating current) and those that use DC (direct current) electricity. In the shielded-electrode method of welding an electric arc is produced between the metal to be welded *(base metal)* and the electrode. This arc melts the electrode metal which is deposited on and fused with the base metal.

Welding Equipment

The arc welding process uses special equipment. The welding student must set up and operate this equipment according to established industrial standards of safe and economical operation.

Arc Welding Machine

Arc welding machines are classified as either AC or DC. DC welding machines may be motor driven generators, figure 1-1, or *rectifier* welders, figure 1-2. A rectifier is a device which converts AC to DC.

Since the DC rectifier welder is most commonly used in shops and in school laboratories, the welding student should concentrate on that type. The DC rectifier furnishes two types of welding current, both of which are utilized for welding and each of which serves specific welding needs in a satisfactory manner.

Direct Current, Straight Polarity (DCSP)

When the parent metal is connected to the positive (+) side of the welder, and the electrode (rod) holder is connected to the negative (–) side of the welder, the circuit is in *straight*

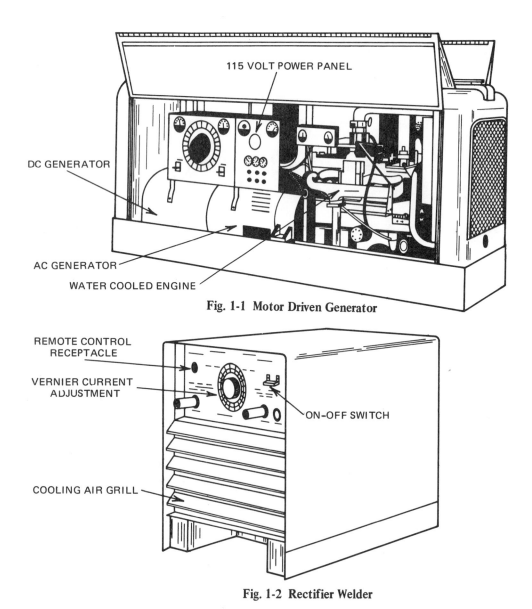

115 VOLT POWER PANEL

DC GENERATOR

AC GENERATOR

WATER COOLED ENGINE

Fig. 1-1 Motor Driven Generator

REMOTE CONTROL RECEPTACLE

VERNIER CURRENT ADJUSTMENT

ON–OFF SWITCH

COOLING AIR GRILL

Fig. 1-2 Rectifier Welder

polarity. With the electrode negative the current travels from the electrode to the base metal, figure 1-3. Two-thirds of the total heat produced is released at the base metal and one-third is released at the electrode. Such concentration of heat at the base metal produces deeper penetration of the weld. DCSP is used with high melting-temperature base metals, for deep penetration, for slow welds and for narrow beads. It is especially recommended for welds which are made *in position* (on the topside of a horizontal surface).

Direct Current, Reverse Polarity

When the parent metal is connected to the negative (–) side of the welder, and the electrode holder is connected to the positive (+) side of the welder, the circuit is called *reverse polarity*. With the electrode positive the current travels from the base metal to the electrode, figure 1-4. Two-thirds of the total heat is released at the electrode and one-third is released at the base metal. This concentration of heat produces shallow penetration,

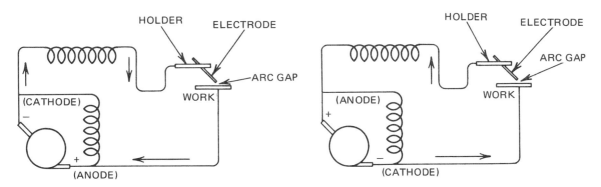

Fig. 1-3 Wiring Diagram, DCSP   Fig. 1-4 Wiring Diagram, DCRP

deposits weld metal at a high rate, and produces fine welds in sheet metal. DCRP is especially helpful for welding *out of position* (welds made in any but the flat position), since the parent metal does not heat up as much and has less tendency to run.

Alternating Current

Most AC welders, figure 1-5, have transformers which step down the voltage and increase the welding current. Electric current furnished by most electric utilities is 60-cycle, alternating current. (The current reverses its direction of flow 120 times per second.) Approximately 50 percent of the heat is released at the parent metal and 50 percent at the electrode. The reversal of current helps to hold down the magnetic field during the welding operation and consequently reduces *arc blow* which sometimes occurs in DC welding. Arc blow is the deflection of the arc from its normal path as a result of magnetic force.

Welding Electrodes

Most welding electrodes are covered with a coating of *flux*. This flux forms a gaseous shield which prevents oxygen from contacting the molten metal. The chemical content of

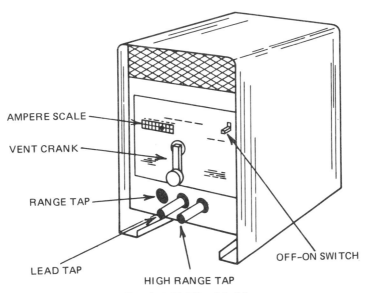

Fig. 1-5 Welding Machine

the coating varies with the manufacturer's specifications. Welding rods are manufactured and coated to fulfill specific welding functions. The welder must learn to identify electrodes and choose the type most suited for the job at hand.

Electrodes are designated with the letter E, followed by four or five digits — E-xxxx(x). The first two (or three) digits indicate the *tensile strength* of the deposited weld, in thousands of pounds per square inch. For example, in the designation E-60xx, the E indicates that it is an electric welding electrode, and the 60 indicates a tensile strength of 60,000 pounds per square inch. The next digit indicates the position of welding for which the electrode is designed. A 1 in this position indicates an all-purpose electrode, or one which can be used in any position; a 2 indicates that it is for flat or horizontal welding; and a 3 indicates that it is intended for flat position welding only. The last digit refers to the operating characteristics, such as coating and polarity. For example, an E-6011 electrode is made of metal with a tensile strength of 60,000 pounds per square inch (60), for use in any position (1), and either AC or DCRP (1).

In addition, rods are generally coded with a color on the end or a colored spot or spots on the covering. Some rods are also designated by a group color, figure 1-6.

Electrode Classification	End Color	Spot Color	Group Color
E 6010	. . . . .	. . . . .	. . . . .
6011	. . . . .	Blue	. . . . .
6012	. . . . .	White	. . . . .
6013	. . . . .	Brown	. . . . .
6014	Red	Brown	. . . . .
6020	. . . . .	Green	. . . . .
6024	. . . . .	Yellow	. . . . .
7010-A1	Blue	White	. . . . .
7011-A1	Blue	Yellow	. . . . .
7016	Blue	Orange	Green
7018	Black	Orange	Green
7020-A1	Blue	Yellow	Silver
8015-B1	White	Brown	Green
8016-B1	White	Black	Green
9016-B3	Brown	Blue	Green
10013-6	Green	Brown	Silver
10015-6	Red	Red	Green
10016-6	Green	Orange	Green
12016-6	Orange	Orange	Green

Fig. 1-6 Color Coding of Electrodes

Leads

The cables used to carry the electric current to the work and back to the welding machine are called *leads*, figure 1-7. Well-built leads of adequate size to carry the current used are essential. Leads which are used a considerable distance from the welding machine must be larger than leads used for jobs close to the machine. The heat which develops in small leads can cause the rubber insulation to overheat and loosen on the copper-wire core of the lead. Two leads are used — one is attached to the electrode holder and the other (the ground) is attached to the work. The leads are subjected to much wear and should be of high quality to ensure long service, figure 1-7.

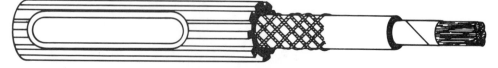

Fig. 1-7 Welding Cable

Electrode Holder

The electrode holder is the part of the arc welding equipment held by the welder, figure 1-8. It is attached to the electrode lead on the welding machine. The "stinger," as it is sometimes called, is a well-insulated handle which is made to withstand the heat from welding.

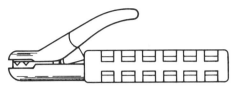

Fig. 1-8 Electrode Holder

Fig. 1-9 Ground Clamp

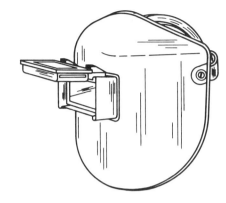

Fig. 1-10 Helmet

Ground Clamp

The ground cable from the machine to the work is generally connected to a spring-loaded clamp which can be easily attached to the work. This is the ground clamp, figure 1-9. In order to do a good job of welding, the ground must be solidly connected to the work.

Helmet

The helmet is generally made of fiber, and formed to cover the front half of the welder's head, figure 1-10. An opening is provided in front of the eyes, and a clear-glass cover lens is installed in the opening. Behind the cover lens is a colored glass which filters the infra-red and ultraviolet rays from the arc. The clear-glass lens is provided to catch the spatter from the welding process which would otherwise adhere to the colored lens. (The clear glass lens is much less expensive than the colored lens.) In addition, the best helmets have a hinged gate in the front which can be raised out of the way when welding is completed. This exposes another clear glass lens which protects the eyes while the weld is cleaned. The use of this type of helmet is recommended, since the entire face remains covered during the weld cleaning process, figure 1-10.

The welding helmet should be examined frequently to insure that no cracks or holes are present which might allow the arc light to leak through. The cover lens and colored lens must be free of cracks or chips for absolute eye protection.

Additional personal equipment should consist of:

- Cap
- Leather gloves
- Leather jacket
- Safety glasses (flash goggles)
- High shoes or boots
- Other wearing apparel should be of a type which is not readily flammable

Arc Welding Safety

Arc welding is not hazardous if a few fundamental safety rules are obeyed. The following points should be observed in addition to the general arc welding safety rules listed in Unit 2:

- Radiation from the arc is dangerous to the eyes. The arc gives off infrared and ultra-violet rays which may burn the eyes and the skin. An arc welder's helmet, with a suitable colored lens, must be worn to keep the rays from the skin and eyes.

- Flying sparks and small globules of molten metal are present most of the time when arc welding. Protective clothing which is not highly flammable, gloves, and high shoes help to protect the welder from burns.

- Avoid striking an arc when other persons are close. Warn others that an arc is to be struck so they may protect their eyes from the arc.

- Fumes given off from the arc and the material being welded may be injurious. Adequate ventilation is required at all times when welding is in process.

- Electric shock is always a possibility. Be sure the floor is dry and wear dry gloves. Use an insulated electrode holder.

- The danger of burns is always present. Do not handle hot metal with the hands. Use tongs or pliers.

SUMMARY

Primary Welding Equipment

- Arc welding machine (either AC or DC)
- Electrode holder and lead
- Ground clamp and lead
- Welding booth, or curtains to protect others from arc flash.
- Metal welding bench
- Welding electrodes

Personal Welding Equipment

- Welding helmet (hood), with clear glass cover lens and No. 10 or No. 11 colored lens
- Cap
- Safety glasses
- Leather jacket
- Leather gloves

Welding Tools

- Pliers
- Combination square

- Chipping hammer

- Ball peen hammer

- 6' steel tape

- 12" adjustable wrench

- Soapstone

- Center punch

- Wire brush

JOB 1: SETTING UP ARC WELDING EQUIPMENT

Equipment:

Arc welding machine
Ground cable and clamp
Electrode holder and cable
Personal hand tools
Gloves
Welding helmet

PROCEDURE	KEY POINTS
1. Attach the ground cable to the post marked "ground" on the welding machine.	1. Use an adjustable wrench.
2. Attach the electrode holder cable to the post marked "electrode" on the machine.	
3. Attach the ground clamp to the welding table and hang up the electrode holder.	3. Be sure the electrode holder is hanging so the metal jaws cannot contact any surface which is grounded to the machine.
4. Adjust the welding helmet to fit by turning the sprocketed wheel on the headband until the helmet is comfortable. Adjust the side buttons on the helmet until the helmet drops slowly.	4. Do not wear gloves when adjusting the headband. Gloves make it difficult to feel the tightness of the adjustment.
5. Be sure the colored lens in the hood is not cracked or chipped. Make sure a clear cover lens is installed and that it is clean.	
6. Turn on the welding machine, then shut it off.	

REVIEW QUESTIONS

1. Why must connections on a welding machine be tight?

2. What is another name for a welding electrode holder?

3. What is another name for a welding helmet?

4. Describe the direction of flow of electricity when using DCRP.

5. Describe the direction of flow of electricity when using DCSP.

6. Describe the direction of flow of electricity when using AC.

7. What welding polarity releases the greatest amount of heat at the base metal?

8. What is the tensile strength of a properly made weld using an E-7014 electrode?

9. What does the digit 1 indicate in the electrode designation E-7014?

10. What are the rays, given off by the arc welding process, which are dangerous to the eyes?

Unit 2   General Arc Welding Safety

OBJECTIVES

After completing this unit the student will be able to:

- demonstrate appropriate safety measures in most common arc welding situations.
- refer to the safety rules governing most arc welding situations.

1. Never use welding machines which are not properly grounded. Ground all power circuits to prevent accidental shock. Stray current can cause a fatal shock. Do not ground the welding machine to any pipes which carry gases or flammable liquids.

2. Check welding cables frequently. Do not overload the capacity of the cable. Do not use cables which have breaks in the insulation. Be certain that all cable connections on the machine are tight. Overloading the cables will cause them to overheat and create a fire hazard.

3. Check the electrode holder frequently. The jaws in the electrode holder should be kept tight. The gripping surface of the jaws must be clean to insure a tight hold on the electrode. Use only fully insulated holders. Never touch two electrodes together when they are connected to two different machines: the operator can receive an electrical shock and the machines can be seriously damaged.

4. The polarity switch on the welder is provided to change the electrode from reverse polarity to straight polarity. The polarity switch must never be operated while the machine is under a welding load. Operate the switch only when welding has stopped. The operator who throws the polarity switch while the welding circuit is in use can be seriously burned.

5. Never weld on containers which have held explosive or combustible materials.

6. Never weld in confined spaces without adequate ventilation, and do not weld near an explosive atmosphere.

7. Do not weld where even small amounts of vapors from solvents are present. Some solvents decompose to form phosgene gas (a deadly poisonous gas).

8. Electricity can be a killer. Before checking any welding machine open the power circuits. The welder should never try to make repairs on a welding machine unless he or she is a fully qualified electrician.

9. Do not touch any exposed or noninsulated parts of cables or clamps.

10. Do not work in a damp area unless insulated from shock. Striking an arc when feet are wet can cause electrical shock.

11. Do not use a cracked or defective helmet. The filter glass in the helmet provides eye

protection from ultraviolet and infrared rays. Cracks in the lens allow these harmful rays to contact the eyes. The use of clean, clear lenses provides a method of cutting down on eyestrain.

12. Never look at an electric arc without eye protection. The intense light and the infrared and ultraviolet rays are harmful to the eyes.

13. Dark colored clothing helps to protect the body from ultraviolet rays.

14. Wear protective clothing when making out-of-position welds.

15. Any equipment which is operated by power can be dangerous. Grinders, drills, sanders, saws, electric welders and all other power tools can cause severe injury or even death.

16. Never do any grinding, weld slagging or heavy hammering without eye and face protection.

17. Do not leave materials or equipment lying on the floor. Clean up the work area. Take care of all safety hazards as soon as they are observed.

18. Do not use defective equipment.

19. Never strike an arc without warning people who are near and liable to see the arc flash.

20. Be alert to the dangers of fire. A small fire is easy to handle; a large fire can be disastrous. Know where the fire extinguishers are and learn how to use them. Combustible material should not be within 35 feet of the welding operation.

21. Wear heavy shoes, preferably those with steel toes.

REVIEW QUESTIONS

Instructions: There are twenty-one general arc welding safety rules listed. For each question in this review write the number of the rules which most closely answer the review questions.

Example: Should welding machines be grounded to pipes?

Rule No. 1 states: "Do not ground the welding machine to any pipes which carry gases or flammable liquids."

1. Can the polarity switch on the welding machine cause a burn hazard to the welder?

2. Can the use of cables which are too small reduce the efficiency of a weld?

3. Can welding on containers which have been used to hold solvents be dangerous?

4. Does the color of the clothes worn by a welder have any effect on safety?

5. Should a welder make temporary repairs on the welding machine so that he can continue welding until an electrician can repair the machine?

6. Is it all right for a welder to watch the arc welding process for a few seconds at a time without eye protection?

7. Should defective equipment be used until new or repaired equipment is available?

8. Do welding machines need to be grounded for safe operation?

9. Can working in a damp area affect the safety of the welder?

10. Can a cracked hood lens have any effect on the welder's eyes?

11. Should a welder know how to use fire extinguishers or are they someone else's responsibility?

12. When two arc welding machines are in operation, can the electrodes be touched together?

13. Is the welder responsible for warning others in the area before striking an arc?

14. Should the welder leave equipment on the floor around the job until the work is finished?

15. Is it safe to handle noninsulated cables and clamps on the welding machine?

16. Is a rule listed which limits the distance that combustible materials must be kept from the welding area?

17. Is good ventilation a part of the safety rules for a welder?

18. In which rule is reference made to the use of grinders and weld slagging?

19. In which rule is reference made to the welding of containers which have held combustible materials?

20. Is out-of-position welding mentioned in the welding safety rules?

Unit 3   Striking an Arc

OBJECTIVES

After completing this unit the student will be able to:

- describe the basic principles of arc welding.
- discuss reasons for welds with poor appearance of lack of penetration.
- successfully strike and maintain an arc, using standard welding electrodes and equipment.

Striking the Arc

When a ground and a wire carrying an electrical charge contact each other, an *arc* (continuous spark) occurs, causing intense heat and bright light. This is the principle on which arc welding operates, except that after the arc is started, the welding electrode is moved a short distance away from the parent metal. This keeps a constant arc and continuous heat. An arc length of 1/8 inch is maintained most of the time. Lengthening the arc by moving the electrode farther away from the parent metal increases the heat and the size of the puddle.

Improper arc length can sometimes be determined by visually inspecting a completed weld. Too short an arc can cause poor *fusion* (fusion is the mixing of the parent metal with the weld metal), *undercutting* (an area where metal is missing), and *porosity* (pinholes). Too long an arc can cause lack of concentration of the heat, excessive splatter, poor *penetration* (penetration is the depth of the weld in the parent metal), and arc action which is not smooth. Holding the arc too short can also cause the electrode to stick to the parent metal.

The arc may be struck by dragging the electrode across the grounded metal much as a match is struck, figure 3-1. If the electrode has a heavy flux coating, it may be necessary to break the coating away from the end of the rod before contact can be made. Another method, and one frequently used, is to bring the rod down onto the plate at a 90-degree angle, then raise it quickly to the correct distance to maintain even heat and penetration, figure 3-2.

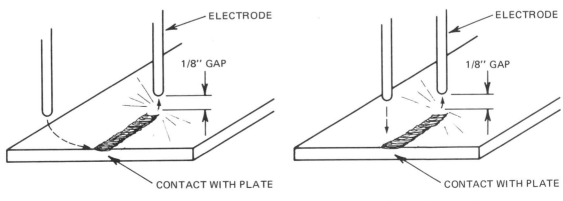

Fig. 3-1

Fig. 3-2 Striking an Arc

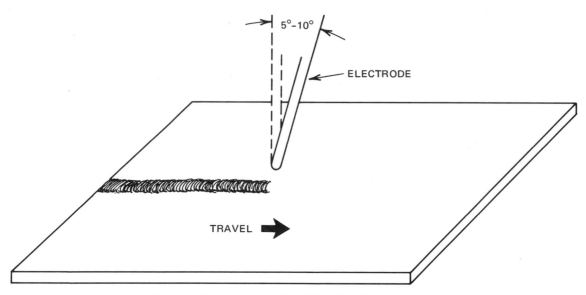

Fig. 3-3 Electrode Angle for Filling Puddle and Welding

After the arc is established, movement across the plate must be made in a steady forward motion, and the arc must be kept a uniform length. Too rapid progress will result in poor penetration. Right-handed welders generally progress from left to right, so that the weld puddle can be seen and the filling (or buildup) of the rod can be controlled. Left-handed welders should work from right to left.

The flux coating on the shielded electrode melts and forms slag over the molten metal. This shields the weld from the action of the gases in the atmosphere. When the weld is completed and cooled, the slag may be removed from the surface of the weld by the use of a chipping hammer.

The heat of the arc melts a crater into the parent metal which must then be filled with the electrode metal. As the electrode is melted into the weld, it is thoroughly mixed with the parent metal. It is not enough to lay a bead on the parent metal. Thorough penetration and fusion between parent metal and electrode metal must take place. As the electrode melts it must be gradually lowered towards the weld to maintain the correct length of the arc.

Electrode

After the arc has been started, the electrode should be held away from the plate to begin the weld. A good rule to follow is to keep the arc at a distance equal to the diameter. of the rod being used. The rod must be held perpendicular to the plate being welded, and tipped slightly (about 10 degrees) in the direction of travel, figure 3-3. The bead width should be approximately twice the diameter of the rod.

JOB 3: STRIKING AND MAINTAINING AN ARC

Equipment and Material:

    Standard AC or DC welding machine
    Helmet
    Gloves

Chipping hammer
Safety glasses
Wire brush
Protective clothing
1/4″ mild steel plate
1/8″, E-6011 or E-6013 electrodes

PROCEDURE	KEY POINTS
1. To form an arc, it is necessary to bring the electrode into contact with the workpiece.	1. Do not turn on the welding machine at this time.
2. Place the metal workpiece in the flat positon.	
3. Install a 1/8-inch electrode in the electrode holder.	
4. Hold the end of the electrode about 1/4 inch above the workpiece, on the left edge of the plate. Practice bringing the electrode down onto the plate, tipping the head forward to bring the helmet down over the face at the same time.	4. Practice this exercise until the electrode strikes the edge of the plate at the same time the helmet drops over the face to cover the eyes.
5. Set the polarity switch on the machine on "Electrode Negative" and the current at 90 amperes.	
6. Turn on the welding machine.	6. Do not make contact between the end of the rod and any grounded material.
7. In the same manner as in No. 4 of this lesson, drop the hood over the face and strike an arc. The electrode should contact the spot the welder was watching before lowering the helmet.	7. **CAUTION: Be very careful not to strike an arc without the colored lens of the hood in front of your eyes. Draw the electrode across the plate with a quick, whipping motion, as if striking a match.**
8. If the electrode sticks to the plate, quickly depress the lever on the electrode holder, releasing the electrode. Break the electrode off of the plate with pliers.	8. **CAUTION: The rod is hot.**
9. Run a bead about 1/2 inch long. Raise the rod to break the arc, then repeat the operation until an arc can be struck and a weld can be run about 2 inches.	9. A weld made at the correct temperature and rate of travel makes a soft, frying sound.

PROCEDURE	KEY POINTS
10. Shift the polarity of the welding machine, alternating between DCSP and DCRP. Practice with both polarities. If AC is available use that also.	10. **CAUTION: Do not change the polarity of a welding machine during an arc. The machine must not have any load on it when the polarity is changed.**
11. Chip the weld beads, using a slag hammer (chipping hammer), brush them with a wire brush, and show them to the instructor for comments.	

REVIEW QUESTIONS

1. What happens when a wire carrying an electrical charge touches a ground?

2. How far away from the metal being welded should the electrode be held for arc welding?

3. List three things which can be observed on a completed weld made with the arc too short.

4. Describe one method of striking an arc with an electrode.

5. What is the reason for the flux coating on the electrode?

6. How should the width of a bead compare with the diameter of the electrode being used?

7. How does the length of the arc affect the amount of heat generated in the welding process?

8. Describe the angle at which the electrode should be held for welding a bead on a flat surface.

9. Define the term penetration, as it is used in welding.

Unit 4   Straight Beads, Flat Position

OBJECTIVES

After completing this unit the student will be able to:

- Identify at least five common welding electrodes by their AWS classification and exhibit knowledge of their specific uses.
- Weld a smooth, even bead using E-6011 electrodes.
- Weld a smooth, even bead using E-6013 electrodes.

Electrode Selection

Welding electrodes are classified according to whether they are to be used with DC reversed polarity (DCRP), DC straight polarity (DCSP), or alternating current (AC). The electrodes used most commonly for mild steel welding are discussed here.

E-6010 indicates an all-position welding rod (flat, vertical, horizontal, and overhead). It performs best when used with DCRP. Deep penetration can be achieved with this electrode which has a thin coating, and which lends itself particularly well to out-of-position welding.

E-6011 also indicates an all-position welding rod. It is particularly suited for use with AC, but it can also be used with DCRP and DCSP. Its thin coating makes it a good electrode for out-of-position work.

E-6012 is another all-position electrode, however, because of its heavier flux coating, it is slightly more difficult to make out-of-position welds with this electrode. It is best suited for use with DCSP or AC.

E-6013 indicates an electrode which is especially suited for deep-penetration welds in the flat position. Because of its heavier coating it is a more difficult electrode for beginners to use than are the E-6011 electrodes. This electrode can be used with all types of polarity.

E-6020 electrodes have a heavy iron powder flux coating. They are used for flat and horizontal welding only. (Notice that the third digit is 2.) These electrodes can be used with DCRP, DCSP, or AC.

E-6030 electrodes also have a heavy iron powder flux coating. As is indicated by the third digit being 3, they are for flat position welding only. These electrodes may be used with DCRP or AC.

Note: E-6020 and E-6030 electrodes are sometimes called drag rods. This is because the welder can run a bead without removing the rod from the parent metal once the arc is struck.

Welding Currents

Generally, the amperage at which the rod runs most readily is indicated by the manufacturer. Differences in rod diameter and in material used for the flux coating require

Diameter of Electrode	Amperage Used					
	E-6010	E-6011	E-6012	E-6013	E-6020	E-6030
1/8″	80-120	80-120	80-130	70-120	100-140	100-140
5/32″	120-160	120-160	120-180	120-170	120-180	120-180
3/16″	140-220	140-220	140-250	140-240	175-250	175-250

Fig. 4-1 Current Setting for Common Electrodes

differences in the current settings used. Figure 4-1 indicates current settings which generally give satisfactory results.

Welding Coupon

The welding *coupon* (or sample) is generally small, therefore the heat from welding tends to concentrate and build up in the plate. After welding is started on the practice plate, the lower ranges of current should be used, and the plate cooled in water frequently. This prevents excessive heat from building up in the coupon.

JOB 4A: STRINGER BEADS IN THE FLAT POSITION, DCRP

Equipment and Material:

Standard AC or DC welding machine
Helmet
Gloves
Chipping hammer
Safety glasses
Wire brush
Necessary protective clothing
1/4″ mild steel plate
1/8″, E-6011 and E-6013 electrodes

PROCEDURE	KEY POINTS
1. Clamp a 1/8-inch, E-6011 electrode in the electrode holder.	
2. Set the welding machine for reverse polarity (electrode positive).	2. Check the manufacturer's chart and set the amperage for the rod being used.
3. Turn on the machine.	3. Be careful that the rod does not come in contact with grounded material.
4. Strike an arc and run a smooth, even straight bead across the sample plate.	4. The welder's face should be covered by the helmet when the arc is struck and while the weld is in progress.
5. Shut off the machine.	

PROCEDURE	KEY POINTS
	CAUTION: The welder must have face and eye protection when chipping welds.
6. Hang up the stinger and remove the sample from the welding jig or table. Use a chipping hammer to remove all the slag from the weld.	6. Open the window in the hood but keep the hood down when chipping or brushing welds. The metal is hot. Handle it with pliers.
7. Brush the weld and the plate with a wire brush and have the instructor check it.	
8. Follow the procedure listed in steps 2 through 7 using a 1/8-inch E-6013 electrode.	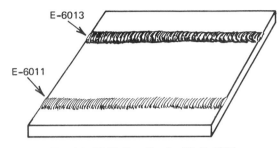
9. Observe the difference in the way the weld is deposited by the two electrodes.	**Fig. 4-2 Weld One Bead with E-6011 and one with E-6013**

JOB 4B: STRINGER BEADS IN THE FLAT POSITION, DCSP

Equipment and Material:

Standard arc welding equipment
Protective clothing
3/16″ or 1/4″ mild steel plate
1/8″ or 5/32″, E-6011 and E-6013 electrodes

PROCEDURE	KEY POINTS
1. Fasten the coupon in a jig or lay it flat on the bench.	
2. Set the welding machine for straight polarity.	2. Check the current setting on a piece of scrap metal.
3. Working in a comfortable position, and beginning at the left side of the plate (if right-handed), deposit a stringer bead with an E-6011 electrode.	3. Center punch the plate if necessary to keep the weld in a straight line. The electrode should be at an angle of 90 degrees with the weld and tipped about 5 degrees in the direction of travel, figure 4-3.

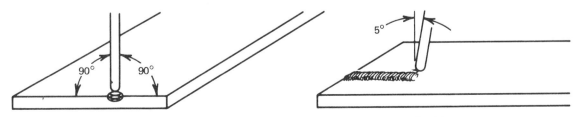

Fig. 4-3 Electrode Angle for Stringer Beads in the Flat Position

PROCEDURE	KEY POINTS
4. Remove the slag and brush the weld.	4. Cool the material in water before attempting to handle it with bare hands.
5. Run another bead 1/2 inch away from and parallel to the first one, using an E-6013 electrode.	
6. Remove the slag, brush the weld, and have it inspected by the instructor.	

JOB 4C: STRINGER BEAD PAD BUILDUP IN THE FLAT POSITION

Equipment and Material:

> Standard arc welding equipment
> Protective clothing
> 3/16" or 1/4" mild steel plate
> 1/8", E-6013 electrodes

PROCEDURE	KEY POINTS
1. Set the welding machine for DCSP.	1. Electrode negative.
2. Run stringer beads across the plate, overlapping each bead about one-fifth, until the plate is completely covered.	2. Right-handed welders should proceed from left to right. Concentrate on achieving penetration into the plate.
3. Slag and brush the welds.	
4. Weld stringer beads across the first layer, overlapping about one-fifth, until the first welds are completely covered, figure 4-4.	4. Concentrate on achieving penetration into the metal deposited on the first weld.

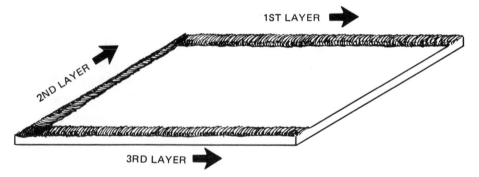

1ST LAYER

2ND LAYER

3RD LAYER

Fig. 4-4 Pass Sequence for Padding

PROCEDURE	KEY POINTS
5. Slag and brush the welds.	
6. Weld a third series of beads across the second layer, overlapping about one-fifth.	6. Concentrate on achieving penetration into the second layer.

PROCEDURE	KEY POINTS
7. Slag and brush the welds. Note: Save this weldment for testing in another unit.	7. The beads should appear smooth and uniform.

REVIEW QUESTIONS

A.   1. What is the purpose of overlapping beads when they are welded close together?

2. Why should a weld be chipped and brushed before another weld is made over it?

3. For what position of welding are electrodes with heavy coatings most suited?

4. What characteristic of an electrode determines the current setting with which it should be used?

5. Describe the proper electrode angle for welding a bead on a flat surface.

B.   Match the characteristics of the electrodes listed on the left with the electrode designations listed on the right.

1. _____   All-position electrode
Heavy coating
Best for AC and DCSP

2. _____   All-position electrode
Thin coating
Best for DCRP

3. _____   Flat-position electrode
Heavy iron-powder coating
Best for DCRP or AC

4. _____   Used for flat and horizontal position
Heavy iron-powder coating
Good for DCRP, DCSP, and AC

5. _____   Flat-position deep-penetration electrode
Heavy coating
Good for DCRP, DCSP, and AC

a. E-6030
b. E-6010
c. E-6012
d. E-6013
e. E-6020

Unit 5   Etching Welds in Mild Steel

OBJECTIVES

After completing this unit the student will be able to:

- list the steps in the etching of welds in mild steel.

- list the safety precautions which should be taken when handling acid.

- etch a mild steel weldment to bring out the penetration.

- describe a method of preserving etched specimens, so that the etching remains visible.

Etching Welds in Mild Steel

If good penetration has taken place in a weld it may be hard to see, because the metal is thoroughly mixed together. A cross section of the weld can be *etched* with acid to make the penetration and other characteristics more visible. In etching, the metal is treated with acid which reacts with the oxides on the surface to make the grain structure of the metal visible.

Before the metal is etched the surface should be smoothed by filing or grinding. Then, a mixture of one part nitric acid and two parts water is applied to the surface with an acid brush.

> **CAUTION:** If acid is not mixed correctly a violent reaction may result, causing acid to be splashed on the skin or in the eyes. Always wear rubber gloves and safety glasses when handling acid. Add the acid to the water. If acid comes in contact with the skin, wash it off immediately with fresh water.

When the acid stops working rinse the weldment in fresh water. The weld penetration will appear darker in color, and any oxides left by the weld will have been removed, exposing blowholes, lack of metal, and porosity. If the etched sample is to be saved for later inspection, a coating of clear plastic or varnish may be applied to it.

JOB 5: ETCHING WELDS IN MILD STEEL

Equipment and Material:

Oxyacetylene cutting equipment or metal-cutting saw
Grinder
File
Rubber gloves
Welded plate from Unit 4
Solution of nitric acid and water
Acid brush

PROCEDURE	KEY POINTS
1. Using the oxyacetylene torch or the saw, make a straight cut through the center of the welded plate.	1. Make as clean a cut as possible to save unnecessary grinding and filing.
2. Cool the cut samples in water, then grind the cut edges smooth. Further smooth the edges with a file, so there are no visible cutting marks.	2. **CAUTION: Wear face protection when grinding.** The smoother the surface is, the better the etching will be. Emery paper can be used for final smoothing.
3. Apply a solution of one part nitric acid and two parts water to the smoothed edges with an acid brush.	3. **CAUTION: Wear rubber gloves and do not breathe the fumes from the etching. The instructor should be present when the acid is being handled.**
4. When the acid stops working, rinse the pieces in fresh water.	
5. Check the weld for penetration and slag inclusions.	5. The penetration shows up as a darkened area on the polished surface. A good weld has no pinholes or slag inclusions and each pass is mixed with the preceding one.

REVIEW QUESTIONS

1. How should metal that is to be etched be prepared?
2. What proportion of acid and water should be used for etching welds?
3. What safety precautions should be observed when mixing acid?
4. What should be done if acid is spilled on the skin?
5. What characteristics of a weld can be observed more easily after etching?
6. What can be done to preserve an etched sample?

Unit 6   Weave Beads, Flat Position

OBJECTIVES

After completing this unit the student will be able to:

- identify a weave bead.
- list three reasons for the use of weave beads.
- describe the process of weaving a bead.
- weave a bead in the flat position.

Weave Bead

Weaving a bead increases the width of the deposit. It also increases the overlap. Weaving is used to widen a bead, to fill undercut at the sides and to assist in slag formation. Weaving is generally recommended for filling poor fitting joints.

A weave bead is deposited by moving the rod back and forth across the surface to be welded. Stringer beads may be run at the edges first. Several different electrode movements may be used, but weaving is generally done in the flat position using a semicircular motion to the left and the right, figure 6-1.

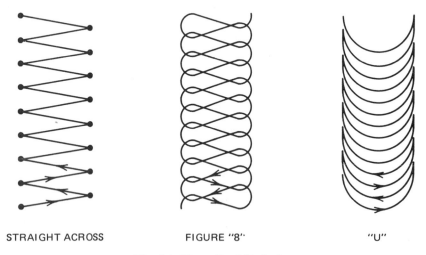

STRAIGHT ACROSS   FIGURE "8"   "U"

Fig. 6-1 Weave Bead Techniques

Flat Position

A weld made on the topside of the parent metal and within 30 degrees of horizontal is called a *flat weld*. The flat position is the most desirable position for welding, since the operator can see the work easily. In many welding shops a device called a *positioner* is used to hold the work, so that it can easily be turned into the flat position. Flat welds can be made successfully with AC, DCRP, or DCSP.

JOB 6: WEAVE BEADS, FLAT POSITION

Equipment and Material:

    Standard arc welding equipment
    Protective clothing
    3/16″ or 1/4″ mild steel plate
    1/8″, E-6010 or E-6011 electrodes

PROCEDURE	KEY POINTS
1. Set the machine for straight polarity.	1. Electrode negative.

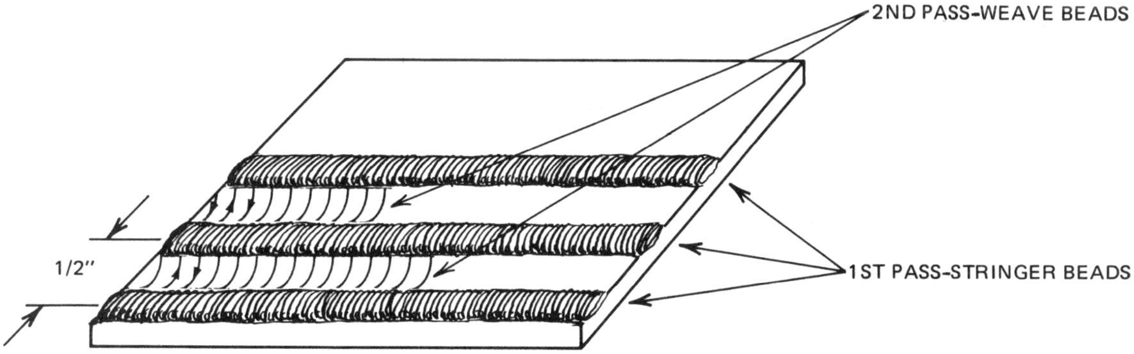

Fig. 6-2 Position of Beads for Weaving

2. Weld stringer beads 1/2 inch apart on a plate, figure 6-2.	
3. Slag and brush the stringer beads.	3. Welds must be cleaned between passes to prevent the slag from being trapped in the welds.
4. Beginning at the left side (left-handed welders begin at the right side) run a weave bead between two stringers, all the way across the plate.	4. The current must be set high enough to make the bead edges flow together.
5. Slag and brush the weld, then have it inspected by the instructor.	
6. Repeat the procedure until the plate has three thicknesses of welded weave beads.	
7. Saw through the center of the plate and smooth the cut edges with a file or grinder. Etch the cut edges with ammonium persulphate or a weak solution of nitric acid.	7. Check the weld for penetration and fusion. No slag holes or slag inclusions should be present.

REVIEW QUESTIONS

1. Draw a sketch of the motion most frequently used for weaving a bead in the flat position.

2. Should stringer beads or weave beads be used to fill poorly fitting joints?

3. Describe the flat welding position.

4. Why is the flat position considered the most desirable position for welding?

5. What is a positioner?

6. Describe two methods which can be used to keep the sides of a weave bead in a straight line.

7. When one bead runs over the top of another, why is it necessary to clean the slag from the first bead?

Unit 7   Fillet Welds, Flat Position

OBJECTIVES

After completing this unit the student will be able to:

- define a fillet and a T weld.
- describe two methods of preparing steel plate for welding.
- list the reasons for chipping and brushing welds.
- make fillet welds in the flat position.

Fillet Weld

The fillet weld, figure 7-1, is the type of weld used most often in industry. A fillet weld is a weld made on two pieces of metal which are joined in any way other than in a flat plane. A *fillet* is a reinforcement, and a weld made in an inside corner is called a *fillet weld.* Fillet welds are sometimes called *T welds,* when the pieces form a 90-degree angle. However, not all fillet welds are T welds, because the pieces may meet at an angle other than 90 degrees.

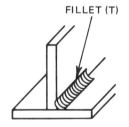
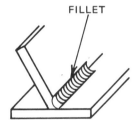
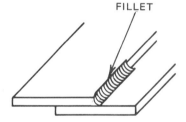

FILLET (T)   FILLET   FILLET

Fig. 7-1 Fillet Welds

Fillet welds should have a balance between the length of the weld and the area they contact on either of the base metal parts. In an attempt to create strength, there is a tendency to build up too much weld when building the fillet. When extra weld metal is used, the cost of the weld is increased and the joint becomes excessively heavy, without increasing the efficiency of the weld.

Preparation for a Fillet Weld

When the pieces to be welded are less than 3/16 inch thick, welding on both sides of the joint should produce a strong joint. When thicker metal is welded the joint must be prepared in such a way that the weld penetration is 100 percent. The pieces may be gapped, as in figure 7-2. Another method is to bevel the edge, figure 7-3 so the weld can penetrate the joint.

Expansion and Contraction

Because metal *expands* (increases in size) as it is heated and *contracts* (decreases in size) as it is cooled, allowances must be made for warpage. The metal should be tack welded in

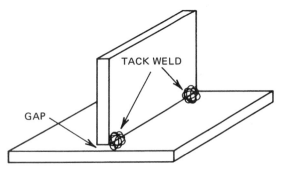

Fig. 7-2 Gapping the Pieces for a Fillet Weld

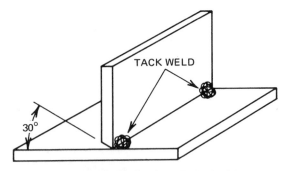

Fig. 7-3 Beveled Edge for a Fillet Weld

position to keep it aligned during the welding. Sometimes the pieces can be tacked in such a way that they are pulled into line by the contraction which takes place as the joint cools. Another method is to clamp the parts in a manner which resists the force of contraction. When neither of these methods can be used, the parts may be welded *intermittently* on each side (short welds alternately on opposite sides of the joint).

JOB 7A: FILLET WELD, FLAT POSITION

Equipment and Material:

> Standard arc welding equipment
> Protective clothing
> 1/4" or 3/16" mild steel plate, 3" x 6"
> 1/8" or 5/32", E-6011 or E-6013 electrodes

PROCEDURE	KEY POINTS
1. Tack weld two pieces of steel plate in position for a 90-degree fillet weld, figure 7-4.	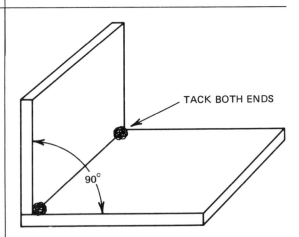 **Fig. 7-4 Setup for Fillet Weld**
2. Set the machine for straight polarity.	
3. Deposit the first bead, from left to right (left-handed welders reverse this) directly in the corner where the two pieces meet.	3. The proper electrode angle is shown in figure 7-5. It is important that the first pass be made carefully. Be certain that full penetration is achieved in the corner. Both plates must be melted together and the rod mixed well.

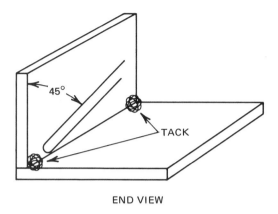

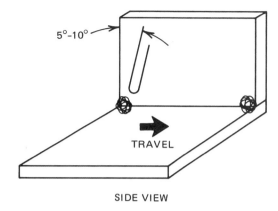

END VIEW   SIDE VIEW

Fig. 7-5 Electrode Angle for a Fillet Weld

4. Chip and brush the weld.

5. Using the weld sequence shown in figure 7-6, make 5 more passes.

6. Cool the weldment in water, then test it by bending the upright plate over the weld, figure 7-7.

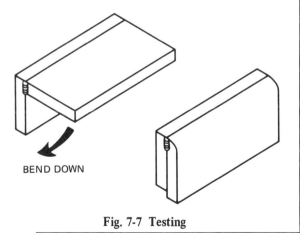

BEND DOWN

Fig. 7-7 Testing

4. Chip and brush each weld before applying the next pass.

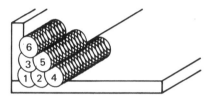

Fig. 7-6 Pass Sequence for a Fillet Weld

6. If the weld breaks, the break should be through the center for the entire length of the weld. The broken metal should be clean and shiny, with no slag inclusions.

JOB 7B: T WELD, FLAT POSITION

Equipment and Material:

Standard arc welding equipment
Protective clothing
Two pieces 1/4" mild steel plate, 4" x 6"
1/8" or 5/32", E-6011 electrodes

PROCEDURE	KEY POINTS
1. Tack weld two pieces of mild steel plate in position for a T weld, figure 7-8.	1. The vertical piece should be beveled 30 degrees, with a grinder.

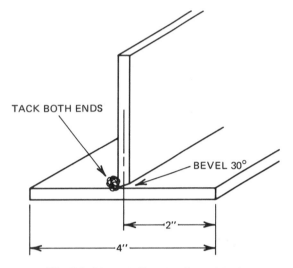

TACK BOTH ENDS

BEVEL 30°

2"

4"

Fig. 7-8 Pieces in Position for a T Weld

2. Set welding machine for DCRP and weld the root pass in the corner where the two plates meet.	2. The root pass must have 100 percent penetration into the corner.
3. Weld 6 passes, chipping and brushing between passes.	3. Care must be taken to avoid undercutting the vertical plate.
4. Cut through the center of the weld and smooth the cut edges with a file or grinder. Etch the weld to observe the penetration.	

REVIEW QUESTIONS

1. Why is it necessary to bevel the edge or leave a gap between the pieces to be welded in a T joint?

2. Why should the pieces be tack welded before a weld is made?

3. What is an intermittent weld?

4. How much penetration is required to produce a welded T joint with full strength?

5. List two reasons why a fillet weld should not be overwelded.

6. Why is it especially important for the first pass on a fillet weld to have good penetration?

7. What is the difference between a T weld and other fillet welds?

8. Describe the proper electrode angle for the first pass on a fillet weld.

Unit 8   Butt Welds, Flat Position

OBJECTIVES

After completing this unit the student will be able to:

- list two methods of preparing steel plate for butt welding.
- describe a land and the reason for its use.
- weld a butt joint in the flat position.

Butt Weld

A butt weld joins the edges of two pieces of metal which are in line with one another. When welded, the two pieces of metal form a flat surface.

For a butt weld to have full strength it must have 100 percent penetration into the parent metal. When 100 percent penetration is achieved, the bottom edges of the plates will be completely fused together. The first pass (or *penetration pass*) is most important, since failure to join the plates completely can cause a break in the metal or the weld.

Preparation for a Butt Weld

Metal which is over 3/16 inch thick must be prepared so that the first pass of the weld will have complete penetration. The pieces to be welded may be gapped to improve penetration. However, if the pieces are over 3/16 inch

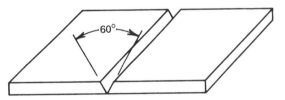

Fig. 8-1 Beveling Plate for a Butt Weld

thick, the edges will melt rapidly and may fill the gap without completely fusing the bottom edges of the joint. If the parts can be welded from both sides this will improve fusion, but often butt welds must be made where the metal is welded from one side only. Such welds can be prepared by beveling the edges of the two plates with a grinder. Usually the edges are beveled about 30 degrees so that when they are fitted together their edges form a V of about 60 degrees, figure 8-1.

Beveling forms a very thin edge, which would burn away rapidly. To prevent this edge from burning away, a *land* (flat nose), figure 8-2, is ground on the edge. This land should be about 1/16 inch thick.

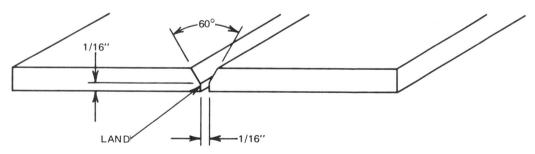

Fig. 8-2 Plate Prepared for a Butt Weld

When the plates are tack welded, a narrow gap is left between them. This helps insure 100 percent penetration of the root pass. After each pass, the weld must be chipped and cleaned, to prevent slag from being trapped in the following passes.

JOB 8A: BUTT WELD, FLAT POSITION

Equipment and Material:

> Standard arc welding equipment
> Protective clothing
> Two pieces 1/4″ mild steel, 5″ x 5″
> 1/8″ and 5/32″, E-6011 electrodes

PROCEDURE	KEY POINTS
1. Grind one edge of each plate to form a 60-degree included angle. Grind a 1/16-inch land at the bottom of the bevel, figure 8-3.	1. The angled edge is called a *bevel*. The squared edge ground at the bottom of the bevel is called the *land*.

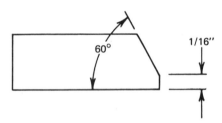

Fig. 8-3

PROCEDURE	KEY POINTS
2. Tack weld the two plates together on a flat plane, so that the bevels fit together to form a V.	2. Leave about a 1/16-inch to 1/8-inch gap between the lands of the plates.
3. Beginning at the left side of the V, weld a pass all the way across the joint with a 1/8-inch, E-6011 electrode.	3. Weld carefully. The bottom edges of the plates must be completely fused.
4. Chip and brush the weld, then examine the root pass. The edges of the two plates should be completely fused together with no unwelded edges.	4. If the root pass fails to completely weld the bottom edges of the plate, cut the weld out and regrind the plate for another weld.
5. When satisfactory penetration of the root pass has been accomplished, fill the remaining V of the weld, using 5/32-inch, E-6011 electrodes.	5. Make as many passes as are necessary to fill the V. The bead should be *convex* (slightly rounded on top), with no low spots.
6. Chip and brush the weld. Note: Save the welded plate for testing later.	

JOB 8B: TESTING A BUTT WELD

Equipment and Material:

Oxyacetylene cutting equipment or metal cutting saw
Grinder
File
Straightedge
Hammer

PROCEDURE	KEY POINTS
1. Grind the crown from the weld so that it is flush with the face of the plate. Grind the back of the plate smooth.	1. Do not overgrind the plates. They should be the same thickness throughout. If the weld is ground below the surface of the plate, it will break first at that point.

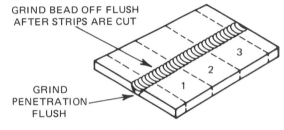

GRIND BEAD OFF FLUSH
AFTER STRIPS ARE CUT

GRIND
PENETRATION
FLUSH

Fig. 8-4

2. Using the straightedge, mark three 1-inch strips across the weld. Leave a small margin at each side of the plate, figure 8-4.	
3. Carefully cut the three strips from the plate.	3. If the cutting torch is used, be careful not to leave gouges in the edges of the strips. Discard the outside margins.
4. Grind and then file the edges of the strips. Work along the length of the strips so that any grinder marks or file marks will be parallel with the length of the specimen.	4. Be careful to keep the width of the strips uniform. Also, any nicks which run across the plate may cause the metal to fail at that point.
5. Following the manufacturer's instructions, place the center strip in a tensile tester and pull until it breaks.	5. The weld should hold more than the parent metal.
6. Bend one of the outside strips over the face of the weld, figure 8-5, and the other outside strip over the root of the weld, figure 8-6.	6. There should be no holes or separation of the bead from the plate anyplace along the weld.

ROOT

FACE

BEND 180°

Fig. 8-5 Bend Over the Face of the Weld

FACE

ROOT

BEND 180°

Fig. 8-6 Bend Over the Root of the Weld

REVIEW QUESTIONS

1. What is a butt weld?

2. How much penetration is required for a butt weld to have full strength?

3. Draw a sketch of the cross section of a piece which has been beveled for a butt weld. Show the proper angle.

4. What is a land?

5. Why is a land necessary on pieces to be butt welded?

6. Why are the pieces to be butt welded gapped?

7. What is the greatest thickness of steel plate which can be successfully butt welded without beveling or gapping the edges?

8. When cutting strips to perform a bend test or tensile test on a weldment, why is it important to file the edges smooth?

9. Where should the specimen break in a tensile test of a weld?

10. When grinding the face of a weld, in preparation for a bend or tensile test, why is it important to keep a uniform cross section?

SECTION 1: WELDS IN THE FLAT POSITION, COMPREHENSIVE REVIEW

A. BUTT WELD IN THE FLAT POSITION

Equipment and Material:

Standard arc welding equipment
Protective clothing
Two pieces of mild steel plate, 1/4" x 4" x 5"
1/8" and 5/32" electrodes (AWS classification to be determined by the student)

PROCEDURE

1. Grind one edge of each plate to form a 60-degree included angle.

2. Grind a 1/16-inch land on each bevel.

3. Tack weld the plates at each end, leaving a gap of about 1/16 inch.

4. Butt weld the two plates in the flat position.

5. Cut three 1-inch strips from the plate, leaving a small margin at each side, as shown in the illustration.

6. Test each strip by bending one over the face of the weld, one over the root of the weld, and pulling one apart in the tensile tester.

7. Have the tested specimens inspected by the instructor.

**Cut Three Strips Across
the Weld for Testing**

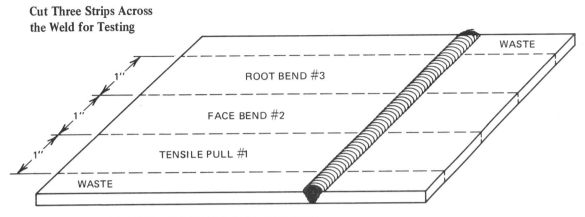

ROOT BEND #3

FACE BEND #2

TENSILE PULL #1

WASTE

WASTE

1"

1"

1"

B.   FILLET WELD IN THE FLAT POSITION

Equipment and Material:

Standard arc welding equipment
Protective clothing
One piece mild steel plate, 3/16" or 1/4" x 6" x 10"
Three pieces of mild steel plate, 3/16" or 1/4" x 3" x 4"
1/8", E-6011 electrodes

PROCEDURE

1. Tack weld the material in the position shown in the illustration.
   Note: If 1/4-inch steel is used it may be necessary to bevel the edge of the upstanding plates.

2. Weld one pass completely around the three plates.

3. Chip and brush the weld.

4. Weld two more passes around the plates to achieve a smooth fillet weld.

5. Chip and brush the welds and examine the appearance of the welds.

6. Cut the weldment to obtain a cross section of each of the three upstanding plates.

7. Smooth the cuts and etch each with acid.

8. Have the etched specimens inspected by the instructor.

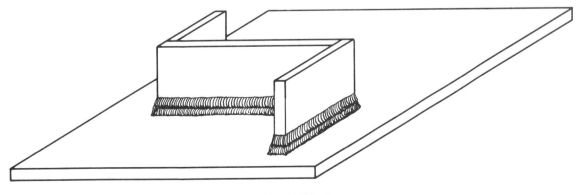

Fillet Weld Review

SECTION 2:
Welds in the Horizontal and Vertical Positions

Unit 9 Stringer Beads, Horizontal Position

OBJECTIVES

After completing this unit the student will be able to:

- describe the horizontal welding position.
- list the problems encountered in horizontal welding and some remedies for these problems.
- weld a bead in the horizontal position.

Horizontal Weld

An *out-of-position* weld is a weld made in any position other than flat. A horizontal weld is a weld made in a horizontal line and against a surface which is approximately vertical.

Horizontal welding presents a problem because gravity works against the welder. The molten metal deposited by the arc has a tendency to sag downwards. This sag must be controlled by rod angle and rod manipulation. Also, care must be taken that an undercut does not develop at the top of the bead. Correct polarity and use of the right electrodes assist in making a good weld.

The electrode should be held at an angle of 90 degrees with the plate being welded, then pointed slightly toward the weld, figure 9-1. If the puddle has a tendency to sag, the electrode can be angled upward slightly (about 20 degrees). In this position, the force of the arc will help hold the molten puddle in place until it has cooled enough to support itself. A slight circular movement of the end of the rod may help the beginning welder maintain bead appearance.

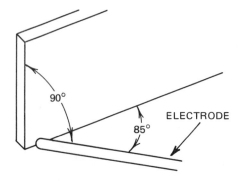

Fig. 9-1 Electrode Angle

JOB 9: STRINGER BEADS, HORIZONTAL POSITION

Equipment and Material:

Standard arc welding equipment
Protective clothing
1/4″ or 3/16″ mild steel plate
E-6011 electrodes

PROCEDURE	KEY POINTS
1. Place a sample plate in position for horizontal welding.	
2. Set the welding machine for reverse polarity.	
3. Beginning at the bottom left corner of the plate, weld a stringer bead across the plate.	3. Hold the welding rod at an angle of 90 degrees with the plate and pointed slightly toward the completed weld. Move at a steady, uniform rate of speed.
4. Chip and brush the weld.	
5. Weld another bead across the plate, just above and slightly overlapping the first bead. The first bead welded will help support the second bead.	5. The second bead should overlap about one-third of the first bead, figure 9-2.
6. Continue welding horizontal beads until the plate has been completely filled with weld, figure 9-3.	6. Be sure to chip and brush each bead before welding the next one.
7. Save this plate for use in unit 15.	

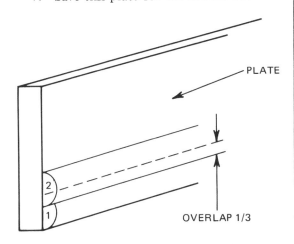

Fig. 9-2 Overlap for Stringer Beads

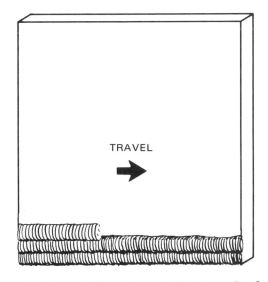

Fig. 9-3 Cover the Entire Plate with Stringer Beads

REVIEW QUESTIONS

1. Describe the horizontal welding position.

2. Explain how the arc can be used to hold the molten puddle in place on a horizontal weld.

3. Describe the best electrode angle for welding a horizontal bead.

4. Why is horizontal welding more difficult than flat welding?

5. When one bead overlaps another, what must be done to prepare the first bead before the second one is welded?

6. What causes undercutting when welding in the horizontal position?

7. How much should beads be overlapped when covering a plate with horizontal beads?

Unit 10   Lap Welds, Horizontal Position

OBJECTIVES

After completing this unit the student will be able to:

- list the reasons why lap welds are avoided where possible.
- describe the difficulties which may be encountered in lap welding.
- weld a lap joint in the horizontal position.

Lap Joint

Lap welds, figure 10-1, are a common application of electric arc welding on mild steel. However, this joint has certain disadvantages which must be recognized. Although the metal may be perfectly lapped, one edge over the other, the space between the overlapped edges is a natural place for corrosion to occur. Moisture can condense in the lap, as a result of heating and cooling. This moisture causes rusting which eventually results in failure of the metal.

Horizontal lap joints are normally easy to weld with good bead appearance. The electrode can be pointed into the corner of the joint and the lower piece of metal acts as a ledge to support the molten metal. Care must be taken to avoid burning away the edge of the lap. Also, if the electrode angle is not correct, the arc flow can wash away the molten metal, causing undercuts. The joint should be welded with constant, uniform progress.

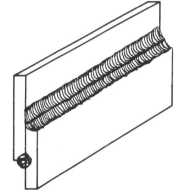

Fig. 10-1 Lap Weld

JOB 10: LAP WELD, HORIZONTAL POSITION

Equipment and Material:

Standard arc welding equipment
Protective clothing
1/4″ or 3/16″ mild steel plate
E-6013 and E-6011 electrodes

PROCEDURE	KEY POINTS
1. Tack weld two pieces of mild steel plate so the edges overlap 1 inch, figure 10-2.	1. Use DCSP and E-6013 electrodes for this weld.

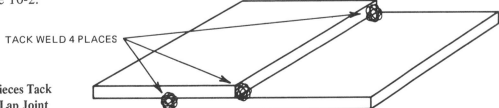

TACK WELD 4 PLACES

Fig. 10-2 Pieces Tack Welded for Lap Joint

PROCEDURE

2. Position the tack welded assembly for a horizontal weld, figure 10-3.

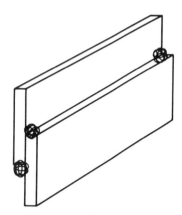

**Fig. 10-3 Lap Joint Positioned
for a Horizontal Weld**

3. Weld the joint using E-6013 electrodes. Chip and brush the weld, then weld two more passes, as shown in figure 10-4.

4. Reverse the plate so that the opposite side can be welded.

5. Weld this side of the joint in three passes using DCRP and E-6011 electrodes.

6. Chip and brush the weld.

7. Cut the lap joint across the center, file or grind the edges and etch them. Inspect the etched edges for penetration. The weld should contain no slag inclusions or pinholes.

KEY POINTS

2. The work may be propped up with a firebrick or held by a positioner.

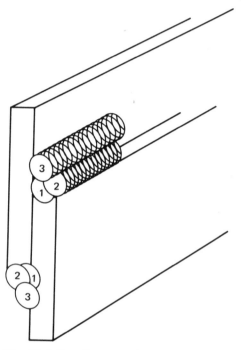

Fig. 10-4 Pass Sequence for a Lap Weld

REVIEW QUESTIONS

1. What are the disadvantages of lap welds, as compared with other joint designs?

2. Why is a lap weld in the horizontal position a relatively easy weld to make?

3. Draw a sketch of a lap weld in the horizontal position. Indicate where the first, second, and third passes should be placed and which is the penetration pass.

4. Describe the difference between welding with E-6011 electrodes and E-6013 electrodes.

5. Describe the difference between welding with DCRP and DCSP.

Unit 11   Butt Welds, Horizontal Position

OBJECTIVES

After completing this unit the student will be able to:

- describe the difficulties encountered in horizontal butt welding.

- prepare steel plate for a butt weld and describe the reasons for this preparation.

- weld a butt joint in the horizontal position.

Horizontal Butt Weld

Welding butt joints in the horizontal position appears to be a relatively simple operation, but it requires careful concentration. While the metal is molten the force of gravity can pull the puddle down, causing the bead to roll over the bottom edge of the weld. This does not leave enough weld metal to fill the top edge of the joint and an undercut results. If the proper electrode angle is used, the pressure of the arc will hold the puddle up and fill the undercut. The electrode should be pointed upward slightly and back into the weld just enough to maintain control over penetration.

Metal Preparation

The preparation of the joint is very important for all butt welds in steel plate. The edges of the metal must be ground to about 60 degrees and a square land must be ground on the edge of the bevel. The two prepared plates should be tacked square with each other with about a 1/16-inch gap between them at the bottom of the bevel.

The gap is important for the penetration of the root pass. If the gap is not the same size throughout the weld, the penetration of the root pass will not be uniform when it comes through the plate. The resulting skips or holes may cause the weld to fail.

The beads should be small enough to be controllable. A horizontal butt weld may require more passes to fill the groove than do welds made in other positions.

JOB 11: BUTT WELD, HORIZONTAL POSITION

Equipment and Material:

Standard arc welding equipment
Protective clothing
Two pieces mild steel plate, 1/4″ x 4″ x 5″
1/8″ and 5/32″, E-6011 electrodes

PROCEDURE	KEY POINTS

1. Grind one edge of each plate to approximately 60 degrees and grind a 1/16-inch land at the bottom of each bevel, figure 11-1.

2. Tack weld the two plates together so that the bevels form a V.

3. Position the weldment for a horizontal weld, figure 11-2.

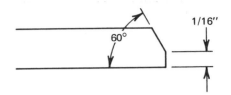

Fig. 11-1

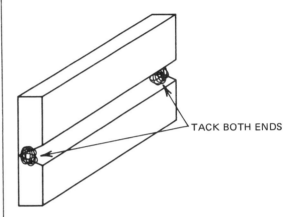

Fig. 11-2 Plates Tacked for a Butt Weld

4. Beginning at the left side, (left-handed welders begin at the right side) run a root pass the length of the joint, using 1/8-inch, E-6011 electrodes.

4. The root pass must completely fuse the edges of the plate on the bottom side.

5. Chip and brush the weld.

5. Examine the weld for complete penetration and fusion of the edges. If complete fusion has not taken place, the weld should be cut out, reground and welded again.

6. When a satisfactory root pass has been made, finish welding the V, using 5/32-inch, E-6011 electrodes.

6. Either two or three more passes will be required to completely fill the V. The *crown* (top) of the bead should be convex.

7. Chip and brush the weld.

8. Test this weld as was done in Unit 8.

REVIEW QUESTIONS

1. Why is undercutting more apt to occur on a butt joint welded in the horizontal position than on one welded in the flat position?

2. How can the electrode be handled to reduce undercutting on a horizontal butt weld?

3. Why is it important that the plates to be butt welded be uniformly gapped?

4. How does the number and size of the beads welded in a horizontal butt joint compare with those welded in other positions?

5. Draw a sketch showing the cross section of a properly welded butt joint in heavy steel plate.

Unit 12 Stringer Beads, Vertical Position

OBJECTIVES

After completing this unit the student will be able to:

- describe the vertical welding position.

- describe the electrode position for vertical welding.

- weld stringer beads in the vertical position.

Vertical Weld

A vertical weld is any weld made in an approximately vertical line. Vertical welding may be difficult for the beginner, but with practice attractive welds with excellent strength can be made.

The electrode should be held at an angle of 90 degrees with the plate, then pointed slightly upward, figure 12-1. Keep the electrode pointed directly at the plate. Allowing it to point slightly left or right will allow the arc to wash the molten metal out of the puddle. If the molten metal has a tendency to flow downward, it may be necessary to flash out of the puddle. Flashing out of the puddle refers to removing the arc from the puddle momentarily, to allow the puddle to cool slightly.

Sometimes a slight up-and-down whipping motion of the electrode is helpful. To use a whipping motion, move the electrode tip up out of the puddle and back down again rapidly. Ordinarily, smooth, even progress up the plate results in a good weld. As the deposited metal cools and solidifies, it forms a shelf which supports the next layer of molten metal. This results in the bead having the appearance of overlapping shingles as the weld progresses up the plate.

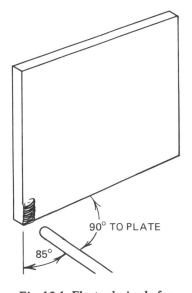

Fig. 12-1 Electrode Angle for Vertical Weld

JOB 12: STRINGER BEADS, VERTICAL POSITION

Equipment and Material:

Standard arc welding equipment
Protective clothing
3/16" or 1/4" mild steel plate
1/8", E-6010 or E-6011 electrodes

PROCEDURE	KEY POINTS
1. Set the machine for reverse polarity.	
2. Position the metal for vertical welding.	2. The work may be tacked to a piece of scrap or held by a positioner.
3. Beginning at the bottom of the plate, strike an arc and weld a bead upward in a straight line to the top of the plate.	3. Keep the rod at an angle of 90 degrees with the plate. Pointing the electrode upward about 5 degrees may help, but if the plate becomes too hot return to the straight-in position. Moving the rod alternately above the puddle and back into it will also help cool the puddle, but do not use this rocking motion unless it is necessary.
4. Chip and brush the weld, then have it checked by the instructor.	
5. Reposition the plate for vertical welding and completely cover the plate with beads. Each bead should overlap the one before it by about one-third, figure 12-2.	5. Clean each pass before the next bead is deposited.

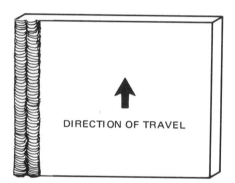

DIRECTION OF TRAVEL

Fig. 12-2 Cover the Entire Plate with Stringer Beads

6. Turn the plate so that a second layer of beads can be run across the first layer. Cover the plate with beads in this direction.	6. Weld from bottom to top. As the welding progresses, remember that a short arc helps cool the puddle and a long arc heats the puddle.
7. Turn the plate again and cover it with a third layer of beads.	
8. Cut through the plate with a torch or saw. Smooth and etch the cut edges to observe the penetration and fusion.	8. The weld metal should be thoroughly mixed with the parent metal and there should be no *porosity* (pinholes) or *slag inclusions* (slag trapped in the weld).

REVIEW QUESTIONS

1. Describe the vertical welding position.

2. Describe the best electrode angle for vertical welding.

3. What is the result if the electrode is pointed too much to the right or left as a vertical bead is being welded?

4. For what reason might the welder flash the arc away from the weld momentarily during a vertical weld?

5. Describe the whipping motion which can be used in vertical welding.

6. What supports the molten puddle during a vertical weld?

7. What are slag inclusions and what causes them?

8. How does the length of the arc affect the heat at the puddle?

Unit 13 Lap Welds, Vertical Position with Weave Beads

OBJECTIVES

After completing this unit the student will be able to:

- describe the difficulties encountered in welding a vertical lap joint.
- describe the procedure for welding a vertical weave bead.
- weld a lap joint in the vertical position.

Vertical Lap Weld

Vertical lap welds may be difficult for a beginner, but with practice they can be mastered. The two plates to be welded must be firmly clamped together before they are tacked. If there is a gap between the two pieces, the edge of the overlapping piece may be melted away, resulting in a poor quality weld.

Electrode Angle

Because a lap weld is made against the flat side of one piece of metal and the edge of another, the electrode angle is important. If the electrode is directed toward the flat surface too much, penetration will not be achieved into the overlapping piece. If the electrode is directed toward the edge of the overlapping piece too much, the edge may be melted away without penetration into the other piece. To achieve good penetration the electrode should be directed slightly more toward the flat surface than toward the edge, and should be angled upward slightly, figure 13-1.

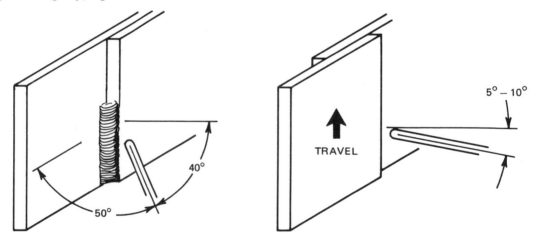

Fig. 13-1 Electrode Angle for a Vertical Lap Weld

Vertical Weave Bead

After the first pass is completed, the remainder of the joint may be filled with a weave bead. The weave bead can be used to fill the joint in one upward operation, figure 13-2.

As with the root pass, the electrode angle is important in welding the weave bead to fill the joint. The heat must be controlled so that there is complete fusion with the parent metal at both sides of the bead. Move the electrode back and forth across the root bead, building layers one upon the other, with uniform speed. The finished weld should have the appearance of a shingled roof, with each layer overlapping the one below it.

JOB 13: LAP WELD, VERTICAL POSITION WITH WEAVE BEAD

Equipment and Material:

    Standard arc welding equipment
    Protective clothing
    3/16" or 1/4" mild steel plate
    E-6010 or E-6011 electrodes

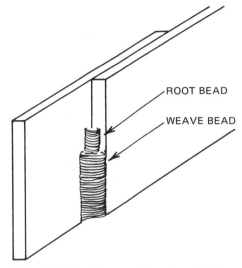

Fig. 13-2 Pass Sequence for a Lap Weld

PROCEDURE	KEY POINTS
1. Set the welding machine for reverse polarity.	
2. Tack weld two pieces of mild steel plate for a lap weld, figure 13-3.	2. Clamp the plates so there is no gap when they are tacked.
3. Place the tacked plates in position for a vertical weld.	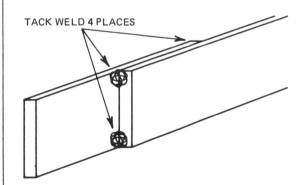 Fig. 13-3 Plates Tack Welded for a Lap Weld
4. Run one bead in the corner, traveling from bottom to top.	4. **CAUTION: Molten metal and sparks will be splattered around the area where vertical welding is being done. Check the nearby area, including the welder's clothing, frequently for fires. Wear protective clothing.**
5. Chip and brush the weld.	

PROCEDURE	KEY POINTS
6. Beginning at the bottom, fill the joint with a weave bead. Move the electrode from side to side, building layers of weld metal, figure 13-4.	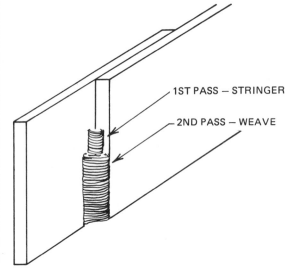 1ST PASS — STRINGER 2ND PASS — WEAVE **Fig. 13-4 Fill the Joint with a Weave Bead**
7. Cut or saw the plate across the weld and etch with nitric acid to check the penetration of the weld.	7. The penetration should be deep into the bottom plate of the lap. The weave bead should be thoroughly mixed with the root pass and no slag holes should appear.

REVIEW QUESTIONS

1. Why is it important to clamp the plates firmly together before tack welding the lap joint?

2. Describe the electrode angle for the first pass on a vertical lap weld.

3. What is the result if the electrode is pointed too much toward the flat surface of the back piece?

4. What is the result if the electrode is pointed too much toward the edge of the overlapping piece?

5. What is the advantage of a weave bead over a series of stringer beads for filling a lap joint after the root pass is welded?

Unit 14 Fillet Welds, Vertical Position

OBJECTIVES

After completing this unit the student will be able to:

- discuss the difficulties which may be encountered when welding in the vertical position.
- describe the proper electrode angle for welding a vertical T joint.
- make a fillet weld in the vertical position.

The vertical fillet weld is one of the joints welded most frequently in industry. The welding student must practice welding this joint to learn to do it well.

The first pass must achieve penetration. There is no way to increase the penetration after the first pass has been welded into the corner. To insure maximum penetration in heavy metal, the corner must be prepared by gapping or grinding the pieces. The corner should be welded carefully and exactly. The electrode should be pointed squarely into the joint and pointed slightly upward, figure 14-1. The welder should observe the puddle and pay particular attention to the amount of penetration which is occurring.

A short arc will make the first pass easier to weld. A long arc increases the chances of too much heat traveling ahead of the weld. The heat, thus generated, melts the metal ahead of the bead and may cause difficulties with penetration and bead appearance.

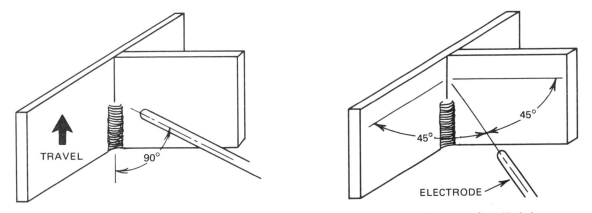

Fig. 14-1 Electrode Angle for Vertical Fillet Weld. The Electrode May be Pointed up Slightly.

JOB 14: FILLET WELD, VERTICAL POSITION

Equipment and Material:

Standard arc welding equipment
Protective clothing
3/16" or 1/4" mild steel plate
1/8" and 5/32", E-6011 electrodes

PROCEDURE	KEY POINTS
1. Tack weld two pieces of mild steel plate for a T weld and position them for a vertical weld.	
2. Using a 1/8-inch electrode, weld one pass from bottom to top.	2. The first pass must achieve good penetration into the corner. Hold the electrode straight into the corner. If the travel is too fast or the arc is too long, poor penetration will result and the bead will have too much crown.

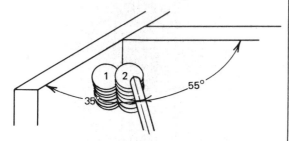

Fig. 14-2 Electrode Angle for Second Pass

PROCEDURE	KEY POINTS
3. Chip and brush the beads and weld two more passes from bottom to top, using 5/32-inch electrodes, figure 14-2.	3. On the second and third passes, the electrode should be pointed into the corner between the previous bead and the steel plate.
4. Chip and brush the weld and examine the appearance of the beads.	
5. Cut through a cross section of the weld, smooth the edges, and etch it to check the penetration.	
6. Tack weld two more plates in position as in step 1.	
7. Weld both sides of the joint, using the procedure outlined in steps 2 and 3. See figure 14-3.	7. A slight weaving motion of the electrode may be necessary to weld the second and third passes.

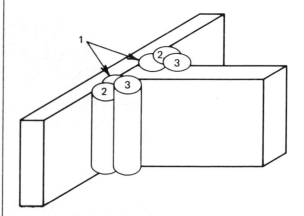

Fig. 14-3 Pass Sequence for Fillet Weld on Both Sides of a T Joint

PROCEDURE	KEY POINTS
8. Chip and brush the weld and examine the appearance.	8. The bead should be smooth and convex, but not with a high crown. There should be no pinholes in the bead.
9. Cut through a cross section of the weld, smooth the edges, and etch it to check the penetration.	

REVIEW QUESTIONS

1. What polarity is best for vertical fillet welding?

2. What is the result if the wrong electrode angle is used when welding the first pass of the fillet weld?

3. What is the advantage of E-6011 electrodes over E-6013 electrodes for vertical fillet welding?

4. Why is the penetration of the first pass especially important on a fillet weld?

5. Describe the proper electrode angle for a vertical fillet weld.

6. What is the effect of holding too long an arc in making a vertical fillet weld?

7. Describe the proper electrode angle for the second pass on a vertical fillet weld.

Unit 15 Padding a Plate with Vertical and Horizontal Beads

OBJECTIVES

After completing this unit the student will be able to:

- describe the pad welding process.
- discuss the importance of carefully chipping and brushing beads when building up surfaces with the electric arc.
- pad a plate to a depth of four beads.

Pad Welding

Pad welding is done by running beads over each other in alternate directions. This procedure is used for building up worn surfaces to their original thickness. Sometimes when the welds are carefully chipped and brushed between passes and there are no slag holes or blowholes, the surface may be machined to give the appearance of the original metal.

In welding the horizontal beads, the sag must be carefully controlled as the first bead across the bottom of the plate is run. As the following beads are welded, each bead is supported by the one directly below. Each bead should overlap the previous bead by about one-third its width.

After the plate is covered with horizontal beads, vertical beads are run from bottom to top, figure 15-1. As with other welding, right-handed welders will find it most convenient to start on the left side of the plate. Each bead must be melted into the preceding bead and those it crosses in the opposite direction.

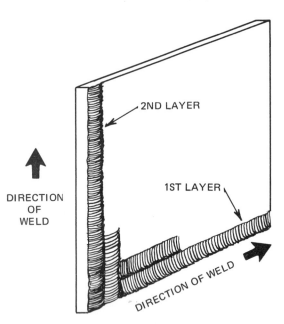

DIRECTION OF WELD

2ND LAYER

1ST LAYER

DIRECTION OF WELD

Fig. 15-1 Pad Buildup by Welding Beads at Right Angles to One Another

JOB 15: PADDING A PLATE WITH VERTICAL AND HORIZONTAL BEADS

Equipment and Material:

Standard arc welding equipment
Protective clothing

Plate welded in Unit 9
E-6011 electrodes

PROCEDURE	KEY POINTS
1. Place the plate welded in Unit 9 in the same position as was used for completing Job 9.	1. The beads welded in Job 9 should be horizontal.
2. Use DCRP for this job.	
3. Beginning in the lower left-hand corner, weld a vertical stringer bead up the side of the plate.	3. This bead should be thoroughly fused into the horizontal stringer beads.
4. Chip and brush the bead.	
5. Continue welding vertical beads, overlapping each about one-third of the width of the preceding bead, until the plate is covered.	5. Chip and brush each bead before welding the next one.
6. Beginning in the lower left-hand corner, weld another complete layer of horizontal stringer beads.	6. Care must be used not to allow the first horizontal bead to sag. Be sure to clean each bead.
7. Weld another layer of vertical beads.	7. The plate should now have four layers of weld metal on the parent metal.
8. Cut the plate in two and etch it to observe the fusion and penetration.	8. If the plate is to be sawed, it must be allowed to cool before sawing.

REVIEW QUESTIONS

1. Why is it important that no slag or blowholes are allowed to accumulate in the weld metal of a buildup job?

2. How can the bead deposited on the first pass of a horizontal weld be controlled so it does not sag?

3. What is an out-of-position weld?

4. Explain why DCRP or AC are better for making out-of-position welds than DCSP.

5. What is pad welding used for in industry?

6. Approximately how much should the beads overlap in pad welding?

Unit 16   Butt Welds, Vertical Position with Weave Beads

OBJECTIVES

After completing this unit the student will be able to:

- describe the preparation of steel plate for a vertical-position weave-bead butt weld.

- explain how the root pass is welded on a vertical butt weld.

- weld a butt joint in the vertical position.

Vertical Butt Weld

The vertical butt weld may be difficult for the beginner, but it can be mastered with practice. When material thicker than 1/8 inch is used, it is usually necessary to grind a bevel on the edge of the material. The two plates should be gapped so that penetration of the molten metal is assured in the first pass. The first pass must be made from bottom to top, with the rod angled slightly upward. If the material is gapped the right amount, a bead will form on the back of the plates and the weld will completely fuse the bottom edges of the plates. After this first pass is completed, the rest of the V joint may be filled by running more vertical passes, or by weaving across the first bead (the *root pass*). The entire joint may be filled in one operation by using a weave bead.

The vertical weave bead must be welded carefully. The heat should be controlled so that the edges of the weave bead thoroughly mix with the plate. The weave bead should penetrate and be completely fused with the root bead. Move the electrode back and forth across the V, building the layers one upon the other, with a uniform speed. The resulting appearance should look something like a shingled roof, with each layer overlapping the one below it.

JOB 16: BUTT WELD, VERTICAL POSITION WITH WEAVE BEAD

Equipment and Material:

Standard arc welding equipment
Protective clothing
Two pieces mild steel plate, 1/4" x 4" x 5"
1/8" and 5/32", E-6011 electrodes

PROCEDURE	KEY POINTS
1. Grind 30 degrees from one edge of each plate and grind a 1/16-inch land on the bottom of each bevel, figure 16-1.	

PROCEDURE	KEY POINTS
2. Tack weld the two plates together so that the bevels form a V.	2. Gap the plates about 1/16 of an inch.
3. Position the weldment for a vertical weld.	

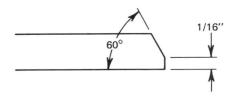

Fig. 16-1

PROCEDURE	KEY POINTS
4. Beginning at the bottom and using 1/8-inch, E-6011 electrodes, weld a root pass up the length of the V, figure 16-2.	4. The root pass must completely fuse the edges of the plate on the bottom side.
5. Chip and brush the weld.	5. Examine the weld for complete penetration and fusion of the bottom edges. If complete fusion has not taken place, the weld should be cut off, reground and welded again.
6. Starting at the bottom of the weld, and using 5/32-inch, E-6011 electrodes, carefully weave a bead across the root bead to the top of the plates, figure 16-3.	6. It is important to keep the electrode pointed straight into the weld. If it is tipped sideways the resulting wash of molten metal may undercut the edges of the weld. The rod may be tipped upward slightly to control the puddle.
7. Chip and brush the weld.	
8. Test this weld as was done in Unit 8.	

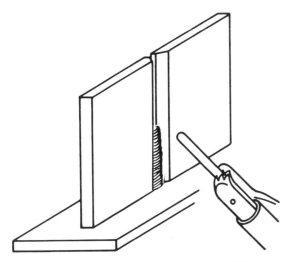

Fig. 16-2 Root Pass for a Butt Weld

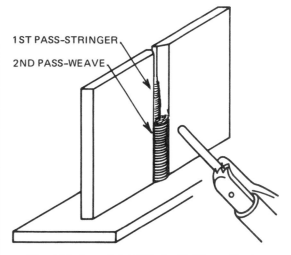

1ST PASS-STRINGER

2ND PASS-WEAVE

Fig. 16-3 Butt Joint Filled with Weave Bead

REVIEW QUESTIONS

1. Why is it necessary to grind a bevel on the edges of plates being butt welded in the vertical position?

2. Why should plates to be butt welded in the vertical position be gapped?

3. Name two types of beads which can be used to fill a vertical butt joint after the root bead is welded.

4. Describe the proper electrode angle for vertical butt welding.

5. What is the result of pointing the electrode to one side or the other on a vertical butt weld?

6. Draw a sketch of the cross section of a beveled butt weld, showing the angle of the beveled edges.

SECTION II: WELDS IN THE HORIZONTAL AND VERTICAL POSITION, COMPREHENSIVE REVIEW

A. LAP WELD IN THE HORIZONTAL POSITION

Equipment and Material

Standard arc welding equipment
Protective clothing
Two pieces of mild steel plate, 1/4" x 4" x 6"
1/8", E-6011 electrodes

PROCEDURE

1. Tack weld the two pieces of steel plate together for a lap weld.

2. Position the weldment for a horizontal weld and weld one side of the joint in a single pass as shown in the illustration.

3. Reposition the plate to weld the opposite side in the horizontal position with one pass.

4. Cut through a cross section of the weld, then smooth and etch the weld. Have the etched weld inspected by the instructor.

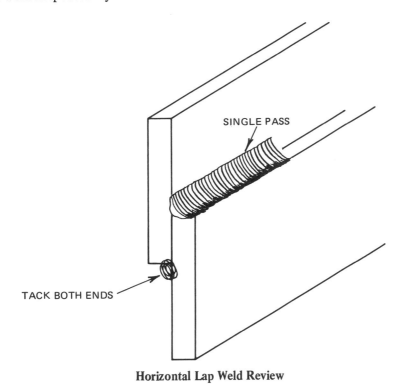

SINGLE PASS

TACK BOTH ENDS

Horizontal Lap Weld Review

B. FILLET WELD IN THE HORIZONTAL POSITION

Equipment and Material:

> Standard arc welding equipment
> Protective clothing
> Two pieces of mild steel plate, 1/4" x 3" x 6"
> 1/8", E-6011 electrodes

PROCEDURE

1. Tack weld the two pieces of steel plate together as shown in the illustration of a fillet weld.

2. Weld the inside corner with one pass in the horizontal position.

3. Weld the outside corner with three passes as shown in the illustration.

4. Cut through a cross section of the weld, then smooth and etch the cut edges. Have the etched weld inspected by the instructor.

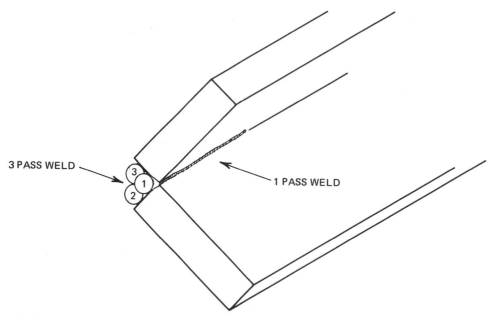

3 PASS WELD

1 PASS WELD

Horizontal Fillet Weld Review

C.   WEAVE BEADS IN THE VERTICAL POSITION

Equipment and Material:

> Standard arc welding equipment
> Protective clothing
> Two pieces of mild steel plate, 1/4″ x 5″ x 6″
> 1/8″, E-6011 electrodes

PROCEDURE

1. Tack weld the two pieces of steel plate for a lap weld as shown in the illustration.

2. Set the welding machine for the correct polarity and amperage.

3. Weld one side of the lap joint in the vertical position with a single stringer bead.

4. Chip and brush the bead and fill the joint with a weave bead in the vertical position.

5. Chip and brush the weld and allow it to cool at room temperature.

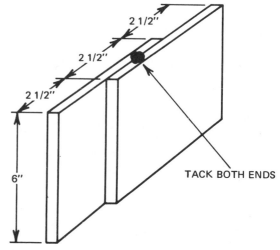

2 1/2″

2 1/2″

2 1/2″

6″

TACK BOTH ENDS

Position of Plates for Vertical Weave Bead Review

6. Test the weld by bending one plate over the bead. The weld should break through the center of the weld for the entire length of the plate.

7. Have the tested specimen inspected by the instructor.

D.   BUTT WELD IN THE VERTICAL POSITION

Equipment and Material:

Standard arc welding equipment
Protective clothing
Two pieces of mild steel plate, 1/4" x 4" x 5"
1/8" and 5/32" electrodes (AWS classification to be determined by the student)

PROCEDURE

1. Grind one 5-inch edge of each plate to form a 60-degree included angle, then grind a 1/16-inch land on each bevel.

2. Tack weld the plates at each end, leaving the proper gap between them.

3. Butt weld the joint in the vertical position.

4. Cut three 1-inch strips from the weldment, as shown in the illustration.

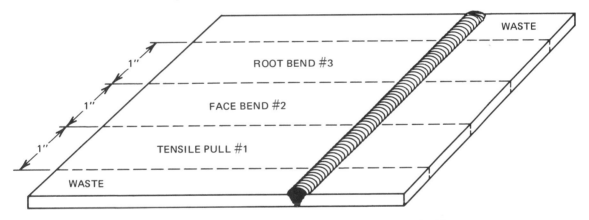

Cut 1-inch Strips Across the Weld to Test the Butt Weld.

5. Test each specimen by bending one over the face of the weld, one over the root of the weld, and pulling one apart in the tensile tester.

6. Have the tested specimens inspected by the instructor.

SECTION 3:
Welds in the Overhead Position

Unit 17   Stringer Beads,
Overhead Position

OBJECTIVES

> After completing this unit the student will be able to:

- describe the overhead welding position.
- list three factors which contribute to the difficulty of making overhead welds.
- describe the differences between DCRP and DCSP welding.
- weld a stringer bead in the overhead position.

Overhead Weld

An overhead weld is one which is made on the underside of the joint and runs in a horizontal line, figure 17-1. Welds which are inclined 45 degrees or less are considered to be in the overhead position.

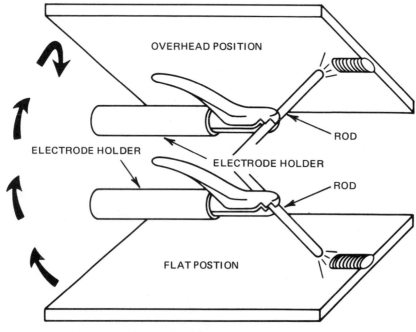

Fig. 17-1

Welding Polarity

Out-of-position welds (welds made in positions other than the flat position) are generally made with DCRP. However, out-of-position welds can be made with DCSP or AC. When the plate is the negative terminal and the electrode is the positive terminal, two-thirds of the welding heat is in the electrode. This means that the coupon does not become as hot and does not have as great a tendency to drip down. AC welding machines release about half of the heat at each terminal, so they are also useful for out-of-position welding.

Difficulties of Overhead Welding

Overhead welding may seem more difficult than other positions because of four factors:

- Welding overhead places the arms and neck in an awkward position and tends to cause muscle cramp.

- Molten metal tends to drip down from the weld because of the force of gravity. If the electrode is not handled correctly and if the welder does not flash off from the weld when the heat becomes too intense, a poor weld results.

- In order to prevent an excessive concentration of heat in one area, the weld must be made with more passes, depositing a smaller bead in each pass. This means that overhead welding takes more time than other welding positions.

- Because of the hazard created by the molten metal dropping from the weld, bulky fireproof clothing must be worn for overhead welding. This bulky clothing slows down the welder and creates an uncomfortable working condition.

Positioner

A fixture which is used to hold material while it is being worked on is called a *jig*. A *positioner* is a special jig for holding material in the desired position for welding. For overhead welding the coupon must be clamped above the welder's head in the positioner.

JOB 17: STRINGER BEADS, OVERHEAD POSITION

Equipment and Material:

Standard arc welding equipment
Protective clothing
Mild steel plate, 1/4" x 5" x 5"
1/8", E-6011 and E-6013 electrodes

PROCEDURE	KEY POINTS
1. Clamp the steel plate in the overhead position.	1. A positioner should be used for this.
2. Use DCRP and an E-6011 electrode.	2. Set the amperage as recommended by the electrode manufacturer.

PROCEDURE	KEY POINTS
3. With the electrode holder in the position shown in figure 17-2, run an overhead stringer bead across the plate.	3. The electrode should be perpendicular to the plate. If it is tipped at all, it should be no more than 5 degrees in the direction of travel.
4. Turn the machine off, hang up the electrode holder and remove the plate from the positioner.	4. The welded plate should be handled with pliers.
5. Clean the weld and have the instructor inspect it.	
6. Repeat the entire procedure, using an E-6013 electrode and straight polarity.	

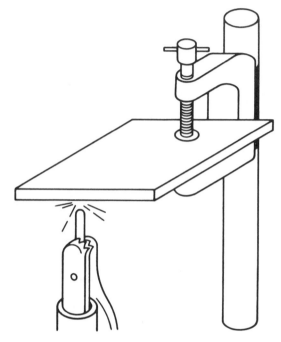

Fig. 17-2 Position of Electrode Holder for Overhead Welding

REVIEW QUESTIONS

1. Describe the overhead welding position.

2. What polarity is generally used for out-of-position welds?

3. Why is this polarity preferred for out-of-position welding?

4. How does the number of passes used to weld a joint in the overhead position compare with the number used to weld the same joint in the flat position?

5. How can the heat developed during overhead welding be controlled?

6. What is a positioner?

7. Why is overhead welding more dangerous than welding in other positions?

Unit 18   Pad Buildup, Overhead Position

OBJECTIVES

After completing this unit the student will be able to:

- explain how the welding machine can be set for the correct current for varying thicknesses of metal and varying electrode diameters.
- describe the electrode angle for welding a pad in the overhead position.
- weld a pad in the overhead position.

Pad Buildup

Building up pads provides excellent practice for the beginning welder. It is also an important industrial operation. As in other positions, overhead padding is done by running stringer beads across the plate, overlapping each by about one-fifth.

The first pass across the plate should be as close as possible to the edge. The electrode should be held nearly perpendicular (at right angles) to the work, figure 18-1. Cleaning each pass is very important. Slag and wire brush each pass when it is completed, before the next pass is welded. This is necessary to prevent slag inclusions and a rough appearance in the next bead.

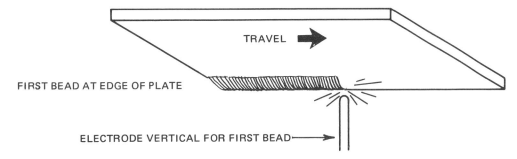

TRAVEL

FIRST BEAD AT EDGE OF PLATE

ELECTRODE VERTICAL FOR FIRST BEAD

Fig. 18-1 Electrode Vertical for First Bead

In all overhead welding, control of the heat is critical. A rocking motion, figure 18-2, may be helpful to control the heat. The rod is rocked forward, out of the weld, momentarily to allow the puddle to cool slightly.

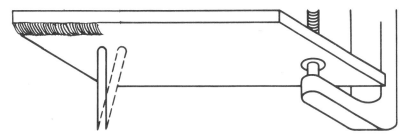

Fig. 18-2 Rocking Motion Used to Control Heat

65

Adjusting Amperage

In order to produce strong welds it is necessary to use the correct amperage setting. The electrode manufacturer's recommendations should be followed for the initial setting, but fine adjustments may be made by the welder. The best amperage can be determined by experimenting on a piece of scrap the same size as the piece to be used for the job. Set the material up as for the welding job, then make a trial weld. Set the current up or down 5 amperes at a time, until the machine is operating exactly as it should. Make it a rule always to test the machine on a piece of scrap before starting any welding job.

Changing Electrode Sizes

When electrodes of different sizes are used the welder must manipulate the rod differently and use varying amounts of heat. As the electrode diameter increases, the electrode movement may be faster, since more weld metal is deposited at a time. The bead buildup will be larger and the width of the bead will increase. Also, larger electrodes require more amperage to produce the greater amount of heat required.

JOB 18: PAD BUILDUP, OVERHEAD POSITION

Equipment and Material:

> Standard arc welding equipment
> Protective clothing
> Two pieces of mild steel plate, 1/4" x 5" x 5"
> 1/8" and 5/32" electrodes

PROCEDURE	KEY POINTS
1. Position the plate for an overhead weld.	
2. Use DCRP and a 1/8-inch electrode.	2. The electrode is positive.
3. Set the amperage according to the manufacturer's recommendations, then readjust it by testing on a piece of scrap metal.	3. Move the amperage indicator up and down five amperes at a time, running a trial bead each time the amperage is changed.
4. Weld a series of overhead beads, overlapping each bead about one-fifth, figure 18-3, until the plate is covered.	

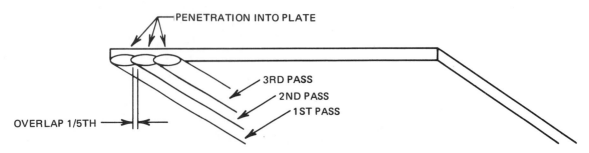

Fig. 18-3 First Layer of Beads for Overhead Padding

PROCEDURE	KEY POINTS
5. Chip and brush the weld and have it inspected by the instructor.	
6. Replace the plate in the positioner and weld another series of beads across the first series, figure 18-4.	
7. Chip and brush the welds and then cut the plate through the center and etch the edges to inspect the fusion and penetration.	7. All welds should be completely fused and have good penetration into the parent metal. There should be no pinholes or slag inclusions.
8. Repeat steps 3 through 7, using 5/32-inch electrodes.	

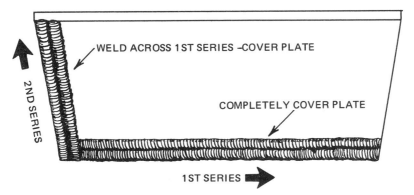

Fig. 18-4 Each Layer is at Right Angles to the One Before.

REVIEW QUESTIONS

1. Why must the welding machine be set for a higher amperage when larger diameter electrodes are used?

2. How does the diameter of the electrode affect the width of a stringer bead?

3. Describe the proper electrode angle for welding an overhead bead.

4. When building up a pad in the overhead position, how much should the beads overlap one another?

5. Why is it important to chip and brush each bead in pad welding?

6. Describe a method of making fine adjustments of the amperage.

Unit 19   Weave Beads,
Overhead Position

OBJECTIVES

After completing this unit the student will be able to:

- describe the technique for weaving beads in the overhead position.

- discuss the differences between welding in the overhead position and welding in other positions.

- weave a bead in the overhead position.

Overhead weave beads are welded in much the same manner as weave beads in other positions. Weave beads are frequently used to reinforce stringer beads, such as those used for a root pass, and to deposit more weld metal in a single pass, such as in pad buildup. The greatest difference between overhead weave beads and those made in other positions is the angle at which the electrode is held. As with other overhead welds, the electrode should be nearly perpendicular to the surface being welded.

There are two techniques the welder may use to insure that the bead is weaved a uniform width throughout. Stringer beads can be deposited at each side of the desired weave bead; then the bead is weaved between these stringer beads. This method helps prevent undercutting at the sides of the weave bead, where less heat is applied. The other method is to mark the path of the weave bead with center-punch marks. With this method it is necessary to pause slightly longer at the sides of the weave to prevent undercutting, figure 19-1.

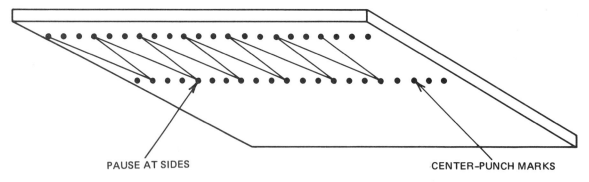

PAUSE AT SIDES   CENTER-PUNCH MARKS

Fig. 19-1 Method of Applying Weave Beads

JOB 19: WEAVE BEAD, OVERHEAD POSITION

Equipment and Material:

Standard arc welding equipment	Center punch
Protective clothing	Mild steel plate, 3/16″ or 1/4″ x 5″ x 5″
Hammer	1/8″, E-6011 electrodes

PROCEDURE	KEY POINTS
1. Position the material for overhead welding.	
2. Weld two stringer beads across the plate, approximately 3/4 inch apart, figure 19-2.	2. The electrode should be held at a 90-degree angle with the plate.

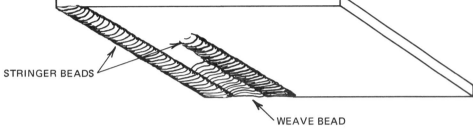

STRINGER BEADS

WEAVE BEAD

Fig. 19-2 Stringer Beads to Guide Welder

3. Weave a bead between the two stringer beads, being careful to fuse the weave bead into the stringer beads.	3. Hold the arc slightly longer at the sides to allow penetration into the stringer beads.
4. Chip and brush the weld, then examine it.	4. The weld should have smooth appearance and good penetration, with no blowholes.

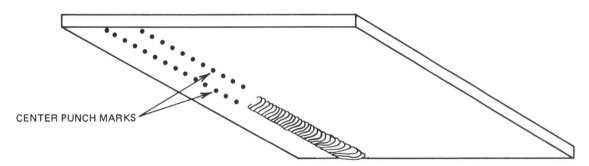

CENTER PUNCH MARKS

Fig. 19-3 Center-punch Marks to Guide Welder

5. On another plate, center punch two lines 3/4 inch apart, figure 19-3. Position this plate for an overhead weld.	
6. Weave a bead between the two rows of center-punch marks.	6. Stay between the center-punch marks, so the edges of the bead are in a straight line.

REVIEW QUESTIONS

1. List two reasons for using weave beads.
2. Describe the electrode angle for welding weave beads in the overhead position.

3. Describe two methods of insuring that weave beads are uniform in width.

4. What is the reason for pausing at the sides of the weave bead?

5. Draw a sketch of the electrode travel for a weave bead, indicating where pausing is required.

Unit 20   AC Welding, Overhead Position

OBJECTIVES

After completing this unit the student will be able to:

- briefly describe what takes place in AC welding.

- explain what a crater is and how to prevent it.

- weld a bead in the overhead position with an AC welding machine.

AC Welding

AC has a definite advantage over DC for welding, because it produces almost no magnetic field. The magnetic field around an arc can cause the arc to whip or wander off its path. This condition is called *arc blow*. Alternating current changes direction 120 times every second (60 cycles per second) which means that the voltage drops to zero 120 times a second. Voltage is the force that causes the current to jump the gap between the electrode and the work. Therefore, there should be no welding taking place at the time the voltage is zero. By covering AC electrodes with a compound *(ionizing agent)* which maintains the electrical path between the electrode and the parent metal, this problem is reduced to a minimum.

Because of the alternations in the direction of current flow, starting the arc is more difficult with AC. However, once the arc is started, good penetration can be achieved and the arc is relatively easy to maintain. The greatest advantage in AC welding is the absence of arc blow, making the arc easy to control. A disadvantage of AC for overhead welding is that there is more of a tendency for the weld to splatter. As with all overhead welding, the welder should be sure to wear protective clothing.

Crater

When a weld is completed and the heat from welding has built up in the metal, the weld has a tendency to melt more rapidly. Because of this rapid melting and the fact that an AC weld ends very abruptly, a *crater* (low spot) may be left at the end of the weld, figure 20-1. This crater is a likely spot for a fracture to begin and should not be left in the weld. To prevent a crater, flash off the end of the weld and then weld back into the crater. This deposits additional weld metal to fill the crater.

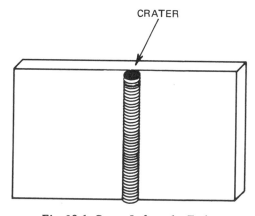

Fig. 20-1 Crater Left at the End of an AC Weld

71

JOB 20: AC WELDING, OVERHEAD POSITION

Equipment and Material:

AC welding machine
Protective clothing
Mild steel plate, 3/16" or 1/4" x 5" x 5"
1/8", E-6010 electrodes

PROCEDURE	KEY POINTS
1. If an AC-DC machine is used, switch the machine to AC.	
2. Test the current setting by welding on a piece of scrap in the overhead position.	2. AC welding generally requires higher current settings than does straight polarity on a DC machine.
3. Clamp the mild steel plate in position for overhead welding, then cover the plate with stringer beads.	3. Remember to chip the weld after each pass. Wear face protection when chipping welds.
4. Cut through the center of the plate and smooth and etch the edges to check penetration.	4. **CAUTION: Wear rubber gloves and eye protection. If acid contacts the skin, wash it off immediately with water.**

REVIEW QUESTIONS

1. What is the advantage of AC welding over DC welding?

2. What is arc blow?

3. The voltage of alternating current drops to zero 120 times every second. What is the reason for this?

4. How are electrodes treated to eliminate problems which would result from AC voltage dropping to zero 120 times a second?

5. What problem is caused by the fact that AC welds end very abruptly?

6. How can the problem mentioned in question 5 be corrected?

7. How does the current setting for AC welding compare with that for DCSP?

Unit 21   Fillet Welds, Overhead Position

OBJECTIVES

After completing this unit the student will be able to:

- explain the importance of a root pass on a fillet weld.
- explain the importance of thorough mixing of the electrode metal as the weld progresses.
- make a fillet weld in the overhead position.

Multiple-Pass Welding

A single-pass weld is one made by running a single bead across the weldment. *Multiple passes* (more than one) are often made to strengthen welds or build up material. It is particularly important for the first pass to have good penetration into the parent metal. The following passes are welded on top of the first pass and do not penetrate the parent metal. If the first pass does not penetrate the parent metal, later passes are of little value to the weld.

Multiple passes should be welded in a sequence that insures that each pass builds on the pass before it, figure 21-1. Each pass of a multiple-pass weld should be thoroughly mixed with the preceding pass and should overlap that pass by about one-fifth. Stirring of the metal as it is welded insures good fusion and eliminates gas pockets and slag inclusions. Using a slight rocking motion stirs the metal.

Polarity

Overhead welding is done most easily by using reverse polarity (electrode positive) and electrodes which are designed for that polarity. AC welding is also used frequently for out-of-position welds. Because of the problem of controlling heat, straight polarity is usually considered the least desirable polarity for out-of-position welding.

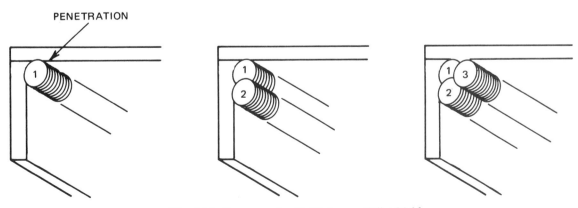

Fig. 21-1 Sequence for Multiple-pass Fillet Welds

JOB 21: FILLET WELD, OVERHEAD POSITION

Equipment and Material:

Standard arc welding equipment
Protective clothing
3/16" or 1/4" mild steel plate
1/8" or 5/32", E-6010 or E-6011 electrodes

PROCEDURE	KEY POINTS
1. Tack weld two pieces of steel plate in position for a fillet weld, figure 21-2.	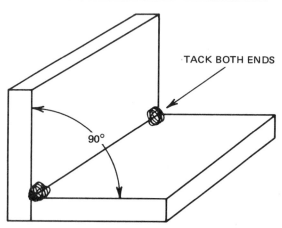
2. Clamp the weldment in position for an overhead fillet weld.	
3. Set the welding machine on DCRP and about 100 to 125 amperes.	
	Fig. 21-2 Plates Tack Welded for a Fillet Weld
4. Using a whipping motion, weld a root pass in the corner between the two plates, figure 21-3.	4. The electrode should be pointed directly into the corner, so it is at an angle of 45 degrees with each plate.
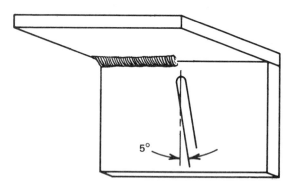	
Fig. 21-3 The Electrode Should be Directed Slightly Toward the Weld.	
5. Chip and brush the weld and have it inspected by the instructor.	
6. Return the weldment to the overhead position.	
7. Weld a second pass, directing the arc at the bottom half of the first bead.	7. Clean each pass before welding the next pass.

PROCEDURE	KEY POINTS
8. Weld a total of 10 passes in the joint as shown in figure 21-4. 9. Cut through a cross section of the weld, then smooth and etch the edges to inspect for penetration and fusion.	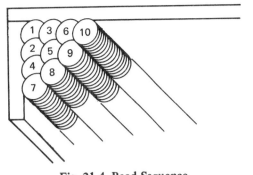 Fig. 21-4 Bead Sequence for Ten-pass Fillet Weld

REVIEW QUESTIONS

1. Why are multiple passes used in making fillet welds?

2. On a multiple-pass weld, which pass penetrates the base metal?

3. What is the best polarity for overhead fillet welding?

4. Describe the proper electrode angle for the first pass on an overhead fillet weld.

5. Describe the electrode angle for the second pass on an overhead fillet weld.

6. Draw a sketch showing the sequence of the beads, if six passes are used to weld a fillet joint.

Unit 22   Butt Welds, Overhead Position

OBJECTIVES

After completing this unit the student will be able to:

- discuss the safety precautions peculiar to overhead welding.
- describe the electrode angle for overhead butt welds.
- weld a butt joint in the overhead position.

Although butt welds made in the overhead position are essentially the same as those made in other positions, several factors become more critical when the weld is done in the overhead position. Some of the factors which become especially important in this position are listed here.

- Molten metal and weld splatter falls on the welder. Protective clothing must be worn at all times.
- The pieces must fit together to form a good joint. Material which is over 3/16 inch thick must be beveled and spaced to insure that the root pass has good penetration.
- A close arc (with the electrode close to the weld) must be used, especially on the root pass, to insure 100 percent penetration.
- DCRP is the easiest polarity for most overhead welding.
- E-6010 and E-6011 electrodes are particularly suited for overhead welding.

Welding Overhead Butt Joints

The plates to be welded should be tacked and solidly held by a jig or positioner in the overhead position. The weldment must not be able to move as the weld progresses.

The electrode should be directed straight into the joint, and tipped about 5 degrees toward the weld, figure 22-1. It may be helpful to use a slight whipping motion with the

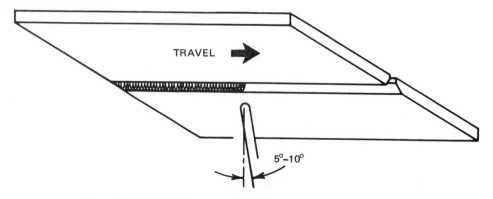

TRAVEL

5°–10°

Fig. 22-1 Electrode Angle for Overhead Butt Weld

the electrode. The bead should be kept uniform in width and the rate of travel should be uniform to produce a weld with good appearance.

On plate which is 1/4 inch thick, or more, do not try to fill the entire joint in one pass. Very heavy beads tend to build up more heat and cause the metal to sag. A series of small stringer beads produces better results. Be sure that each pass is well fused with the one before it and clean all slag from each bead before welding the next one. Weld enough stringer beads to completely fill the joint. The finished weld should be slightly convex, with a smooth crown above the surface of the parent metal.

JOB 22: BUTT WELD, OVERHEAD POSITION

Equipment and Material:

Standard arc welding equipment
Protective clothing
Two pieces of mild steel plate, 1/4″ x 4″ x 5″
1/8″, E-6010 or E-6011 electrodes

PROCEDURE

1. Grind one edge of each plate to form an included angle of 60 degrees and grind a 1/16-inch land on the bottom of each bevel, figure 22-2.

2. Tack weld the two plates for a butt weld.

3. Fasten the weldment in position for an overhead weld.

4. Weld a root pass the length of the joint, figure 22-3.

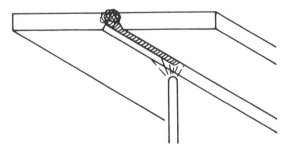

Fig. 22-3 Root Pass in an Overhead Butt Weld

5. Chip and brush the root pass, then examine it for complete penetration and fusion on the reverse edge.

KEY POINTS

1. The bevel must be ground in a straight line, so the plates will fit together well.

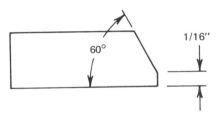

Fig. 22-2

4. The root pass must completely fuse the edges of the plates on the reverse of the joint. Right-handed welders should weld from left to right; left-handed welders should weld from right to left.

5. If complete fusion has not taken place, the weld should be cut out, reground, and welded again.

PROCEDURE	KEY POINTS
6. When a satisfactory root pass has been welded, fill the remainder of the joint with stringer beads.	6. Hold a close arc and be sure that fusion is taking place between the plate, the puddle, and the previously welded bead. Chip and brush the weld between each stringer bead. The finished weld should have a slightly crowned surface.
7. Test this weld as was done in Unit 8.	

REVIEW QUESTIONS

1. How can the best arc length for overhead welding be described?
2. Why is overhead welding more dangerous than welding in other positions?
3. Why must material which is over 3/16 inch thick be beveled for a butt weld?
4. List two AWS classifications of electrodes which are well-suited for overhead welding.
5. Describe the proper electrode angle for an overhead butt weld.
6. Why are very large beads undesirable for overhead welding?

SECTION 3: WELDS IN THE OVERHEAD POSITION, COMPREHENSIVE REVIEW

A.   PAD BUILDUP IN THE OVERHEAD POSITION

Equipment and Material:

Standard arc welding equipment
Protective clothing
Mild steel plate, 3/16" or 1/4" x 5" x 5"
5/32" electrodes (AWS classification to be determined by the student)

PROCEDURE

1. Position the plate for overhead welding.
2. Set the welding machine on DCRP and adjust the current as required.
3. Cover the plate with stringer beads, being sure to clean each bead before welding the next one.
4. Cover this first layer of weld metal with a second layer of stringer beads running at right angles to the first.
5. Weld a third layer in the same direction as the first.
6. Cut through the center of the plate, smooth the cut edges, and etch both pieces. Have the etched pieces inspected by the instructor.

B.   BUTT WELD IN THE OVERHEAD POSITION

Equipment and Material:

> Standard arc welding equipment
> Protective clothing
> Two pieces of mild steel plate, 1/4" x 4" x 5"
> 1/8" and 5/32" electrodes (AWS classification to be determined by the student)

PROCEDURE

1. Grind one of the 5-inch edges of each plate to form a 60-degree included angle.

2. Grind a 1/16-inch land on each bevel.

3. Tack weld the two pieces with a 1/16-inch root opening.

4. Position the weldment for an overhead weld.

5. Weld the root pass, using 1/8-inch electrodes.

6. Chip and brush the weld, then fill the joint with 5/32-inch electrodes.

7. Cut three 1-inch strips from the weldment, as shown in the illustration.

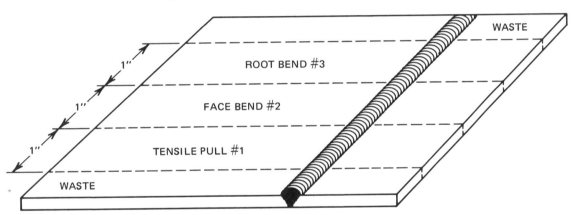

Cut 1-inch Strips Across the Weld to Test Butt Weld

8. Test each strip by bending one over the face of the weld, one over the root of the weld, and pulling one apart in the tensile tester.

9. Have the tested specimens inspected by the instructor.

SECTION 4:
Welding Mild Steel Sheet Metal

Unit 23 Welding Mild Steel Sheet Metal, Vertical-Down

OBJECTIVES

After completing this unit the student will be able to:

- describe vertical-down welding.
- describe two electrode angles that can be used with vertical-down welding.
- weld a bead vertical-down in sheet metal.

Sheet Metal Welding

Welding sheet metal requires practice in controlling the puddle and the speed of welding. While it might seem that the amperage should be reduced to weld the thinner metal, the opposite is true. 16-gage sheet metal can be very successfully welded with the same amperage as that used for welding steel plate.

Very little of the electrode is actually deposited on the parent metal. The heat from the electrode is used to fuse the pieces of parent metal together. By using a high rate of travel, the arc does not deposit much of the electrode.

Vertical-down Welding

The easiest way to weld sheet metal is usually by welding from the top to the bottom. This is called *vertical-down* welding. In *vertical-up* (bottom-to-top) welding the rate of travel is slower. Vertical-up welding allows more time for the metal to heat up and results in holes being melted through the joint.

A   B

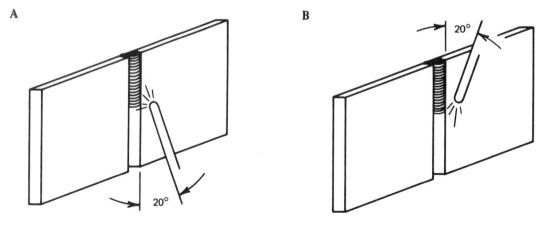

Fig. 23-1 Two Electrode Positions for Vertical-down Welding

There are two electrode angles which may be used for vertical-down welding. The electrode may be angled up into the weld at about a 20-degree angle, figure 23-1A, or it may be angled down at an angle of about 20 degrees, figure 23-1B.

Direct current reverse polarity (DCRP) may be used with E-6010 or E-6011 electrodes to produce good welds. Expert welders can also use other polarities and other electrodes to make good welds. Alternating current (AC) does an excellent job of welding on light-gage metal.

JOB 23: WELDING MILD STEEL SHEET METAL, VERTICAL-DOWN

Equipment and Material:

    Standard arc welding equipment
    Protective clothing
    Four pieces of 16-gage mild steel, 4″ x 6″
    1/8″, E-6011 and E-6013 electrodes

PROCEDURE	KEY POINTS
1. Place a 16-gage steel strip in position for vertical welding.	1. A piece of scrap may be tacked to the back or it may be held in a jig.
2. Set the welding machine on DCRP and adjust the amperage.	2. The amperage should be adjusted the same as it is for welding mild steel plate.
3. Using an E-6011 electrode, weld a stringer bead vertical-down.	3. The electrode should be at an angle of about 20 degrees with the plate. The rate of travel should be much faster than it is for vertical-up welding.
4. Weld a stringer bead vertical-down on each of the other three strips. Use both types of electrode for DCRP, then use both for AC — four beads in all.	4. The current can be adjusted for the speed of travel. Faster travel allows for higher current settings.
5. Chip and brush the beads and have the instructor examine them.	5. The beads should have smooth appearance, with good penetration.

REVIEW QUESTIONS

1. What is vertical-down welding?

2. How does the amperage required to weld sheet metal compare with that required for steel plate?

3. How does the amount of electrode metal deposited in vertical-down welding of sheet metal compare with that for other types of welding?

4. Why is vertical-down welding recommended for sheet metal?

5. Describe two electrode angles which can be used with vertical-down welding of sheet metal.

6. As the rate of travel is increased, what adjustment should be made to the current setting on the welding machine?

Unit 24 Outside Corner in Sheet Metal, Vertical-Down

OBJECTIVES

After completing this unit the student will be able to:

- explain why less electrode metal is used for vertical-down welding sheet metal.

- describe the procedure for welding outside corners in sheet metal.

- vertical-down weld an outside corner in sheet metal.

Fabrication plants require welders to have the ability to quickly weld a leakproof seam in mild steel sheet metal. This is frequently done by vertical-down welding. With little practice, a beginning welder can successfully weld sheet metal corners rapidly and efficiently.

Outside corner welds require very little electrode metal. The heat from the arc melts the sheet metal at the corner. The molten metal fuses the corner together with 100 percent penetration. These welds require little or no reinforcing of the weld, because the corner is thicker than either of the pieces of parent metal.

Speed of travel down the joint must be controlled to produce good bead appearance. The arc must be held close, so the melting process can be controlled. The melting process must be controlled to prevent holes from being melted through the joint.

Any polarity, DCRP, DCSP, or AC, may be used successfully for vertical-down welding of outside corners. However, the arc is more difficult to control using DCSP.

JOB 24: OUTSIDE CORNER IN SHEET METAL, VERTICAL-DOWN

Equipment and Material:

Standard arc welding equipment
Protective clothing
Eight pieces of 16-gage mild steel, 2" x 6"
1/8", E-6011 and E-6013 electrodes

PROCEDURE	KEY POINTS
1. Tack weld two pieces of sheet metal every 1 1/2 inches to form a corner joint, figure 24-1.	1. The two pieces must fit together tightly. If small gaps are left in the joint, the arc will melt through the corner.
2. Position the assembly for a vertical weld.	

PROCEDURE	KEY POINTS

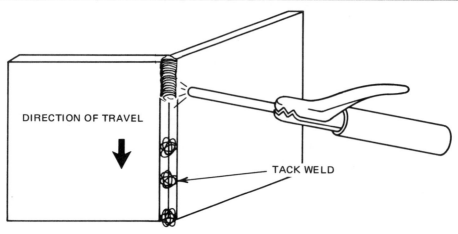

DIRECTION OF TRAVEL

TACK WELD

Fig. 24-1 Outside Corner Tacked for Down Welding

3. Weld the joint vertical-down, using an E-6011 electrode and DCRP.

3. The electrode angle should be about the same as that used for running a bead vertical-down, 20 degrees from the parent metal, figure 24-2. Hold a short arc.

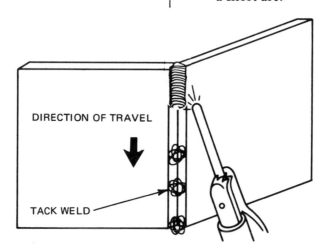

DIRECTION OF TRAVEL

TACK WELD

Fig. 24-2 Electrode Angle for Welding an Outside Corner Vertical-Down

4. Weld three more corner joints. The following combinations of polarity and electrodes are to be used:
   - E-6011 with DCRP
   - E-6011 with AC
   - E-6013 with DCRP
   - E-6013 with AC

5. Chip and brush all welds and have them inspected by the instructor.

5. The joints should be thoroughly fused, with no porosity.

REVIEW QUESTIONS

1. Why is it not necessary to add a large amount of filler rod for vertical-down welding of corner joints?

2. How much penetration should be achieved on a corner weld?

3. What polarities are effective for welding corner joints?

4. Describe the best arc length for welding outside corners in sheet metal.

5. Describe the electrode angle for vertical-down welding an outside corner in sheet metal.

6. Why is it especially important that the pieces of an outside corner weld in sheet metal fit well?

Unit 25　T Welds in Mild Steel Sheet Metal, Vertical-Down

OBJECTIVES

After completing this unit the student will be able to:

- discuss the importance of the rate of travel in welding T joints vertical-down.

- describe the joint preparation necessary for welding a T joint vertical-down.

- vertical-down weld a T joint in mild steel sheet metal.

Since the T joint is a commonly used joint design, especially in sheet metal fabrication, the welder must master it. Smooth, steady welding and knowledge of heat transfer are important to produce good T welds.

Because the weld is made in a corner, more heat is required than on some welds, such as an outside corner weld.

The arc must be held close, and speed down the corner must be regulated to control the amount of penetration in the corner. Tipping the rod to one side or the other will develop more heat on the side toward which the rod is pointed. The plate which is exposed to the heat has a tendency to melt through.

The sheet metal pieces must fit tightly. It may be necessary to tack them in several places in order to prevent one piece from warping away from the other due to expansion and contraction.

DCRP and either E-6010 or E-6011 electrodes allow for maximum speed and uniformity on vertical-down T welds in sheet metal. DCSP and E-6013 electrodes can also be used effectively, as can AC welding.

JOB 25:　T WELD MILD STEEL SHEET METAL, VERTICAL-DOWN

Equipment and Material:

Standard arc welding equipment
Protective clothing
Eight pieces of 16-gage mild steel, 2" x 4"
1/8", E-6011 and E-6013 electrodes

PROCEDURE	KEY POINTS
1. Tack weld the sheet metal pieces to make four weldments with T joints, figure 25-1. Tack the joint at 2-inch intervals.	1. It is very important that the joint fits tightly, with no visible crack between the two pieces. If the joint does not fit well, welding will be nearly impossible.

PROCEDURE	KEY POINTS

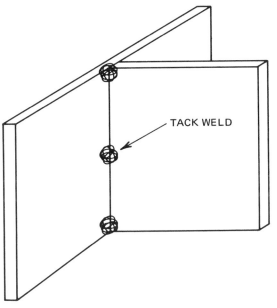

Fig. 25-1 Setup for T Weld

2. Weld all four joints from top to bottom, using all combinations of DCRP, AC, E-6011, and E-6013.

2. The electrode should be pointed slightly in the direction of travel. Use a slight rocking motion to help control the heat.

3. Chip and brush the welds and have them inspected by the instructor.

3. The beads should have a smooth appearance and be somewhat concave, with good penetration.

REVIEW QUESTIONS

1. Why is the length of the arc critical when welding sheet metal?

2. Describe the proper electrode angle for welding T joints in sheet metal.

3. Why do joints in sheet metal require more tack welds than joints in steel plate?

4. Why is it especially important for the pieces to fit well for a T weld in sheet metal?

5. What polarity is best for welding T joints in sheet metal?

6. What electrode classifications are best for T welds in sheet metal?

Unit 26 Mild Steel Sheet Metal, Flat Position

OBJECTIVES

After completing this unit the student will be able to:

- discuss the control of heat when welding sheet metal in the flat position.
- explain centerline shrinkage and its control.
- weld a bead on sheet metal in the flat position.

Sheet metal may be more difficult to weld in the flat position than in other positions, because the heat is not readily absorbed by the base metal. With sufficient practice, the student welder should master this weld.

If the metal to be welded is placed on a firebrick or metal tabletop, more heat will be absorbed and the rate of travel will need to be slower. If the metal is supported between two firebricks or in a jig, so there is no backing, heat builds up rapidly and holes may melt through the base metal. Experience is the most valuable guide to producing good welds in sheet metal.

The electrode should be pointed about 20 degrees into the weld. This tends to force the puddle toward the end of the weld and helps prevent the weld from melting through the base metal.

If the weld progresses too rapidly a condition known as *centerline shrinkage* can result. The metal is welded in a hot, expanded condition and as it cools the weld shrinks. This causes the center of the weld to be concave.

DCRP with E-6010 electrodes proves the most satisfactory for this work. The thin coating on the electrode does not build up on the weld, so the welder can see the work while it is in progress.

JOB 26: MILD STEEL SHEET METAL, FLAT POSITION

Equipment and Material:

Standard arc welding equipment
Protective clothing
Four pieces of 16-gage mild steel, 4" x 6"
1/8", E-6011 and E-6013 electrodes

PROCEDURE	KEY POINTS
1. Place a piece of sheet metal in the flat position.	
2. Set the welding machine for 100 amperes.	

PROCEDURE	KEY POINTS

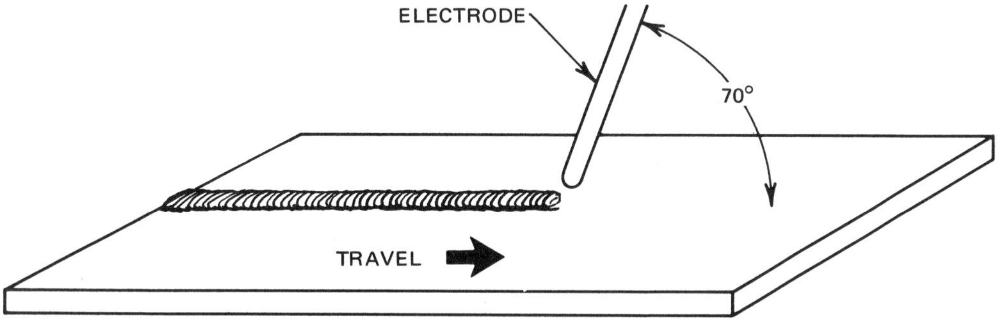

Fig. 26-1 Electrode Angle for Sheet Metal in Flat Position

3. Weld one bead on each of the four pieces of sheet metal. Use all combinations of DCRP, AC, E-6011, and E-6013.	3. The electrode should point about 20 degrees into the weld, figure 26-1. Hold a medium length arc. If the rate of travel is too slow, the weld will have a concave surface.
4. Chip and brush the welds and have them inspected by the instructor.	4. Practice these welds until smooth, uniform appearance and 90 to 100 percent penetration results.

REVIEW QUESTIONS

1. Describe the proper electrode angle for welding mild steel sheet metal in the flat position.

2. What is centerline shrinkage?

3. What causes centerline shrinkage?

4. What is the difference between welding sheet metal which is placed directly on a firebrick surface and that which is suspended between two bricks?

5. What polarity is best for welding mild steel sheet metal in the flat position?

6. What electrode classification is best for welding sheet metal in the flat position?

Unit 27   Outside Corner Weld in Sheet Metal, Flat Position

OBJECTIVES

After completing this unit the student will be able to:

- explain the procedure for welding an outside corner in sheet metal in the flat position.
- describe the electrode angle for welding outside corners in sheet metal in the flat position.
- weld an outside corner in sheet metal in the flat position.

In industry, welders are frequently called upon to weld outside corners in sheet metal. This is a difficult weld to make in the flat position, because the thin edge created by the corner has a tendency to melt away rapidly. With practice, the beginning welder should be able to produce satisfactory welds in this position.

It is extremely important that the pieces to be joined fit together well. It is necessary to tack weld the joint securely to prevent it from being distorted by expansion and contraction. Solid tack welds also keep the joint closed tightly, preventing excessive melting of the edges.

The joint should be welded with a uniform rate of travel, fast enough to prevent it from melting through and slow enough to prevent centerline shrinkage. The electrode should be pointed about 20 degrees into the completed weld, as in other sheet metal welding.

The current should be set high enough to insure that the molten metal flows evenly. If the current is set too low, the bead will be built up without penetrating the base metal.

JOB 27: OUTSIDE CORNER WELD IN SHEET METAL, FLAT POSITION

Equipment and Material:

Standard arc welding equipment	Eight pieces of 16-gage mild steel, 4″ x 6″
Protective clothing	1/8″, E-6011 and E-6013 electrodes

PROCEDURE	KEY POINTS
1. Tack the sheet metal pieces together, forming four corner joints and place them in the flat position, figure 27-1.	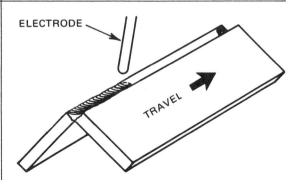

Fig. 27-1 Outside Corner Weld in the Flat Position

PROCEDURE	KEY POINTS
2. Weld each of the four joints. Use all combinations of DCRP, AC, E-6011, and E-6013.	2. The rate of travel determines the amount of buildup on the weld.
3. Chip and brush the welds and have them inspected by the instructor.	The amperage must be high enough to insure good wetting action at the edges of the bead.

REVIEW QUESTIONS

1. To weld an outside corner joint in sheet metal, how wide a gap should be left between the pieces?

2. What is the result if the rate of travel is too slow when welding an outside corner in the flat position?

3. What is the result if the rate of travel is too fast when welding an outside corner in the flat position?

4. Describe the proper electrode angle for welding outside corner joints in sheet metal in the flat position.

5. What is the result if the current is set too low when welding outside corner joints in sheet metal in the flat position?

SECTION 4 WELDING MILD STEEL SHEET METAL, COMPREHENSIVE REVIEW

A. T WELD IN MILD STEEL SHEET METAL

Equipment and Material:

Standard arc welding equipment
Protective clothing
Two pieces of 16-gage mild steel, 4″ x 6″
1/8″, E-6011 electrodes

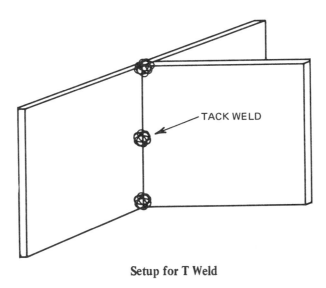

TACK WELD

Setup for T Weld

PROCEDURE

1. Tack weld the two pieces of sheet metal for a T joint as shown in the illustration.

2. Place the weldment in position for a vertical weld.

3. Weld the joint from top to bottom.

4. Chip and brush the weld.

5. Bend the upright piece over the bead to check the penetration.

6. Have the tested specimen inspected by the instructor.

B. OUTSIDE CORNER WELD IN MILD STEEL SHEET METAL

Equipment and Material:

Standard arc welding equipment
Protective clothing
Two pieces of 16-gage mild steel, 4″ x 6″
1/8″, E-6011 electrodes

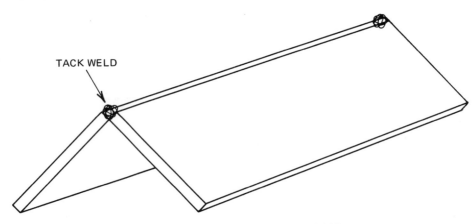

TACK WELD

Setup for Outside Corner Weld

PROCEDURE

1. Tack weld the two pieces of sheet metal for a corner joint as shown in the illustration.

2. Place the weldment in position for a vertical weld.

3. Weld the joint from top to bottom.

4. Chip and brush the weld.

5. Bend the pieces over the face of the weld to check the penetration.

6. Have the tested specimen inspected by the instructor.

Final Review Projects

These projects are included to give the welding student additional practice in electric arc welding. The student should read and interpret the drawings, select the proper electrodes, and adjust the welding machine for the suitable polarity and amperage to complete the project. When the project has been completely welded, it should be allowed to cool, be cleaned, and presented to the instructor for evaluation.

Material Required for Project 1

30 inches of 1/2" round stock
3" x 3" x 1/4" angle iron, 5" long

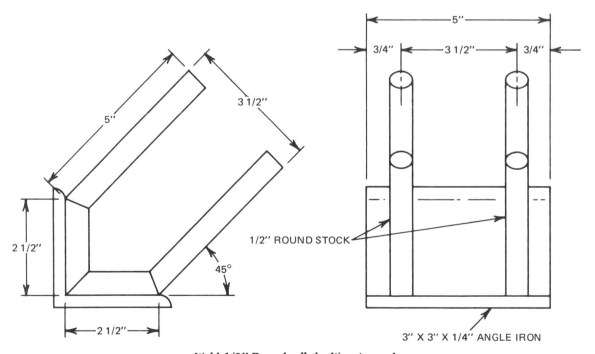

Weld 1/2" Rounds all the Way Around

Material Required for Project 2

Mild steel plate, 1/4" x 4" x 13"

3" x 3" x 1/4" angle iron, 8" long

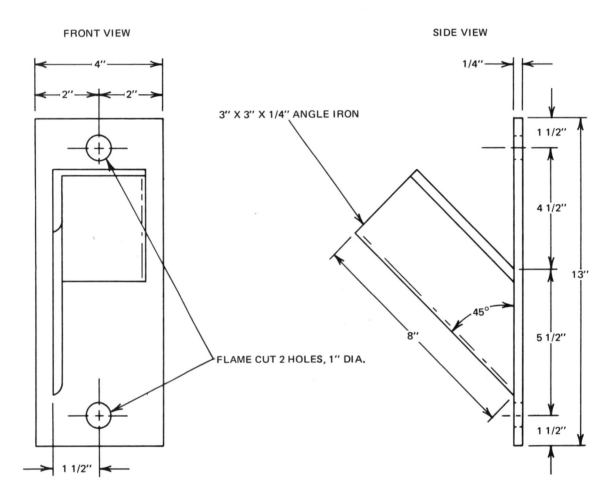

FRONT VIEW

SIDE VIEW

3" X 3" X 1/4" ANGLE IRON

FLAME CUT 2 HOLES, 1" DIA.

Weld Angle Iron to Plate all the Way Around

Glossary

Arc Blow: Deflection of an electric arc from the normal path because of magnetic forces.

Arc Voltage: The voltage across the welding arc.

Axis of the Weld: An imaginary line drawn through the length of a weld.

Base Metal: The metal to be welded.

Bead: The line of metal deposited by welding.

Bevel: An edge which has been ground at an angle.

Blowhole: A gas pocket in a weld.

Butt Joint: A joint between two pieces which are in line with one another.

Butt Weld: A weld joining the edges of two pieces of metal which are in line with one another.

Concave Weld: A weld having a concave face (curved inward).

Convex Weld: A weld having a convex face (curved outward).

Corner Joint: A joint in which the pieces form an L shape.

Cover Glass or Lens: Any clear, transparent material used to protect the filter lens of a welding helmet.

Crater: A depression at the end of a weld bead.

Crown: The buildup of the bead above the thickness of the parent metal.

Drop-Through: The sagging of molten metal through a joint caused by overheating.

Electrode: Part of the welding circuit which conducts the current between the electrode holder and the arc. (Sometimes called *welding rod.*)

Electrode Holder: The device which holds the electrode for welding.

Face of the Weld: The surface of the weld on the side from which the welding is accomplished.

Filler Metal: The metal added to a weld by the electrode.

Fillet Weld: An inside corner weld made where two pieces of metal join at any angle. When one plate forms a 90-degree angle with the other, it is often called a T weld.

Filter Lens: Colored glass used in the welding helmet to stop harmful rays given off by the arc.

Flat Position: The position of a weld made on the topside of the parent metal and within 30 degrees of horizontal.

Fusion: Melting together of filler metal and base metal.

Gas Pocket: A hole in the weld caused by trapped gas in the molten metal (a blowhole).

Groove: Any opening provided for a groove weld.

Groove Angle: The included angle of the two parts to be joined.

Groove Weld: Any weld made in a groove between two pieces to be welded. These welds may include the square groove, the single- and double-V, the single- and double-U, the single- and double-J, and the flare-bevel and flare-V.

Helmet: A device used to protect the face, eyes and neck from the heat and harmful rays of the arc welding process. (Also called a hood.)

Horizontal Position: A weld made within 45 degrees of horizontal and against a vertical surface.

Incomplete Fusion: A cold lap.

Joint: The place where welding is to be done on two or more pieces.

Jig: Any device used for holding material while work is being done. (A holding fixture.)

Land: A small square edge ground at 90 degrees with the bottom edge of a bevel. It is also referred to as the "nose" of a bevel.

Lap Weld: A weld made on two pieces of metal with the edge of one overlapping the other.

Overhead Position: The position of a weld made on the underside of the parent metal and within 45 degrees of horizontal.

Parent Metal: The metal being welded.

Penetration: The depth the welding puddle melts into the parent metal.

Porosity: Gas pockets or pinholes in a weld.

Positioner: A device for holding parts to be welded in the desirable position.

Positions of Welding: Flat, vertical, horizontal and overhead.

Reverse Polarity: The setup of direct current arc welding wherein the work is the negative terminal and the electrode is the positive terminal.

Root Opening: The separation between the pieces to be welded at the bottom of the joint.

Spatter: Metal particles which do not melt into the molten puddle and are thrown off by welding.

Straight Polarity: The setup of direct current arc welding wherein the work is the positive terminal and the electrode is the negative terminal.

Stringer Bead: A welded bead, run in a straight line, with no side-to-side motion.

Stick Electrode: A filler metal electrode which has a metal wire core, usually with a covering to protect the molten metal from the atmosphere.

Tack Weld: A weld, generally of small size, made to hold parts together and in line until final welding is done.

T Joint: A joint in which the two pieces of metal are located at approximately right angles to each other.

Undercut: The thinned, metal section resulting when fluid metal drops from the vertical surface of a weld.

V Grinding: Grinding the edges to pieces to be welded so that a V is formed when they are put together.

Vertical-Down Welding: Welding from top to bottom in the vertical position.

Vertical Position: The position of a weld which is made on a vertical surface with the bead perpendicular to the ground or floor.

Vertical-Up Welding: Welding from bottom to top in the vertical position.

Weave Bead: A type of weld where the bead is made by a series of side-to-side movements.

Weldment: An assemble of two or more parts which have been welded together.

WELDING PROCEDURES MIG & TIG

by Frank R. Schell
Bill Matlock

Preface

MIG (metallic inert gas) and TIG (tungsten inert gas) welding are the two most important techniques added to the field of welding since the introduction of arc welding. The development of MIG and TIG welding in the early 1940s makes them two of the more recent and technical welding methods used today. The use of an inert-gas shield to protect the weld zone produces so many desirable advantages, that these processes have become extremely popular. Skilled welders who are competent in the use of inert-gas welding equipment are in great demand.

WELDING PROCEDURES: MIG & TIG is designed to help the student acquire the ability to operate inert-gas welding equipment in a variety of positions and situations. The text is divided into eleven sections, each of which relates to various processes, materials, or positions. These sections are subdivided into units which are introduced by clearly stated behavioral objectives that cover a particular joint design or welding situation. After the stated objectives, each unit contains a concise discussion of any pertinent information being introduced for the first time. This part of the unit contains such things as new terms, safety information, and other related information.

Also included in the units is a specific welding procedure to be completed by the student. These jobs are designed to give the student an opportunity to "learn by doing." Each procedure is supplemented with key points which relate to the procedural steps. Each unit is completed with unit-end questions which the student may use to review some of the highlights of the unit or to test newly-obtained knowledge without the aid of the instructor.

All new or technical terms are explained as they are introduced throughout the text. In addition, the textbook includes a glossary which serves as a ready reference so that the student may locate the definition of a term quickly.

Frank Schell has been a journeyman welder for twenty-six years. He is currently Curriculum Development Coordinator and Professor of Welding at The College of Southern Idaho. He is a member of The American Vocational Association and The Idaho Vocational Association. Mr. Schell is also the author of several other pieces of instructional material, including WELDING PROCEDURES: OXYACETYLENE and WELDING PROCEDURES: ELECTRIC ARC.

Bill Matlock is currently an Associate Professor of Welding Technology and other related subjects at The College of Southern Idaho, of which he is a graduate. Mr. Matlock has also written several books on exercises in welding which are presently used throughout the state of Idaho.

Contents

Unit 1 Inert-Gas Welding Safety

OBJECTIVES

After completing this unit the student will be able to:

- list the hazards created by inert-gas welding.
- demonstrate the accepted safety practices of inert-gas welding.
- list the personal welding equipment required to protect the welder.

Inert-Gas Welding Safety

As with any industrial operation, hazards that could cause serious health problems or bodily injuries are present. Welding operators need to be fully informed of these dangers so that they may avoid or correct unsafe practices that exist in the work area. It is every welder's responsibility to consider other workers' safety as well as his or her own. Many industrial accidents that shorten careers could have been avoided by using common sense on the job. Professional welding is a safe industry as long as safety rules and regulations are strictly followed.

Hazards:

- Radiation from the arc of inert-gas welding is dangerous to the eyes and skin. The ultraviolet rays given off during inert-gas welding are more exposed and about twice the strength of the rays given off during stick-arc operations; therefore, a darker filter lens is needed for eye protection.

- Flying sparks and small drops of molten metal are present most of the time when inert-gas welding. Protective clothing which is not highly flammable, gloves, and high work shoes help to protect the welder from burns.

- Avoid striking an arc when others are nearby. Warn co-workers when an arc is to be struck so they may protect their eyes from the arc.

- Fumes given off from the arc and the material being welded may be toxic. These fumes escape when welding on materials treated with rust inhibitors, such as zinc galvanizing, cadmium, and lead. Adequate ventilation is required at all times when welding operations are in progress.

- Electric shock is always a possibility. Be sure the floor is dry where you are standing. Use insulated welding guns. Wear dry gloves.

- The danger of burns is always present. Do not handle hot metal with your hands, use tongs or pliers.

Personal Inert-Gas Welding Equipment

The *helmet* (hood) is generally made of fiber reinforced plastic, and formed to cover the front half of the head. An opening is provided in front of the eyes and a clear glass cover lens is installed in the opening. Behind this is a colored glass which filters the infrared and ultraviolet rays from the arc. The clear glass lens is provided to catch the spatters from the welding process which would otherwise adhere to the colored lens. (The clear glass is much cheaper than the colored lens.) In addition, the best type of helmet has a hinged gate in the front which can be raised out of the way when welding is completed. This exposes another clear glass lens which protects the eyes while the weld is being cleaned. The use of this type of helmet is recommended, since the entire face remains covered during any weld cleaning process, figure 1-1. The welding

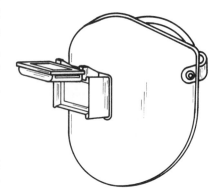

Fig. 1-1 Helmet

helmet or hood should be examined frequently to ensure that no cracks or holes are present which might allow the arc light to leak through. The colored lens and cover lens must be free of cracks or chips for absolute eye protection.

Additional personal equipment should consist of:

- Cap
- Leather Gloves
- Leather Jacket or Apron
- Safety Glasses
- High Shoes or Boots
- Other clothing should be of a type which is not highly flammable.

General Inert-Gas Welding Safety Rules

1. Never use welding machines which are not properly grounded.
   Ground all power circuits to prevent accidental shock. A stray current can create a fatal shock. Do not ground the welding machine to any pipe lines which carry gases or flammable liquids.

2. Check welding cables frequently.
   Check the ground cable and clamp often. Do not use cables which have breaks in the insulation. Be certain all cable connections are tight at the machine and ground clamp. Overloading of cables will cause them to overheat and create a fire hazard; efficiency of the weld is reduced when the welding cable is too small to carry the current.

3. Check the welding gun or torch frequently.
   The contact tip of a MIG gun should be tight and centered in the nozzle. The nozzle needs to be cleaned frequently, due to the spatter created in MIG welding. When TIG welding, the tungsten electrode should be tight and free from contamination. The

inert-gas nozzle must be tight on the torch body. Use only fully insulated guns and torches. Check the gas transfer hoses occasionally for leaks.

4. Never weld on containers which have held explosive or combustible materials.

5. Never weld in confined spaces without adequate ventilation and do not weld near an explosive atmosphere.

6. Do not weld where even minute amounts of vapors from solvents are present. Some solvents decompose to form *phosgene gas,* a deadly poisonous gas, when exposed to the heat created during welding.

7. Check for fire hazards. Combustible material should be at least 35 feet from the welding operations.

8. Electricity can be a killer. Before checking any welding machine, open the power circuits to stop the current. Repairs on a welding machine should only be made by a fully qualified electrician.

9. Do not touch any exposed, or noninsulated parts of cables, clamps, or other electrical parts.

10. Do not work in a damp area. Striking the arc can cause electrical shock if a worker's feet are wet.

11. Do not use a cracked or defective helmet. The filter glass in the helmet provides eye and face protection from ultraviolet and infrared rays. Cracks in the lens allow these harmful rays to contact the eyes. The rays given off by the inert-gas processes are approximately two times more harmful than the stick-arc rays, due to the action of the flux coating. The inert-gas processes are used on reflective surfaces such as aluminum or stainless steel which intensifies the rays.

12. Never look at the electric arc without eye protection. Blindness can result.

13. Wear heavy shoes, preferably with steel toes. Dark colored clothing helps to protect the body from ultraviolet rays.

14. Always wear protective clothing when making out-of-position welds. Do not wear ragged or frayed clothing and button the flaps on shirt pockets. Do not wear cuffs on trousers. Wear a cap when working in the overhead position.

15. Any equipment which is operated by power is potentially dangerous. Grinders, drills, sanders, saws, electric welders, and all other power tools can cause serious injury or death. Never do any grinding, weld slagging, or heavy hammering without eye and face protection.

16. Do not leave materials or equipment lying on the floor. Clean up the work area. Take care of all safety hazards as soon as they are observed.

17. Do not use defective equipment.

18. Never strike an arc without warning people who are nearby and might be exposed to the arc flash.

19. Be alert to the dangers of fire.

Fires can be disastrous. Know where the fire extinguishers are and learn how to use them. Saving time is important in the case of fire.

20. In case of equipment malfunction, notify the person in authority. Never, under any circumstances, operate defective equipment.

REVIEW QUESTIONS

Instructions: There are twenty general inert-gas welding safety rules listed. For each question in this review write the number of the rule which most closely answers the question.

Example: Should welding machines be grounded to pipe lines?

Rule #1 states: "Do not ground the welding machine to any pipe lines which carry gases or flammable liquids."

1. Should all power circuits be grounded?

2. When about to strike an arc, is the welder responsible for warning nearby workers?

3. The harmful rays given off by the inert-gas welding process are ultra-violet and infrared. Are they liable to be more harmful from inert-gas welding than from the stick-arc welding process?

4. Should a welder know how to use fire extinguishers, or are they someone else's responsibility?

5. How far should combustible materials be from the operator when welding?

6. Is good ventilation a part of the safety rules for a welder?

7. Can the welder watch the arc without eye protection if he limits the watching to a few seconds at a time?

8. Does the color of the clothes worn have any effect on the welder's safety?

9. Should a welder leave equipment on the floor around a job until after the job is completed?

10. Should defective equipment be used until new or repaired equipment is provided?

11. Does the size of the ground cable have any effect on the welding process?

12. Can welding on containers which have held solvents be dangerous?

13. Can a cracked hood lens have any effect on the welder's eyes?

14. Is it necessary to clean the nozzle of the MIG welding gun more than once during a job?

15. Is the use of power tools considered dangerous?

16. Can working in a damp area affect the safety of a welder?

17. Should the welder make repairs on a welding machine?

18. Is there any rule listed which pertains to making out-of-position welds?

19. In which rule is a reference made to the welding of containers which have held combustibles?

20. In which rule is reference made to the slagging of welds?

SECTION 1:
Fundamentals of MIG Welding

Unit 2 Setting Up MIG Welding Equipment

OBJECTIVES

After completing this unit the student will be able to:

- summarize and explain the theory on which MIG welding is based.

- list six advantages of MIG welding over conventional stick-arc welding.

- name and explain the function of the main components of a MIG welding outfit.

- define the uses of shielding gases.

- explain the AWS classification of MIG welding wires.

- successfully set up MIG welding equipment in preparation for welding.

- contrast and explain the difference between short-arc and spray-arc, MIG welding.

The MIG Process

MIG (metal inert gas) welding is a semiautomatic, manually operated welding process. The MIG process is more properly referred to as gas metal-arc welding (GMAW) by the American Welding Society (AWS), although MIG welding is the term most commonly used. The AWS is the professional organization which sets standards in the field of welding. An electric arc is maintained between the work and the electrode as in stick-arc welding (shielded metal-arc welding). In this case, a shielding gas is introduced to exclude the atmosphere from the weld zone. The absence of oxygen and other gases from the weld zone prohibits oxidation of the heated metal. Therefore, little or no slag forms on the completed weldment.

The electrode is a consumable wire which is usually connected to the positive terminal. The workpiece is grounded by connecting it to the negative terminal, figure 2-1. This type of polarity is known as direct current, reverse polarity. The electrode is automatically fed into the weld zone and then melted by the electric arc.

The *welding gun* (sometimes called the head or torch) is hand held and guided along the joint to be welded by the operator. The automatically fed filler wire is forced through the torch by an electrical machine. As the wire is fed through, the shielding gas is released to protect the weld from the atmosphere. Although oxygen presents the greatest problems, hydrogen and nitrogen also produce negative reactions when allowed to contact heated metals.

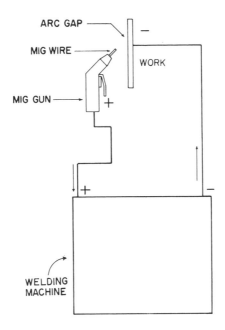

Fig. 2-1 In typical MIG welding, the base metal is connected to the negative terminal and the gun is connected to the positive terminal.

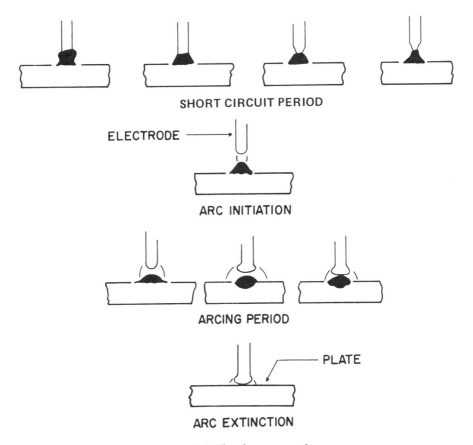

Fig. 2-2 The short-arc cycle

There are two basic methods of transferring the metal from the wire to the weld puddle. The *short-arc* method employs a continuous cycle. As shown in figure 2-2, page 7, the initial stage of the short-arc cycle is the contact of the wire to the work. This contact creates a short between the wire and the work for a fraction of a moment. This short produces enough heat to melt the wire. The electrical current then *pinches off* the molten drop. This recreates the arc used to shape the puddle. During this time, the wire is once again approaching the work to make contact and begin a new cycle.

The *spray-arc* method uses a high enough current to melt the wire before it contacts the work. The push of the current forces the molten droplets across the arc into the weld zone in the form of a spray, figure 2-3.

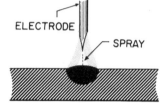

Fig. 2-3 The spray-arc process

The MIG process is a fairly recent development, but the fast production and easy application have rapidly made it popular in the welding industry. Many operations done with the stick-arc process in the past are now done by the MIG process. A successful welder must be competent in MIG welding.

Advantages of the MIG Process

MIG welding produces strong, high-quality welds at high speeds without the use of coated electrodes. Since cleaning agents and fluxes are not used, the amount of cleaning is minimal after the weld is completed. In MIG welding approximately 95 percent of the consumable electrode wire is deposited in the weld puddle. This method reduces the waste of electrode stubs in contrast to the stick-arc process.

Labor time is reduced, because the consumable wire is continuously fed into the weld. This means the weld does not need to be stopped and restarted to change the electrodes as in the stick-arc process. Because the weld does not need to be stopped and restarted, slag inclusions, cold laps, shallow penetration, crater cracks, and poor fusion are greatly reduced. Large gaps which may be found in both new construction and repair work can be easily filled by bridging using the short-arc, MIG process.

The MIG welding process is effective in all welding positions and it can be mastered by any student who has gained the proper knowledge and experience.

MIG Welding Equipment

The MIG welding process is highly specialized and uses expensive equipment. The student must set up and operate the machines and attachments according to the rules and regulations established by the welding industry. When these procedures are followed, safe and economical welding can be done.

MIG welding equipment consists of four primary parts, figure 2-4.

- Power Source
- Wire-Feed Unit
- Welding Gun
- Shielding-Gas Supply

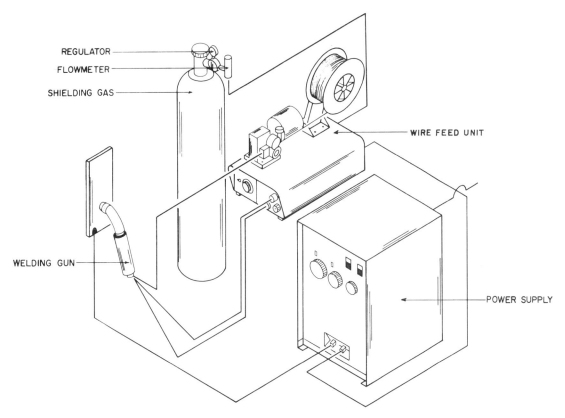

REGULATOR

FLOWMETER

SHIELDING GAS

WIRE FEED UNIT

WELDING GUN

POWER SUPPLY

Fig. 2-4 The basic components of a MIG welding outfit

Power Source. Most MIG welding operations are completed with a *rectifier* or *motor generator,* that supplies direct, constant-voltage, welding current, figure 2-5, page 10. Although the voltage is constant, it is also variable, within certain limits, from its original setting. As the welding operator moves the welding gun closer to, or farther away from the work, the power supply unit automatically increases or decreases the voltage output. Increasing or decreasing the voltage controls the burn-off rate of the consumable wire fed into the weld zone.

Voltage Controls. MIG welding machines have controls on the front panel of the power source. The voltage setting is very important for the smooth operation of the welding process. Voltage output and wire speed determines whether the short-arc or spray-arc process is used. Short-arc welding requires a low initial voltage setting and low wire speed. Spray-arc welding requires a high initial voltage setting and high wire speed.

The voltage and the wire speed must be adjusted properly for each welding process. If the wire speed is high and the voltage is low, the consumable wire will hit the work before it is melted at the proper point to maintain the arc, figure 2-6, page 10. If the wire speed is low and the voltage is high, the consumable wire will melt close to the contact tip and it will be impossible to maintain the proper arc length, figure 2-7, page 10.

Proper adjustment of these two controls will result in sound welds of high quality.

Inductance Controls. The inductance control regulates the *wetness* (fluidity) or *stiffness* (rigidity) of the weld puddle, by increasing or decreasing the time the arc is on.

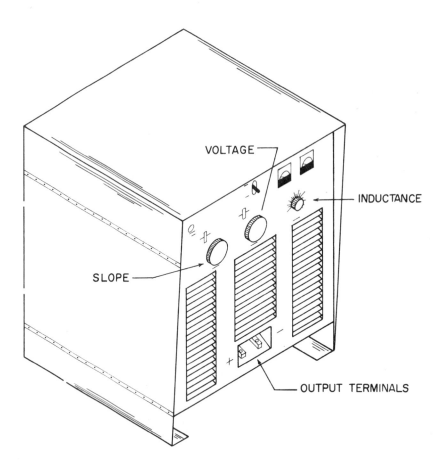

Fig. 2-5 MIG welding power source

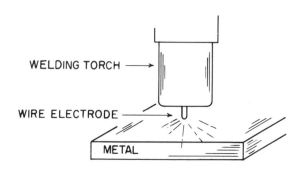

Fig. 2-6 If the wire speed is set too high and the voltage is set too low, the wire will strike the work.

Fig. 2-7 If the wire speed is set too low and the voltage is set too high, the wire will pinch off close to the contact tip.

Out-of-position welds (those made in any position other than flat) can be made by using low inductance input, which decreases the on time of the arc. This decrease results in a colder, stiffer weld puddle. Higher inductance input increases the on time of the arc, resulting in a hotter, more fluid weld puddle. This more fluid puddle is suitable for welding in the flat position. Some constant-voltage power sources have built-in, fixed inductance which cannot be adjusted.

Slope Controls. The slope control regulates the relationship between output voltage and amperage. This controls the pinch force exerted on the consumable wire and decreases the current surge as the initial arc is struck.

Controlling the slope increases versatility, fluidity, and smoothness in the weld puddle. Correct slope control will greatly reduce the amount of spatter while increasing efficiency. When little or no slope is in the welding circuit, the short circuit current goes up to a higher level and creates spatter. When too much slope is used, the consumable wire will carry the full current and a short circuit does not clear itself so that the wire may pile up on the metal being welded.

Cables. Large-diameter, flexible cables carry the electrical current to and from the welding operation. Welding cables are well insulated with rubber and a woven reinforcement. The cables are subject to wear and should be checked frequently for breaks in the insulation. The cable must be large enough for the amount of current it carries, figure 2-8.

Ground Clamp. The cable from the machine to the work is connected to a spring-loaded clamp which can be easily attached to the work, figure 2-9. Clamps should be heavy enough to carry the required currents. The ground clamp capacity should correspond with the cable capacity.

Wire Feed Unit. The wire feed unit, figure 2-10, page 12, and the *gas-flow solenoid* are automatically engaged as the gun power trigger is pressed. The wire feed unit pulls the wire from a spool and pushes it into the conduit tube and through the contact tip into the weld puddle.

At the same time as the wire starts to move, the gas solenoid is activated to release flow of shielding gas. The gas solenoid is mounted in the wire feed unit. The knob which controls the speed of the wire feed is located on the front panel of the wire feed unit.

LEAD NUMBER	CURRENT CAPACITY 0' TO 50'
4/0	600 AMPS.
3/0	500 AMPS.
2/0	400 AMPS.
1/0	300 AMPS.
1	250 AMPS.
2	200 AMPS.
3	150 AMPS.
4	125 AMPS.

Fig. 2-8 Sizes for cables up to 50 feet long

GROUND CLAMP

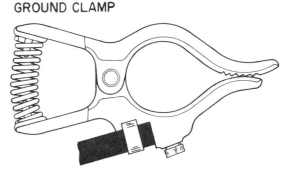

Fig. 2-9 The ground cable is attached to the ground clamp, which can be easily attached to the work.

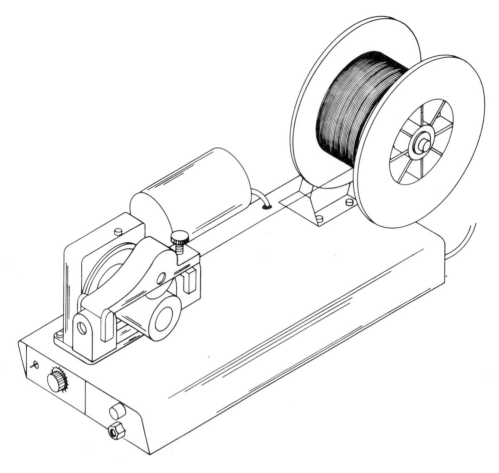

Fig. 2-10 Wire feed unit

Wire speed increase or decrease plays a double role in MIG welding, because the amperage of the arc is increased or decreased as the wire speed changes. The wire feed units may be mounted directly on the power source, a traveling track, or positioned away from the welding machine.

MIG Welding Gun. MIG welding guns, figure 2-11, are either air– or water-cooled. Guns which are used with very high amperages are normally water-cooled. The welding gun supplies the wire, shielding gas, and welding current to the weld zone.

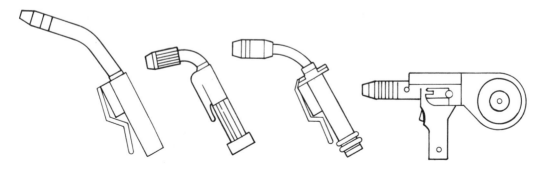

Fig. 2-11 Various styles of MIG welding guns

Most MIG welding guns are of the push-type, which utilizes drive rolls positioned inside the wire feed unit. Pull-type guns are used for small diameter hard wires, and for soft wires, such as aluminum or magnesium. Both types of guns have a trigger which is pressed to start the wire feed, gas flow, and the arc. When the trigger is released, all three operations stop immediately.

Guns are available with many different contact tip and nozzle designs. The *contact tip* delivers the electrical current from the cable to the wire electrode. New tips have smooth round holes which are slightly larger than the wire. The passing of the wire will wear the hole larger allowing spatter to enter and stick to the tip, figure 2-12.

The flow of the shielding gas is directed by the *gas nozzle* (or gas cup). Most MIG welding nozzles are made from copper or its alloys. Nozzles are held over the weld zone and catch a good deal of spatter. The spatter buildup can be decreased through the use of spatter-proof sprays or liquids. Nevertheless, MIG nozzles must be cleaned frequently, or the spatter buildup will slow down the flow of the inert gas used to shield the weld zone. To clean the nozzle, use a hollow reamer of the proper size or a small screwdriver.

Shielding Gas Equipment. The shielding gas equipment stores the gas safely, reduces the high pressure in the *cylinder* (or bottle) to a safe working pressure, and delivers the gas to the weld zone. This equipment includes gas cylinders, regulators, valves, and hoses.

Carbon dioxide (CO_2) is the shielding gas that is used most often, because it is economical. Carbon dioxide is normally bottled in its liquid state in fifty-pound cylinders. Truly *inert* gases (those that do not react chemically with other elements), such as argon, helium, and mixtures of argon and helium, are normally bottled in their gaseous state. The gas cylinder provides a means of storing and handling the gas safely.

A combination regulator and flowmeter delivers the gas to the gas *solenoid* (electrically controlled valve) located at the wire feed unit. The gas is delivered through special hoses and fittings. The same type of hoses and fittings deliver the gas from the wire feed mechanism to the gun. As the gas is discharged from the gun, it pushes the atmospheric elements away from the weld zone.

Carbon Dioxide. Carbon dioxide is often used as a shielding gas for welding mild steel. When carbon dioxide is exposed to an electric arc, carbon monoxide (a poisonous gas) and oxygen are given off. Some of the released oxygen is transformed into ozone (also a poisonous gas). The presence of these two by-products forces the welder to operate in a well-ventilated area.

To compensate for the oxygen which is present in the weld zone when carbon dioxide is used, a consumable wire electrode with deoxidizing agents should be used. The most common deoxidizers added to the wire are silicon, aluminum, titanium, manganese, and vanadium.

When carbon dioxide is produced according to the proper standards, many of the undesirable characteristics of other shielding gases are eliminated. Carbon dioxide produces broad, deep penetration patterns and good bead contour. Under-cutting is less of a problem with carbon dioxide than with argon.

CONTACT TIPS

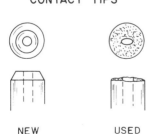

NEW   USED

Fig. 2-12 Contact tips become obstructed with spatter build-up during use.

The main disadvantage of carbon dioxide is the violent action in the arc zone. This is due to the high electrical resistance of the gas. The arc may spatter, but for most operations, this is not an important factor.

Argon and Oxygen. A mixture of argon and oxygen is used as a shielding gas for welding mild steel and stainless steel. Argon gas is an important element in the spray-arc method of MIG welding. When welding on mild steel using argon, the bead contour and penetration may be of poor quality.

To eliminate the problems encountered with pure argon, 1 to 5 percent oxygen may be added to the shielding gas. The addition of oxygen to the mixture widens the penetration zone, improves the bead contour, and reduces undercutting at the edges of the welds, figure 2-13.

The main consideration for welding nonferrous metal (metal containing no iron) is absolute protection from atmospheric contamination. If the alloying elements of nonferrous metal become oxidized, the metal will most likely have reduced corrosion resistance.

MIG Welding Wire Electrodes

The wire for consumable wire electrodes must be clean, smooth, and constant in diameter, so that the wire feed unit can push it into the weld zone without difficulty. It must be evenly wound on an expendable spool. Wire for consumable electrodes is available in spools of one pound to sixty pounds.

For successful MIG welding it is important to match the filler metal (wire electrode) with the base metal. Using the correct wire with the proper shielding gas results in completed welds that meet or surpass their requirements.

The American Welding Society (AWS) specifies the chemical composition of solid bare wire electrodes. The society also designates the numbering system used to classify MIG welding wire electrodes. Figure 2-14 shows how the classification numbers are assigned to MIG wire electrodes.

The two wires best suited for learning situations are AWS numbers E 70S-3 and E 70S-4. These two wire electrodes can be used with argon and oxygen, argon and carbon dioxide, or carbon dioxide as the shielding gas for short-arc operations. Spray-arc operations are possible with both electrodes, using argon and oxygen as the shielding gas.

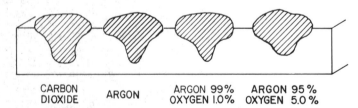

CARBON DIOXIDE   ARGON   ARGON 99% OXYGEN 1.0%   ARGON 95% OXYGEN 5.0%

Fig. 2-13 Penetration and bead appearance can be varied by using different shielding gases.

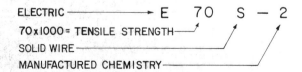

ELECTRIC ⟶ E 70 S — 2
70 x 1000 = TENSILE STRENGTH
SOLID WIRE
MANUFACTURED CHEMISTRY

Fig. 2-14 AWS classification for gas shielded solid wire

E 70S-3 wire electrodes produce good welds in killed and semikilled steel using all three gas shields. *Killed* steel has most of the atmospheric gas removed in the manufacturing process by deoxidizers. *Semikilled* steel is similar to killed steel, except that the atmospheric gases are deoxidized to a lesser extent. Rimmed steel should be welded using argon and oxygen or argon and carbon dioxide as a shield. *Rimmed* steel is manufactured with very little deoxidation. This produces a steel that is almost pure on the outer surface.

The weld puddle of E 70S-4 wire produces a suitable wetting action. The bead is generally flatter and wider than those made with E 70S-3 electrodes. E 70S-4 wire is used in the construction of ships, piping systems, broiler pressure vessels, and general steel structures.

JOB 1: SETTING UP MIG WELDING EQUIPMENT

Equipment and Materials:

> Direct current, constant voltage MIG welding machine
> Wire feed unit
> Roll of E 70S-4 MIG wire (.045 diameter)
> MIG welding gun, complete with contact tip
> Contact centering device
> Nozzle
> Power cable and gas hose
> Inert-gas regulator with flowmeter
> Hose and fitting from the regulator to the wire feed unit
> Cylinders of argon and oxygen

PROCEDURE	KEY POINTS
1. Attach the ground clamp to the ground cable.	1. Use an adjustable wrench.
2. Attach the ground cable to the negative (–) post on the welding machine.	2. Some MIG welding machines do not have a polarity switch on the front panel. To change polarity on these welders, the ground and electrode cables must be switched from the negative (–) post to the positive (+) post.
3. Attach the ground clamp to the work or bench.	
4. Connect the power cable, contractor cable, wire conduit, and gas hose to the wire feed unit and the welding gun.	
5. Insert the contact tip centering device, contact tip, and nozzle into the MIG gun.	

PROCEDURE	KEY POINTS
6. Attach the electrode cable to the positive (+) post of the welding machine.	6. Since the power source must be connected to the electrical supply according to the electrical code, make certain the power source is properly grounded.
7. Connect the electrode cable to the wire feed unit.	
8. Secure a cylinder of argon and oxygen (1 to 5 percent) next to the welding machine.	8. **CAUTION: All high-pressure gas cylinders should be fastened tightly in the upright position while in use. When moving cylinders, always put the valve cap on the cylinder and tighten it securely.**
9. Crack the cylinder valve by opening it slightly and closing it immediately.	9. This allows a small amount of the escaping gas to remove any foreign particles which may have accumulated.
	CAUTION: Do not stand in front of the cylinder when cracking the valve. Do not point the valve toward a wall since the force of the gas may cause particles on the wall to be blown into the operator's eyes.
10. Attach the inert-gas regulator and flowmeter to the cylinder.	
11. Connect the inert-gas hose to the regulator and wire feed unit.	11. Inert gas used for MIG welding is very expensive. Check for and correct leaks in inert-gas fitting, connections, and hoses. Gas leaks may be found with soap and water mix. Escaping gas will make bubbles where leaks are occurring.
12. Install the electrode wire on the wire feed unit.	
13. By hand, feed the wire into the drive rollers of the wire feed unit.	

REVIEW QUESTIONS

Answer the following questions briefly.

1. What are the four primary parts of MIG welding equipment?

2. What is the function of the constant voltage power supply?

3. Describe the flow of electricity in direct current, reverse polarity, MIG welding.

4. What is the relative voltage and wire speed used for short-arc welding?

5. Why is low inductance input preferred for out-of-position welds?

6. What is the function of the slope control?

7. How is the wire feed activated for striking an arc?

8. What is the function of the shielding gas?

9. What are the two toxic gases which are released when welding with carbon dioxide as the shielding gas?

10. What desirable features are added to the argon gas shield when oxygen is mixed with it?

11. What is killed steel?

12. What is rimmed steel?

Unit 3 Striking the Arc and Adjusting the MIG Welding Machine

OBJECTIVES

After completing this unit the student will be able to:

- explain the process of striking an arc with the MIG machine.
- identify direction of travel as it applies to MIG welding.
- define penetration.
- successfully set up and adjust the MIG machine, strike and maintain an arc, and shut off the machine.

Striking the Arc

An arc occurs when the ground and the wire carrying an electrical charge are brought close enough together to close an electrical circuit. The arc jumps the air gap between the wire and the ground, producing ultraviolet and infrared rays and large amounts of heat. These rays are very harmful to the eyes and must be filtered out with a filter lens. The rays made by the MIG welding process are stronger than those of stick-arc welding because the arc is more open and sometimes the welding is done on bright metal like stainless steel or aluminum.

The MIG welding arc is struck by holding the nozzle above the work and pressing the trigger of the MIG gun. When the trigger is pressed, three things start:

- the wire feed unit
- the gas flow
- the electric current

When the wire touches the work, the electric circuit is closed and the arc flashes. To stop the arc, the gun trigger is released and the arc is broken, the wire feed unit stops and the gas flow is shut off.

After the arc is struck, the machine automatically maintains the arc length. The gun nozzle should be about 1/2" above the metal when the welding is being done, figure 3-1.

MIG welding amperage (or heat) is set by the wire speed. If the inches of wire which go through the cable in a minute's time (IPM = inches per minute) are fed through the wire feed faster, the welding heat is increased. If the inches per minute are slowed down, the heat is not as great. When the amperage is varied by changing the speed at which the wire is added to the puddle, the voltage has to be adjusted so the pinch-off of the wire is correct.

When the amperage and voltage are correct, welding can begin. The bead should be run across the plate in a steady forward motion and the gun-nozzle distance above the plate

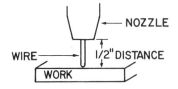

Fig. 3-1 The nozzle should be held 1/2 inch from the work.

should remain constant. If the welding is done too fast, the weld bead will not form properly and the weld will not be deep enough in the plate (penetration). There may be holes inside of the weld because the fast speed will not let the gas come out of the nozzle fast enough to cover the puddle. If the welding is done too slowly, excess heat will build up in the weld zone. This will cause the bead to become very large or possibly burn through the metal.

Direction of Travel

The welding direction of travel is usually made in a left-to-right motion so the weld puddle can be watched and the buildup of the bead can be kept the same. In MIG welding, the bead is clean inside the gas shield and there is not much time lost cleaning the weld.

Penetration

When the arc is struck on the plate, a crater or melted spot is made which is filled with the filler wire. This process of making the crater and filling it up is called *penetration*. The wire and the plate (base metal) must be well mixed for a satisfactory weld.

MIG Gun Angle

After the arc is struck, the nozzle of the gun should be held above the plate at a height and angle that will make a good bead. For a flat weld stringer bead, the nozzle should be straight up-and-down and tipped so that it is about 10 degrees in the direction the welder is making the weld.

Forehand Welding

When the gun is held so that the nozzle is pointed in the direction of travel, it is called *forehand welding,* figure 3-2(a).

Backhand Welding

When the gun is held so that the nozzle is pointed away from the direction of travel, it is called *backhand welding.* Both of the motions are used for different welds, figure 3-2(b).

Purging

When the gas is turned on to start welding, it is a good practice to let some of the gas go through the lines so that all the lines have only the pure gas coming through. This is called *purging the lines.*

Fig. 3-2(a) Forehand **Fig. 3-2(b) Backhand**
The two techniques of welding in a certain direction are forehand and backhand.

Bleeding

After welding is finished and the gas is turned off, the remaining gas should be released from the lines by pressing the trigger on the gun. This is called *bleeding the lines.*

JOB 3: STRIKING THE ARC AND ADJUSTING THE MIG WELDING MACHINE

Equipment and Materials:

Standard MIG welding equipment with E 70S-4 .045 diameter wire (figure 2-4, page 9)
Personal hand tools and protective clothing and gear
3/8" x 6" x 6" mild steel plate

PROCEDURE	KEY POINTS
1. Check all electrical systems and gas system fittings.	1. Welding should be done in a well-ventilated area.
2. Connect the ground clamp to the work.	
3. Turn the wire speed control dial to "off," or "zero." Set the slope or inductance dial halfway up the scale.	3. The slope or inductance controls increase or decrease the wetness of the puddle. Set the control(s) halfway up the scale until welding has started, and then make adjustments as necessary.
4. Turn the voltage control dial to off, or zero.	
5. Turn the flowmeter dial to zero.	
6. Open the gas cylinder slowly and let the pressure build up, then open it fully.	6. Always stand to one side when opening regulators on high pressure cylinders.
7. Turn on the welding machine.	
8. Turn on the wire feed unit.	
9. Turn the flowmeter control counter-clockwise to set the gas flow.	9. Set the gas flow for 40 cubic feet per hour (CFH = cubic feet per hour).
10. Purge the gas system by pressing and releasing the gun trigger.	10. Inert gases are expensive. Do not waste gas.
11. Check the wire and drive rollers to make sure they are clean.	
12. Insert the wire through the drive rollers and into the conduit guide. Remove the contact tip.	12. The wire drive rollers are grooved to fit the size of wire being used. When the wire size is changed the rollers and contact tips must be changed to fit it, figure 3-3, page 22.

PROCEDURE	KEY POINTS

Fig. 3-3   The drive rollers must be of the proper size to match the diameter of the wire.

13. Set the wire feed control dial on half speed.	13. When the wire speed and voltage are set high, *spray arcing* will take place.
14. Set the voltage control dial on 23 volts.	
15. Set the wire feed drive rollers so there is some tension on the wire.	
16. Press the trigger on the gun until the wire passes through the nozzle.	16. The contact tip must be the right size for the wire used.
17. Insert the contact tip.	17. The contact tip must be removed when feeding the wire through the gun. This prevents the wire from hitting or catching on the contact tip.
18. Put on the helmet and gloves.	
19. Lower the helmet over the face and press the gun trigger to make the wire contact the plate.	19. The MIG machine will not strike an arc until the gun trigger is pressed.
20. Run a bead 3″ long–then release the trigger.	
21. Examine the bead.	21. Adjust the machine to correct any faults which appear in the weld.
22. Run more short beads on the plate.	22. Make any adjustments necessary. There are two ways to correct defects: (1) adjust the voltage; (2) adjust the wire feed.
23. Stop welding by releasing the gun trigger.	
24. Turn the wire speed dial to zero.	
25. Turn the voltage dial to zero.	
26. Turn the flowmeter dial to zero.	
27. Bleed the gas lines by pressing the gun trigger.	

PROCEDURE	KEY POINTS
28. Turn off the wire feed unit.	
29. Turn off the welding machine.	
30. Hang up the welding gun.	
31. Disconnect the ground clamp and hang it up.	
32. Clean up the work area.	
33. Submit the weld beads to the instructor for comments and further instructions.	

REVIEW QUESTIONS

Answer the following questions briefly.

1. What causes an electric arc?

2. What is amperage?

3. What happens when the MIG weld is made too rapidly across the workpiece?

4. What is penetration?

5. Are wire feed rollers designed to fit a certain size of wire or are they universal?

6. What conditions must exist to have a spray arc?

7. What two methods are listed for correcting the welding arc on the MIG welding machine?

Unit 4 Running a Bead, Spray Transfer, Flat Position

OBJECTIVES

After completing this unit, the student will be able to:

- name some of the special ways to find out which welding wire is to be used.
- know the general descriptions and uses of E 70S-2, E 70S-3, E 70S-4, E 70S-5, E 70S-6 welding wire.
- successfully weld stringer beads, using the spray-transfer method of MIG welding.

Welding Wire Selection

The welder must know the type of base metal to be welded before the proper wire can be chosen. Other factors may also have an effect on wire choice:

- Joint design
- Thickness of the metal to be welded
- The welding position
- The shielding gas to be used

After the type of metal to be welded is determined, a manufacturer's chart can be used to select the right wire.

The wires used for welding carbon steels are numbered by the American Welding Society (AWS). Figure 4-1 shows two of the wires and their makeup.

E 70S-2

This wire has a high content of deoxidizers. A deoxidizer is an element added to the filler wire to help keep the metal from reacting with oxygen. E 70S-2 is made to make high quality welds in all types of low-carbon steels. It works well on new or old, rusty steel. This wire can be used in all positions of welding using argon/oxygen, argon/carbon dioxide, or carbon-dioxide gas.

E 70S-3

This wire is very popular. E 70S-3 wire will mix well with all kinds of mild steel using carbon dioxide, argon/oxygen, and argon/carbon dioxide. If it is used on rimmed steels,

AWS CLASSIFICATION	CARBON			MANGANESE			SILICON			SULFUR			PHOSPHORUS		
	AWS	A	B	AWS	A	B	AWS	A	B	AWS	A	B	AWS	A	B
E 70S-3	.06-.15	.09	.11	.90-1.40	1.15	1.20	.45-.70	.50	.50	.035	.025		.025	.017	
E 70S-6	.07-.15	.10	.11	1.40-1.85	1.65	1.65	.80-1.15	.97	1.12	.035	.02		.025	.01	

Fig. 4-1 This chart shows the percentage amount of each of the deoxidizers in the two wires listed.

the best welds can be made with argon/oxygen and argon/carbon dioxide gases. The bead will have a flat surface contour. If out-of-position welds are made, it should be used in a small diameter wire and should be welded in single passes.

E 70S-4

This wire has a larger percentage of deoxidizers than either the -2 or -3 wires. The weld is effective on all mild steels. If high strength welds are required, E 70S-4 can be used with any of the shielding gases.

E 70S-5

E 70S-5 has aluminum in the deoxidizers to help make a stiffer puddle. It must be used only in the flat position if the spray-arc transfer method is used. All three shielding gases can be used with E 70S-5 and it will weld all kinds of mild steel.

E 70S-6

E 70S-6 wire has a larger amount of deoxidizers than any of the wires mentioned above. This wire produces very high-quality welds and can be used for high-speed welding with argon/oxygen with the oxygen content at 5 percent or greater. There is no aluminum in this rod, so it can be used for the short-arc process by using carbon dioxide or argon/carbon dioxide for the shielding gas.

Welding Pass

When a bead is run across a plate it is called a *pass*. Out-of-position welds (any weld except flat welds) are generally made with several small passes.

Stringer Bead

A pass run across the plate being welded in a straight line, with no side-to-side motion, is a *stringer bead*.

JOB 4: RUNNING A BEAD, SPRAY TRANSFER, FLAT POSITION

Equipment and Materials:

Standard MIG welding equipment
Personal hand tools
Protective clothing and gear
3/4" x 6" x 6" mild steel plate
E 70S-5 wire electrode
Argon/oxygen shielding gas

PROCEDURE	KEY POINTS
1. Turn on the ventilating system.	1. Proper ventilation is required for MIG welding. The fumes made by the electric arc passing through the gas shield are toxic.

PROCEDURE	KEY POINTS
2. Turn on the power source.	
3. Turn on the wire feed unit.	
4. Open the cylinder valve on the gas system.	
5. Purge the gas system.	5. Make sure the wire feed speed control is set on zero so that the wire does not feed out while the system is purging.
6. Set: Voltage 　　Wire feed 　　Slope inductance	6. All settings should be made according to the wire manufacturer's recommendations.
7. Position the plate in place for welding in the flat position.	
8. With the welding helmet over the face, strike an arc. Run a bead across the plate, using a backhand motion, figure 4-2.	8. Side-cutting pliers are used to cut wire when necessary.
9. Adjust the machine if necessary. Run another bead across the plate each time adjustments are made until the bead is satisfactory.	 **Fig. 4-2 Running a bead**
10. Run the stringer beads 1/2″ apart all across the plate.	10. Turn the wire feed speed up and down to see the difference in the bead appearance and in the operation of the machine.
11. Turn off the MIG machine.	
12. Hang up the MIG gun.	
13. Consult the welding instructor for comments for improving weld quality.	*Note:* Jobs in later units will not give the basic operations needed to set up the machine. The student should become familiar with the procedures followed in setting up and closing down the MIG machine.

REVIEW QUESTIONS

Answer the following questions briefly.

1. Name three gases, or combinations of gas, most generally used for MIG welding.

2. What is a deoxidizer?

3. In your opinion, what are the reasons that E 70S-5 electrode wire was selected for Job #4?

Unit 5 Four Layer Stringer Bead Pad Buildup, Spray Transfer, Flat Position

OBJECTIVES

After completing this unit the student will be able to:

- list the characteristics of a metal.

- explain the methods used to designate groups of steel.

- define low-carbon steel, medium-carbon steel, high-carbon steel, and very high-carbon steel by carbon percentage.

- list the uses of the four carbon classes of steel.

- successfully weld a pad buildup using the spray transfer technique of MIG welding.

MIG Welding Currents

The wire feed unit dial sets the amount of amperes (current) in the arc zone.

Each size and type of wire will have a low and a high current amount in order to make effective welds, figure 5-1.

Each welder chooses the current or heat he or she likes best. Current settings have a wide range and welders can set the machine so that it makes the best welds for their style of welding.

The time it takes for the wire to burn off (called the *burn-off rate*), the bead formation, the depth of penetration and the speed of travel are all affected by the amount of current at the weld zone. When the current is changed, there is a change in the arc at the weld puddle. When the welding current is increased, the wire melts off faster, and the penetration is deeper. The amount of weld deposited and the travel speed of welding will also change. When welding current is decreased, all of the above decrease.

MIG CURRENT RANGE	
WELDING CURRENT RANGE	WIRE ELECTRODE DIAMETER
370 TO 600 AMPS	1/8″
200 TO 550 AMPS	3/32″
160 TO 370 AMPS	1/16″
100 TO 220 AMPS	.045
70 TO 160 AMPS	.035
50 TO 140 AMPS	.030

Fig. 5-1 Welding current ranges for various diameter electrodes.

If the current is set below the ranges shown in figure 5-1, the burn-off rate slows down and poor beads, poor fusion, and shallow penetration will result. If the current is set above the ranges shown, the burn-off rate is too fast and undercutting, burn-through, poor beads, and ruined contact tips can occur.

The wrong current settings can ruin the advantage of MIG welding. MIG welding is used because it makes high-quality welds at high rates of speed.

Wrong settings can also affect the base metal being welded, since they can lower the strength of the metal. Improper current settings can ruin a weld by creating porosity.

Steel Classes and Identification

It is important for a welder to have knowledge of the classes of metals.

A definition of metal includes the following: A class of chemical elements which are: solid at normal temperatures, opaque, a conductor of heat and electricity, will reflect when polished, and will expand and contract in reaction to heat or cold.

The different classes of metal have been identified by giving them numbers, according to their chemical composition. The numbers used here are called S.A.E. numbers because they have been put together by the Society of Automotive Engineers. The first number in the group of numbers used means the type of steel, the second number shows the alloy in the steel, and the last two or three numbers show the carbon content, figure 5-2.

Carbon Steels	1	xxx	**Corrosion and Heat-resisting Steels** . .	30	xxx
Plain carbon	10	xx			
Free cutting (screw stock)	11	xx	**Molybdenum Steels**	4	xxx
Free cutting, manganese	x13	xx	Chromium	41	xx
High manganese	T13	xx	Chromium nickel	43	xx
Nickel Steels	2	xxx	Nickel . . . . 46 xx and 48		xx
.50% nickel	20	xx	**Chromium Steel**	5	xxx
1.50% nickel	21	xx	Low chromium	51	xx
3.50% nickel	23	xx	Medium chromium	52	xxx
5.00% nickel	25	xx	Corrosion and heat resisting	51	xxx
Nickel Chromium Steel	3	xxx			
1.25% nickel; 60% chromium . . .	31	xx	**Chromium Vanadium Steel**	6	xxx
1.75% nickel; 1.00% chromium . .	32	xx	**Tungsten Steel** . . . . 7 xxx and 7		xxxx
3.50% nickel; 1.50% chromium . .	33	xx			
3.00% nickel; 80% chromium . . .	34	xx	**Silicon Manganese Steel**	9	xxx

Fig. 5-2 The S.A.E. system of numbering steels

On the chart, notice that the number 1 shows carbon steel, the number 2 shows nickel steel, and so forth. If a steel is S.A.E. number 1240, the chart would tell that it is in the carbon steel group, is alloyed with nickel, and the carbon in it is 0.40 of 1 percent.

Most of the welding is done on the carbon steel group. There are four main parts in this group, all named by the amount of carbon in them, figure 5-3.

• Low carbon steel has a carbon content of 0.05 to 0.30 percent. This steel is called mild steel. It is tough, easily formed, and welds well.

• Medium carbon steel has a carbon content of 0.30 to 0.45 percent. This steel is strong, hard, and not easily welded. When the amount of carbon is increased, the steel is brittle and easily broken in the weld area.

• High carbon steel has a carbon content of 0.45 to 0.75 percent. This steel can be heat treated. To weld it the operator must have special welding wire. Generally, it must be preheated to be weldable.

• Very high carbon steel, carbon content 0.75 to 1.75 percent. This steel is very hard to weld.

CARBON GROUP	CARBON RANGE BY PERCENTAGE	GENERAL USES OF THE GROUP
LOW CARBON	0.05-0.15	GENERAL FASTENING ITEMS USED IN THE BUILDING TRADES
	*0.15-0.30	STRUCTURAL STEELS
MEDIUM CARBON	0.30-0.45	AUTOMOTIVE INDUSTRY, INTERNAL COMBUSION ENGINE PARTS
HIGH CARBON	0.45-0.60	CONSTRUCTION EQUIPMENT, BLADES & BUCKETS ON EARTH MOVING EQUIPMENT
	0.60-0.75	SPRING STEELS AND TOOLS
VERY HIGH CARBON	0.75-1.50	USED EXTENSIVELY IN THE TOOL INDUSTRY

*GROUP THAT WILL BE OF GREATEST IMPORTANCE TO WELDERS.

Fig. 5-3 Carbon steels and their uses

JOB 5: FOUR LAYER STRINGER BEAD PAD BUILDUP, SPRAY TRANSFER, FLAT POSITION

Equipment and Materials:

Standard MIG welding equipment
Personal hand tools
Protective clothing and gear

3/4" x 6" x 6" mild steel plate
E 70S-5 wire
Argon/oxygen mix gas

PROCEDURE	KEY POINTS
1. Set up machine for spray-arc welding.	1. Position the plate material for flat position.
2. Run a stringer bead all the way across the plate in the center of the plate.	2. Follow the bead sequence in figure 5-4.

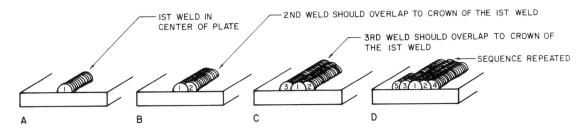

Fig. 5-4 Sequence for padding a plate

3. Continue to run stringer beads until the plate is covered.	3. The small piece of plate will become very hot. Dip it in a cooling tank when necessary. **CAUTION: Use pliers to pick it up and when dipping it in the tank, push it under the water quickly, to avoid steam burns.**

PROCEDURE	KEY POINTS
4. Show the completed first layer to the welding instructor.	
5. Turn the sample around so the beads will cross the plate in the opposite direction. Weld the pad full of beads.	
6. Repeat the welding, turning the metal until four layers have been welded.	
7. Check with the instructor for comments and instructions.	

REVIEW QUESTIONS

Answer the following questions briefly.

1. What is the function of the wire feed unit control dial?

2. What is the welding current range for 1/8-inch diameter electrode wire?

3. What is the welding current range for an .035-inch diameter electrode wire?

4. Four operational functions are affected by the amount of current present in the arc zone. What are they?

5. What are the four functions which are affected by a welding current increase?

6. If the current settings are too low, what happens to the burn-off rate?

7. What characteristics of the base metal may be affected by wrong current settings?

8. List the characteristics of a metal.

9. How does the Society of Automotive Engineers indicate the carbon content, alloy, and type of steel by the use of numbers?

10. If a steel was designated as "1025 S.A.E." what would be its carbon content?

11. What is the low carbon content limit of mild steel? The high carbon content limit?

Unit 6 Fillet Weld, Spray Transfer, Flat Position

OBJECTIVES

After completing this unit the student will be able to:

- define a fillet weld.
- define and contrast convex and concave beads.
- define arc blow.
- list several methods of controlling arc blow.
- successfully MIG weld a fillet joint using the spray-arc method of metal transfer.

The Fillet Weld

A fillet weld is defined as an inside corner weld made within any angle formed by the pieces to be welded. Fillet welds are one of the major weld designs. A fillet weld is triangular in shape and used in T, corner, or lap joints. One or more beads may be used to weld a fillet joint.

The first pass on the fillet weld should be directed into the corner where the plates meet. It must be a deep penetration pass, with good fusion. If the corner is not fused in the first pass, it is impossible to reach into the corner again for full penetration.

Convex and Concave Beads

A convex bead is a built-up bead. The center of the bead is higher than the edges. In general, a convex bead should be used for all joints. The dome of a convex bead should be slightly higher than the surface of the base metal.

A concave bead has a center section which is lower than its edges. This type of bead is shallow in the middle which weakens the joint.

Arc Blow

A direct current sometimes causes arc blow. Arc blow is the wandering of the arc from the path of the weld bead. This movement is created because of the magnetic force produced by the flow of electricity. The magnetic force pulls the arc off course, which makes it difficult to keep the molten puddle in place.

Arc blow is a greater problem in MIG welding than in conventional stick-arc welding. This is a result of the bare electrode used in MIG welding. The flux coating on the stick-arc welding rods helps to control the magnetic force which creates arc blow. The bare wire used in MIG welding has no coating to help control arc blow.

MIG welding with bare electrodes can cause arc blow in all types of joints. Corner welds and butt welds are the most susceptible joints to this problem. These joints, having

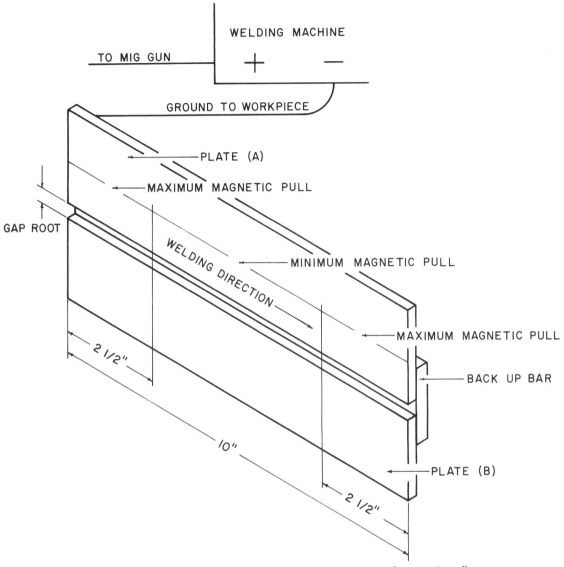

Fig. 6-1 The ends of the work have the largest amount of magnetic pull.

two pieces of metal separated by a groove, can cause different magnetic forces which pull against each other, figure 6-1.

Controlling Arc Blow

Arc blow can be controlled to some extent by any of these methods:

- Directing the angle of the MIG gun towards the right place.
- Changing the motion being used. If forehand motion is being used, change to backhand.
- Decreasing the arc length.
- Increasing or decreasing the wire feed.
- Changing the location of the ground clamp.

The student must be careful that the adjustment of these controls does not affect the quality of the weld.

JOB 6: FILLET WELD, SPRAY TRANSFER, FLAT POSITION

Equipment and Materials:

Standard MIG welding equipment
Personal hand tools
Protective clothing and gear
1 piece, 3/4" x 6" x 6" plate
2 pieces, 3/4" x 3" x 6" plate
E 70S-5 wire electrode
Argon/oxygen mix shielding gas

PROCEDURE	KEY POINTS
1. Turn on and adjust the MIG machine and other equipment for spray-arc welding.	
2. Tack weld the plates as shown, figure 6-2.	

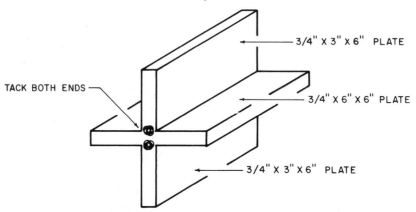

Fig. 6-2 Proper location of tack welds

3. Position the plates for welding as shown, figure 6-3.

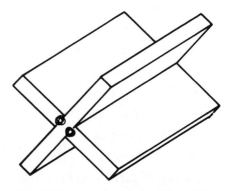

Fig. 6-3 Proper positioning for welding

PROCEDURE   **KEY POINTS**

4. This is a 3-pass weld sequence, figure 6-4.

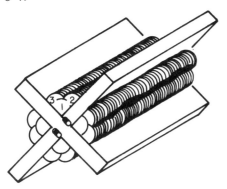

Fig. 6-4 The beads are run in sequence.

5. The angle of the wire electrode should be approximately one-half that of the fillet joint. The gun should also be tilted approximately 10 degrees to facilitate handling of the gun and the view of the weld bead.

6. Travel speed should be appropriate. Welds should not be overwelded (convex) or underwelded (concave), figure 6-5.

4. Use the bead sequence shown. Cool the plate when it becomes overheated. Rotate the joint until all four sides have been welded in the flat position.

6. Slow travel speed increases deposits and chances of cold laps and undercutting. A travel speed that is too fast may decrease metal deposits, produce concave beads, and weaken the joint.

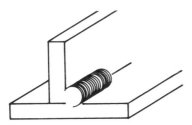 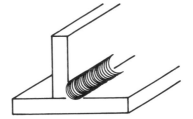

Fig. 6-5 Slow travel speed may result in a convex bead. Fast travel speed may result in a concave bead.

7. Complete the welds, shut down the welding station, and hang up the gun.

8. Submit the welds to the instructor for comments and further instructions.

REVIEW QUESTIONS

Answer the following questions briefly.

1. Define arc blow.

2. Why do bare wire electrodes have a tendency to produce more arc blow than regular stick electrodes?

3. List five methods of controlling arc blow.

4. What is a 3-pass weld sequence?

Unit 7 Butt Weld with Backup Bar, Spray Transfer, Flat Position

OBJECTIVES

After completing this unit the student will be able to:

- define a butt joint.
- name and define five types of groove welds.
- explain the use of a backup bar.
- define a starting tab and a run-off tab.
- successfully complete a butt weld using a backup bar and the spray-arc method of metal transfer.

Butt Weld

A butt weld is defined as a weld joining two pieces of metal with their edges in a straight plane. Generally, metals up to 3/16" thick can be welded with complete penetration without edge preparation. This type of weld is known as a square butt weld, figure 7-1. A square butt weld can be welded from one side or both sides and can be of either the closed or open design.

Groove Welds

Metal over 3/16" thick must be prepared so that 100 percent penetration can be made on the joint. Two methods are used:

- Gapping the metal, or leaving a space between the pieces to be welded;
- Grinding the edges of the pieces at an angle so that there is room at the bottom of the plates for the first pass to gain complete penetration.

Several different designs are used to prepare the edges of the plates, figure 7-1, page 40.

Root Opening

When plates are to be welded together, they should have a space left between them so that the first weld will penetrate through the plate. This space is called a root opening.

Land

Grinding the edge of a piece of metal to a slope makes the bottom edge very thin. The thin edge will burn off quickly and cause *drop-through* (excessive penetration). To keep this from happening, the bottom edge of the bevels are ground square to make it thicker. Such an edge is called a land, figure 7-2, page 40.

Backup Bar

A deep groove weld is usually made with the aid of a backup bar. A backup bar is a plate positioned behind the groove to help support the weld zone. A backup bar is a narrow piece of metal tack welded on the bottom of the plates, directly below the groove. The weld is started on this backup bar so that the bead is flowing smoothly by the time the weld starts in the groove. By welding the first pass on the back-up bar, the lower edges of the plates are fused together with 100 percent penetration. The weld is ended on the backup bar to eliminate craters or holes which might be at the end of the weld.

The backup bar is cut off the welded plates after the weld is finished. After the backup bar is cut off, the bead on the top side of the plates is ground or machined off before testing is done on the weld, figure 7-3.

Starting Tab and Run-Off Tab

The part of the backup bar which extends past the plates to be welded on the starting end of the weld is called the starting tab. The bar which extends past the finishing end of the weld is called the run-off tab.

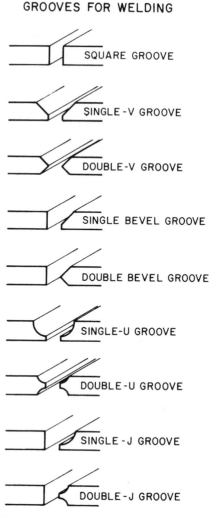

GROOVES FOR WELDING

Fig. 7-1 **Various types of groove joints**

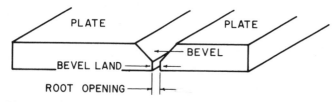

Fig. 7-2 **The flat section of an edge prepared for welding is a land.**

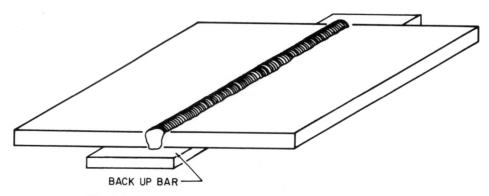

Fig. 7-3 **The backup bar is positioned under the root of the weld.**

JOB 7: BUTT WELD WITH BACKUP BAR, SPRAY TRANSFER, FLAT POSITION

Equipment and Materials:

Standard MIG welding equipment
Personal hand tools
Protective clothing and gear
2 pieces, 3/4" x 4" x 6" plate
1 piece, 3/8" x 2" x 9" strip
E 70S-5 wire electrode
Argon/oxygen shielding gas

PROCEDURE	KEY POINTS
1. Set up the machine for spray-arc welding and tack weld the beveled plates as shown in figure 7-4.	1. Backup bars must fit tightly on the plates. Any gap increases the chance of defective welds.

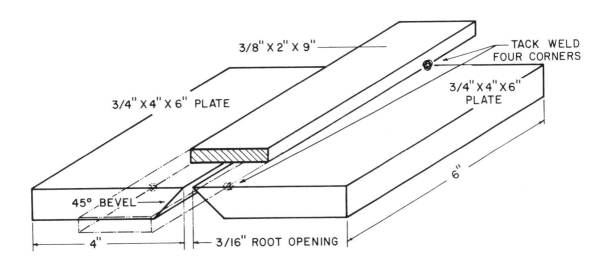

Fig. 7-4 The backup bar is tack welded in place.

2. After tacking the backup bar on the plates, turn the joint over so that the groove is on top.

3. The backup bar extends past the plate sides. This overlap is called (a) the starting tab, and (b) the run-off tab. The backup bar extensions are used when 100 percent welded samples are required.

PROCEDURE	**KEY POINTS**
4. Begin welding on the sample following the bead sequence shown in figure 7-5.	4. If more fill is needed than is received with 6 passes, add more beads to the groove. The bead must be built up above the surface of the base metal.

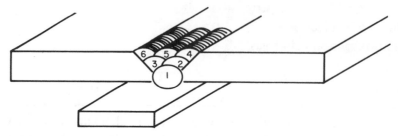

Fig. 7-5 The running of beads must be done in the proper sequence.

5. Submit the weld to the instructor for comments and instructions.	
6. Save this weld for testing in Unit 8.	6. When welds are to be tested they should not be cooled in water. Cooling makes the weld weak and brittle.

REVIEW QUESTIONS

Answer the following questions briefly.

1. Define a square butt weld.

2. Define a groove weld.

3. Explain the need for a starting tab and a run-off tab.

4. How is a backup bar removed from a weld specimen?

Unit 8 Testing Welds

OBJECTIVES

After completing this unit the student will be able to:

- list three types of weld tests.
- list the procedure used to perform a guided-bend test.
- successfully perform a guided-bend test of the butt weld sample from Unit 7.

Types of Weld Tests

In order to make sure the welder can complete welds of good quality, three kinds of tests may be used.

- Visual tests
- X-ray tests
- Destructive tests

Visual tests are made by simply looking at the weld to judge it by the appearance. It is not a good test. Sometimes the weld may look smooth and strong, but it may contain slag holes or porosity and the penetration may be poor.

An x-ray test is a nondestructive test used to examine the internal structure of a weld. Porosity and inclusions can be easily spotted using the x-ray photograph.

A destructive test is one in which the metal is bent or torn apart. This kind of test destroys the metal for further use, but does prove if the weld is strong enough to hold the base metal.

In most cases where a welder is seeking a job, welds will have to pass a destructive test to prove that the welder can apply sound, strong welds in any welding position.

Guided-Bend Test. One of the most commonly used destructive welding tests is the guided-bend test. In this test, a butt weld is made by the operator and cut into strips, figure 8-1, page 44. These strips are bent in a machine called a guided-bend test machine, figure 8-2, page 45. The strips are bent 180 degrees against the face of the weld and the root of the weld to see if the weld separates from the base metal. The face of the weld is formed on the surface of the side that the welder works on. The root of the weld is at the bottom and directly opposite the face. A good weld will hold more than the material that was welded.

Acid Etching Test. The nitric acid etching test is employed to determine penetration qualities of a welded sample. By applying the etching solution the depth of penetration of a weld can be visually checked. The nitric acid solution etches rapidly and when used on a polished surface will show the refined weld zone as well as the base metal.

Nitric acid solution is mixed by adding 1 part nitric acid to three parts water, by volume. For safety reasons, acids must be handled with extreme caution. Nitric acid will cause severe burns and bad stains.

CAUTION: **The nitric acid must always be poured into the water when the solution is mixed.**

To conduct the test, a welded sample must be cut to show the complete cross-sectional area. The face of the cut is then filed and polished to a smooth surface. The prepared surface of the sample may be etched by applying the mixed solution to the face of the sample with a glass stirring rod, at room temperature. All tests should be conducted in a well ventilated area.

After the test sample has been etched it should be immediately rinsed in clear hot water. Remove the excess water and dip the etched surface in ethyl alcohol.

Remove the sample from the alcohol bath and dry it in a warm air blast.

The weld penetration into the base metal will show as a dark area. The etched surface may be preserved by coating it with a thin, clear lacquer. Destructive testing of this nature is a good means of determining the abilities of the welder.

JOB 8: TESTING WELDS

Equipment and Materials:

Carbon-arc or oxyacetylene cutting equipment   Guided-bend test machine
Personal hand tools   Butt weld sample from Job #7
Protective clothing and gear

PROCEDURE	KEY POINTS
1. Use the butt weld sample made in Job #7.	
2. Remove the backup bar with the carbon-arc or oxyacetylene cutting process.	2. The weld sample must not be damaged or destroyed during this process.
3. Lay out and section the sample as shown in figure 8-1.	

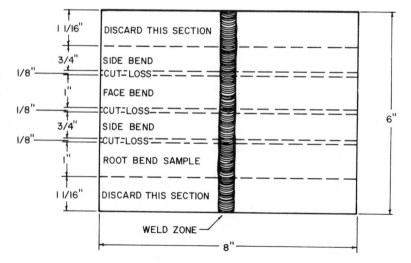

Fig. 8-1 Layout of butt weld sample used to produce strips needed for guided-bend test

PROCEDURE	**KEY POINTS**
4. Bend the root bend strip in the guided-bend machine, figure 8-2.	GUIDED BEND TEST 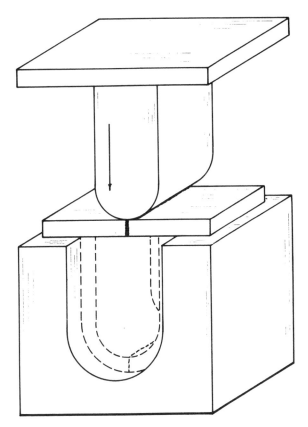
5. Bend the face bend strip in the guided-bend machine.	
6. Bend one side bend (root facing out) in the guided-bend machine.	
7. Bend the other side bend (root facing in) in the guided-bend machine.	**Fig. 8-2 Guided-bend test machine**
8. Inspect the bent samples for flaws.	8. There should be no separation of weld from the base metal and no pin holes or slag inclusion.
9. If the weld samples are defective, construct a new butt weld sample as in Job #7 and repeat Job #8.	
10. When the bent samples are satisfactory, submit them to the instructor for comments and further instructions.	

REVIEW QUESTIONS

Answer the following questions briefly.

1. Why is a visual check of a weld not a good determining factor for evaluating the weld?

2. Name a nondestructive test used to examine the internal structure of a weld.

3. What type of test may a job applicant have to pass to gain a job in the welding trade?

4. Name two types of destructive tests and list the procedure used to conduct each one.

SECTION 1: REVIEW A

OBJECTIVE:

- Produce a fillet weld of proper penetration and quality using standard MIG welding equipment and procedures. Subject the sample to a visual test.

Equipment and Materials:

Standard MIG welding equipment
Personal hand tools
Protective clothing and gear
1 piece, 3/4" x 6" x 6" plate

2 pieces, 3/4" x 3" x 6" plate
E 70S-5 wire electrode
Argon/oxygen mix shielding gas

PROCEDURE

1. Set up and adjust the MIG welding equipment for spray-arc welding.

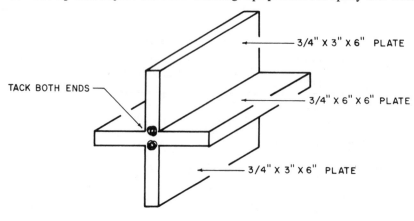

3/4" X 3" X 6" PLATE

TACK BOTH ENDS

3/4" X 6" X 6" PLATE

3/4" X 3" X 6" PLATE

Fig. R1-1 Proper placement of the plates and the tack welds

2. Position and tack weld the plates in the proper place, figure R1-1.

3. Arrange the plates for flat position welding and run a three bead sequence as shown in figure R1-2.

4. Run three more beads in the proper sequence directly opposite those just completed, figure R1-3.

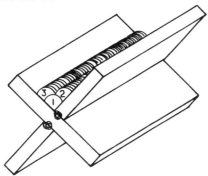 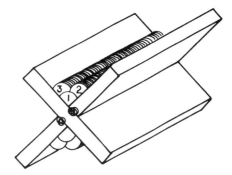

Fig. R1-2 Proper position and sequence of beads for flat position

Fig. R1-3 Proper position of second sequence of beads

5. Submit the welds to the instructor for comments and suggestions.

6. Complete the sample as shown in figure R1-4.

7. Submit the welds to the instructor for comments and instructions.

Fig. R1-4 Completed weld sample

SECTION 1: REVIEW B

OBJECTIVE:

• Produce a butt weld of proper penetration and quality using standard MIG welding equipment and procedures. Subject a sample butt weld to the guided-bend test.

Equipment and Materials:

Standard MIG welding equipment
Carbon-arc or oxyacetylene cutting equipment
Personal hand tools
Protective clothing and gear
Guided-bend test machine
2 pieces, 3/4" x 4" x 6" plate
1 piece, 3/8" x 2" x 9" strip
E 70S-5 wire electrode
Argon/oxygen mix shielding gas

PROCEDURE

1. Grind the edge of one plate to form a single-V groove weld.

2. Tack weld the backup bar in place as shown in figure R1-5 and turn the sample over.

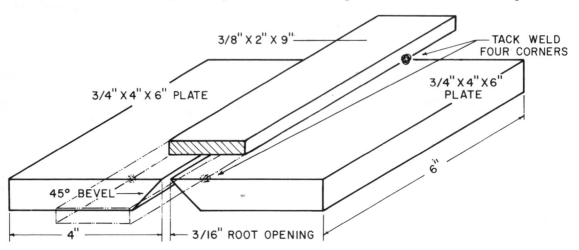

Fig. R1-5 Proper position of the plates, the backup bar and the tack welds

3. Run enough beads in sequence to fill the groove, figure R1-6.

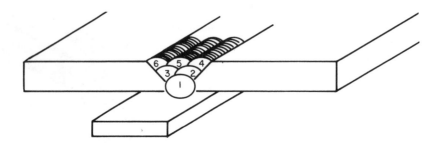

Fig. R1-6 The proper sequence for running the beads

4. Remove the backup bar using the carbon-arc or oxyacetylene cutting process.

5. Lay out and cut the sample into sections as shown in figure R1-7.

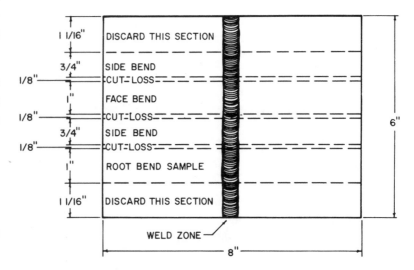

Fig. R1-7 The layout of the test sections

6. Perform a guided-bend test on the face, the root, and the sides of the various strips.

7. Submit the bent samples to the instructor for comments and further instructions.

SECTION 2:
Welds in the Flat Position, Short-Arc Technique

Unit 9 Striking the Arc and Adjusting the Machine, Short Arc, Flat Position

OBJECTIVES

After completing this unit the student will be able to:

- explain the application of short-arc MIG welding.
- name the three general areas of MIG welding defects.
- list eleven common MIG welding problems, their causes, and possible ways to correct them.
- discuss and apply efficient cleaning methods to the MIG welding gun.
- list the negative results caused by spatter buildup.
- successfully set up the equipment and adjust it so that satisfactory beads can be run on mild steel plate in the flat position using the short-arc method of metal transfer.

Short-Arc Welding

The short-arc process requires less amperage than spray-arc. The machine must be set to lower amperages. Follow the wire manufacturer's recommendations for setting the machine. When welding with short-arc, the wire is fed into the front of the puddle.

Troubleshooting for MIG Operations

With the correct materials (a match up of ferrous or nonferrous metals and ferrous or nonferrous filler wire), gas coverage, welding conditions and techniques the MIG process will yield high-quality welds.

As with any other welding process, when any of these conditions are incorrect, welds of poor quality will result. Most of the bad welds are caused by wrong machine settings. When the reasons for poor welds are known, corrections are easily made.

Troubleshooting for MIG can be grouped into three probable causes:

- Electrical
- Mechanical
- Welding Procedures

The chart, figure 9-1, has been included to help the welder identify the causes of welding troubles. It lists some of the corrections to make so the welder saves time and effort while welding. The chart covers only the most common causes of wrong operations.

DEFECT	CAUSE	CORRECTION
1. Wire feed unit stops feeding wire while welding.	1. Electrical fuse blown in primary power, power source or wire feed unit.	Have qualified electrician check electric power circuits.
	2. Control relays of MIG gun defective.	Replace defective relay switches.
	3. Loss of wire feed roller tension.	Inspect and replace all defective feed drive rollers.
2. Wire electrode feeds but arcing does not occur.	1. Electrical fuse blown in primary power, power source or wire feed unit.	Have qualified electrician check electric power circuits.
3. Weld porosity.	1. Dirty base metal.	Proper joint preparation.
	2. Faulty gas system.	Check out and replace defective components of the gas system.
	3. Faulty torch assembly.	Replace the contact tip centering device.
	4. Improper torch manipulation.	Use recommended torch manipulation.
	5. Spatter-clogged nozzle.	Clean the nozzle.
4. The wire electrode stubs into the workpiece.	1. Excessive slope.	Adjust slope.
	2. Low voltage.	Increase voltage.
	3. High wire feed.	Decrease wire feed.
5. Excessive spatter.	1. Excessive or insufficient gas flow.	Adjust gas flow.
	2. Insufficient slope setting.	Adjust slope.
	3. Excessive voltage.	Adjust voltage.
	4. Wrong shielding gas.	Use recommended gas coverage.
	5. Wrong electrode.	Use recommended wire electrodes.
6. Undercut at sides of weld puddle. (Called "wagon tracks.")	1. Wrong shielding gas.	Use recommended gas coverage.
	2. Wrong torch manipulation.	Use recommended torch motions.
7. Bad arc starts.	1. Wrong voltage.	Adjust voltage.
	2. Wrong slope.	Adjust slope.
	3. Wrong inductance.	Adjust inductance.
	4. Glass slag.	Chip and brush welds.
8. Fluid puddle.	1. Wrong inductance.	Adjust inductance.
	2. Wrong electrode.	Use recommended wire electrodes.
	3. Wrong shielding gas.	Use recommended gas coverage.
9. Rigid puddle.	1. Wrong inductance.	Adjust inductance.
	2. Wrong electrode.	Use recommended wire electrodes.
	3. Wrong shielding gas.	Use recommended gas coverage.
10. Convex deposit.	1. Wrong voltage or wire feed.	Adjust controls.
	2. Wrong inductance.	Adjust inductance.
	3. Wrong torch manipulation.	Use recommended torch motion.
	4. Wrong travel speed.	Increase travel speed.
11. Concave deposit.	1. Wrong voltage or inductance.	Adjust controls.
	2. Wrong torch manipulation.	Use recommended torch motion.
	3. Wrong travel speed.	Decrease torch travel speed.

Fig. 9-1

Care of the MIG Welding Gun and Accessories

All welding processes require proper care and usage of the tools used. Correct handling of MIG welding equipment prolongs the life of the nozzles, contact tips, wire conduit, wire feed drive, rollers, (and so forth).

A welder who is not used to handling a MIG gun may push the gun nozzle into the weld puddle. Many gun nozzles and contact tips may be melted in this manner, and they will have to be thrown away. Nozzles must be kept free of spatter. If too much spatter builds up, it may interfere with the gas flow through the nozzle causing unsatisfactory welds.

Spatter should not be cleaned from the nozzle by tapping the gun on any solid object. The gun, nozzle, and contact tip are made from soft materials. Tapping to remove the spatter can distort or damage the threads used to position the parts in the gun. If the nozzle loses shape, the gas flow may be pushed to one side and not cover the weld puddle. Also, if the threads are damaged in the nozzle, it may have to be thrown away. The high temperature insulation in the gun can be easily broken.

Spatter on the contact tip increases friction on the wire and slows down the feed speed. This has a bad effect on the welds. Worn tips will also make the wire feed change.

The drive rollers and contact tip are made in sizes to fit the wire. They must be changed when the wire is changed in size.

JOB 9: STRIKING THE ARC AND ADJUSTING THE MACHINE, SHORT ARC, FLAT POSITION

Equipment and Materials:

Standard MIG welding equipment	3/8" x 5" x 6" mild steel plate
Personal hand tools	Carbon dioxide shielding gas
Protective clothing and gear	E 70S-3 or E 70S-4, .035" wire electrode

PROCEDURE	KEY POINTS
1. Install E 70S-3 or E 70S-4, .035" wire on the MIG machine wire feed. Either of these wires should be used for short-arc application.	1. The drive rollers and contact tip must be changed for the small diameter wire. When loading new wire on the machine, leave the contact tip out until the wire passes through the gun, then install the contact tip in place.
2. Turn on the welding machine, the wire feed unit, and the gas system. Feed the wire through the nozzle.	2. Insert the contact tip.
3. Set the gas flow for 25 cubic feet per hour (25 CFH) and make later adjustments as necessary for good welding results.	3. Too much gas flow wastes gas. Also, extra gas will make a swirl around the weld zone, which may cause porosity in the weld. CO_2 gas may cause some slag on the weld and chipping and brushing may be necessary.

PROCEDURE	KEY POINTS
4. Set the wire feed control dial on half speed and make necessary adjustments later as needed.	
5. Set the voltage control dial on 19 volts. Make necessary adjustments later as needed.	
6. Adjust the wire feed drive roller tension.	6. The wire feed tension should not be too tight or the wire will be deformed.
7. Begin welding beads, using the short-arc process. Run beads 3″ long and adjust the machine for proper operation.	

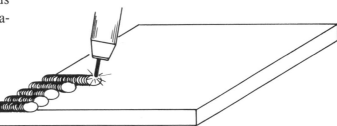

Fig. 9-2 **Running beads using the short-arc process**

8. Run several beads, readjusting the machine as necessary. At the end of the weld, weld back into the crater, to fill the crater and prevent crater cracks.	8. Short-arc welding uses less voltage and wire feed speed than spray-arc. A frying, crackling sound will be heard when the proper setting is made.
9. When the welds are of high quality, turn off the machine and shut off all the systems. Hang up the gun.	9. When the welding station is shut down, all the controls should be returned to zero. At the start of each project the controls will have to be adjusted for correct operations.

REVIEW QUESTIONS

Answer the following questions briefly.

1. When troubleshooting defective welds in the MIG process, what three general categories may the faults be grouped in?

2. If the wire feed unit stops feeding wire when MIG welding, what are three possible causes?

3. When excessive spatter occurs while MIG welding, what five possible reasons should be looked for?

4. What is the condition of a weld which is "wagon tracked"?

5. If the puddle of a MIG weld is too fluid, what are three possible reasons to suspect as the faults?

6. Describe a good method of setting the MIG machine before welding is attempted on a project.

7. What negative effects can result if the MIG gun is pushed into the weld puddle?

8. Why is it not good practice to tap the MIG nozzle on the welding bench to clean out spatter?

9. What effect can a distorted nozzle have on the gas flow during MIG welding?

10. What is the diameter of the wire called for in Job #9?

11. Is it necessary to change the drive rollers when the wire size in the MIG process is changed?

Unit 10 Running a Bead, Short Arc, Flat Position

OBJECTIVES

After completing this unit the student will be able to:

- identify and describe three common faults found in the MIG welding process.
- understand and correct weld failures caused by porosity, poor penetration, and excessive penetration.
- successfully MIG weld stringer beads on mild steel plate using the short-arc method of metal transfer.

Fusion

When the MIG wire and the base metal are melted together and mixed, the process is called fusion. The more fusion (mixing) there is between the wire and the metal, the better the weld.

Defects in Penetration

Welds should be of good quality throughout the melted sections. If this is not the case, generally one of two defects has occurred:

- Shallow penetration
- Excessive penetration

Shallow penetration is caused by either too much or too little welding current and an improper gun angle. When MIG welding, the heat output is increased by using more wire speed in the weld zone. If the wire feed is too fast, and the heat is increased by it, the welder has to move along the joint too fast, and weld penetration is lost. Correct this by lowering the wire feed inches per minute (IPM). MIG guns should be angled so that the arc is held in the leading edge of the puddle. The proper root openings will ensure penetration on open butt welds, figure 10-1.

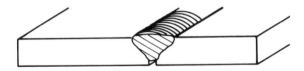

Fig. 10-1 Shallow penetration

Excessive penetration (burn through), figure 10-2, page 57, is caused by too much heat in the weld zone, too slow a travel speed, or improper joint preparation. The welding heat can be lowered by slowing down the inches per minute (IPM) of wire feed into the

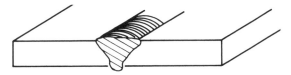

Fig. 10-2 Excessive penetration

molten puddle. If the welding current is correct, faster travel along the joint will eliminate the burn through caused by slow travel speed.

JOB 10: RUNNING A BEAD, SHORT ARC, FLAT POSITION

Equipment and Materials:

Standard MIG welding equipment
Personal hand tools
Protective clothing and gear
3/8″ x 5″ x 6″ mild steel plate
Carbon dioxide shielding gas
E 70S-5 wire electrode

PROCEDURE	KEY POINTS
1. Turn on the welding machine, wire feed unit and the gas supply.	
2. Use side-cutting pliers and cut off 1/8″ from the end of the wire.	2. When welding stops, the end of the wire will be round. Cut the round end off, it may make it hard to start the next arc.
3. Set the gas flow.	3. Sometimes if the gas flow from CO_2 bottles is more than 25 CFH (cubic feet per hour) the regulators will freeze. If this happens, regulator heaters must be used.
4. Begin welding stringer beads. Continue welding and making adjustments until correct results are achieved.	4. The welder should be able at this point to adjust the voltage, wire feed, and slope/inductance for proper welding.
5. Turn off the machine, shut down all the systems and hang up the welding gun and ground cable.	
6. Submit the completed welds to the instructor for comments and instructions.	

REVIEW QUESTIONS

Answer the following questions briefly.

1. What are the three most commonly encountered weld defects in MIG welding?

2. What is shallow penetration?

3. How may shallow penetration be corrected?

4. What is excessive penetration?

5. How may excessive penetration be corrected?

Unit 11 Four Layer Stringer Bead Pad Buildup, Short Arc, Flat Position

OBJECTIVES

After completing this unit the student will be able to:

- define porosity, undercutting, crater cracks, and whiskering.

- list corrective action which may eliminate weld defects.

- adjust the MIG equipment and successfully complete a four layer stringer bead pad buildup.

Weld Defects

Four of the major weld defects are:

- Porosity
- Undercutting
- Weld Cracking
- Whiskering

Porosity. The presence of holes in a weld is called porosity. The holes may be on top of the weld where they can be seen, or inside the weld. If they cannot be seen, x-rays are sometimes used to check the internal structure of the weld. Porosity can be caused by not aiming the MIG gun properly so that the shielding gas is over the weld puddle. Porosity can be kept out of the weld by using the correct forward speed and gun angle, figure 11-1.

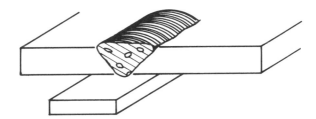

Fig. 11-1 Porosity is the result of small holes in the weld.

Undercutting. Undercutting is the melting away of base metal which is not replaced by the fusion of the base metal and wire in the arc zone. During MIG welding, incorrect travel speed, wrong arc voltage, or fast wire speeds are the major reasons for undercutting. Corrections can be made by changing one or all three of these factors, figure 11-2, page 60.

Weld Cracking. Weld cracks are broken down into two groups, called (1) hot cracks, and (2) cold cracks. Cracking is caused by stresses which are greater than the strength of the base metal. Hot cracks, generally crater cracks, are formed when the weld is very hot. When the crater at the end of the weld cools, cracks sometimes show up because the crater

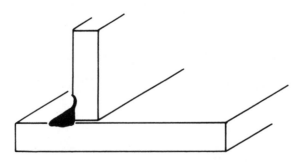

Fig. 11-2 Undercutting of the vertical member

is thinner than the rest of the hot metal. Hot cracks can be eliminated if the welder backs the arc up enough to fill the crater with molten metal when finishing the weld, figure 11-3.

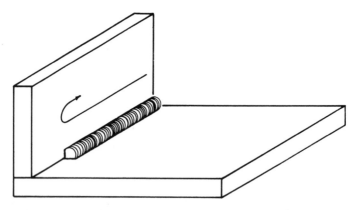

Fig. 11-3 Fill the crater by reversing direction to avoid cracks.

Cold cracks can run the length of the weld. These flaws are caused by putting too much load on the weld. They can be corrected by using more weld deposit so the weld will support more weight, figure 11-4.

CRACKING

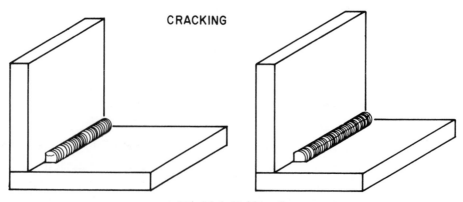

Fig. 11-4 Weld cracks

Whiskers. Sometimes when MIG welding, a defect never found in other arc welding processes appears. It is known as whiskering. Careless use of the MIG gun causes the wire to leave the leading edge of the puddle and go through the root opening. When it is through the root opening it makes contact and cuts off a piece of wire which is fused to the underside of the plate. The fault is cured by slowing down the travel speed and keeping the wire in contact with the base metal.

JOB 11: FOUR LAYERS STRINGER BEAD PAD BUILDUP, SHORT ARC, FLAT POSITION

Equipment and Materials:

Standard MIG welding equipment
Personal hand tools
Protective clothing and gear
3/8" x 5" x 6" mild steel plate
Carbon dioxide shielding gas
E 70S-5 wire electrode

PROCEDURE	KEY POINTS
1. Turn on the welding machine, the wire feed unit and the shielding gas system.	
2. Set the gas flow at 25 CFH and adjust as necessary when welding.	
3. Weld a stringer bead down the middle of the plate, full length. Follow the diagram in figure 11-5 for later welds.	3. After bead #1 has been completed, bead #2 should cover the edge of weld #1, up to 1/3 of the bead. Bead #3 should cover the toe of weld #1 on the left side, up to 1/3 of the bead width. Each following bead should be welded in the same way.

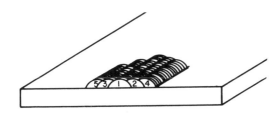

Fig. 11-5 Proper sequence for Job 11

PROCEDURE	KEY POINTS
4. Cool the sample between the weld passes.	4. Small weld plates get a large amount of heat input and cannot lose the heat. It is necessary to cool the sample often.
5. Adjust the machine if weld defects are found.	
6. Turn the pad 90 degrees so the next layer of welds will cross the first layer and continue welding until the pad has been turned enough to make four layers of weld, figure 11-6, page 62.	

PROCEDURE   **KEY POINTS**

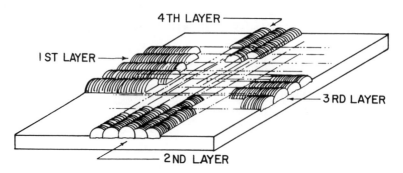

Fig. 11-6 The pad is turned 90° between each fall layer.

7. Submit the pad to the instructor for comments and approval to proceed to Unit 12.

REVIEW QUESTIONS

Answer the following questions briefly.

1. Name three weld defects common to fillet and root bead welds.

2. Define undercutting.

3. How may porosity in welds be avoided?

4. What are the reasons for undercutting?

5. Name three methods of eliminating undercutting from the weld.

6. What are the two groups of centerline cracking?

7. What is a crater crack?

8. How can it be eliminated?

9. What is a cold crack?

10. How can it be eliminated?

11. What are whiskers in MIG welding?

12. How is whiskering controlled?

Unit 12 Fillet Weld, Short Arc, Flat Position

OBJECTIVES

After completing this unit the student will be able to:

- identify and describe a lap joint, a T joint, and a corner joint.

- explain the necessity for grooving a T joint.

- successfully weld a fillet weld using the short-arc method of metal transfer, in the flat position.

Lap Joints

Lap joints are used because there is very little joint preparation. The joint is one in which a plate is placed on top of another, forming a joint that requires a fillet weld bead. Lap welds may be of the single or double joint design, figure 12-1.

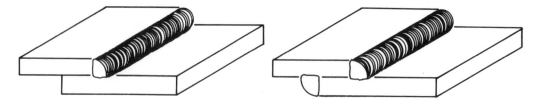

Fig. 12-1 A. Single lap fillet weld   B. Double lap fillet weld

T Joints

When T joints are used, little plate preparation is necessary. One plate is positioned 90 degrees on top of the other forming a joint which requires a fillet weld. For added strength, T joints are sometimes prepared as a single or double-bevel grooved joint, figures 12-2 and 12-3. This allows for 100 percent penetration if the plates are 1/8″ thick or more.

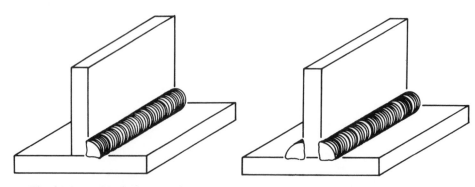

Fig. 12-2 A. Single T joint fillet weld   B. Double T joint fillet weld

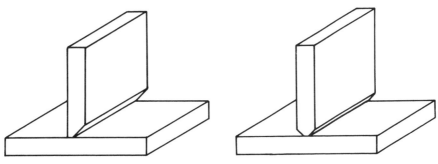

Fig. 12-3 A. Singel-bevel T joint groove   B. Double-bevel T joint groove

Corner Joints

Corner joints are used when designed into the job. Strength is added by the use of groove joints. The joint is made by putting one edge of a plate at a 90 degree angle to the edge of another plate. Corner joints can be further broken down into inside corner joints and outside corner joints. The inside corner joint is placed within the 90 degree angle, figure 12-4(a). The outside corner joint is placed on the 270 degree angle side of the joint. If both sides are to be welded, a bevel must be ground on one of the plates, figure 12-4(b).

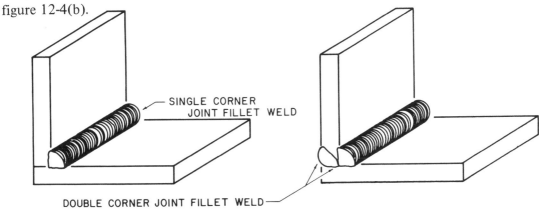

SINGLE CORNER
JOINT FILLET WELD

DOUBLE CORNER JOINT FILLET WELD

Fig. 12-4 Corner joints
A. Single inside corner joint fillet weld   B. Double inside and outside corner joint fillet weld

Leg Size

Many times the welder has to make the fillet weld legs a certain size. Leg size is measured from the corner to the edge of the weld in both directions, figure 12-5. The size of the load that the joint will carry is related to the size of the weld. The weld may be made with more than one pass to increase its size.

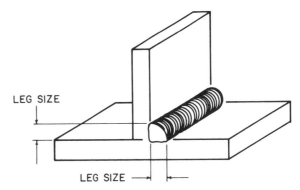

LEG SIZE

LEG SIZE

Fig. 12-5 Leg size of a fillet weld is measured from the corner to the edge of the weld.

JOB 12: FILLET WELD, SHORT ARC, FLAT POSITION

Equipment and Materials:

Standard MIG welding equipment
Personal hand tools
Protective clothing and gear
2 pieces, 3/8" x 5" x 6" mild steel plate
2 pieces, 3/8" x 2 1/2" x 6" mild steel plate.
Carbon dioxide shielding gas
E 70S-5 wire electrode

PROCEDURE	KEY POINTS

1. Position the plates for flat welding and tack weld both samples, figure 12-6.

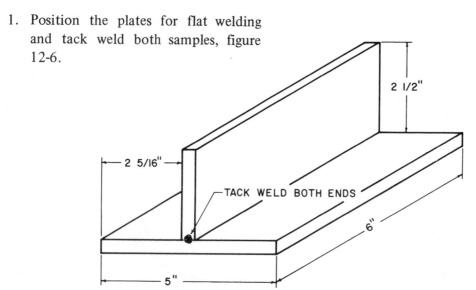

Fig. 12-6 Proper position for plates and tack welds

2. Weld one side of one sample, following the bead sequence in figure 12-7, using the forehand motion.

    2. Check the weld for any defects.

Fig. 12-7 Proper bead sequence for a fillet weld

3. Weld the other side of the joint, using the same bead sequence, with a backhand motion.

    3. Machine adjustments may be necessary because of added weld and change of travel direction.

PROCEDURE	KEY POINTS
4. Weld the other sample.	4. Follow the same procedures.
5. Shut down all systems.	
6. Submit the sample to the instructor for comments and instructions.	

REVIEW QUESTIONS

Answer the following questions briefly.

1. Define a lap weld.

2. What is a T joint?

3. Why is it sometimes necessary to groove a T joint?

4. How is the fillet weld leg size specified?

Unit 13 Butt Weld with Open Root, Short Arc, Flat Position

OBJECTIVES

After completing this unit, the student will be able to:

- list the characteristics of the different types of groove joints used for welding.
- explain the increased costs of metal preparation for welding.
- successfully weld an open butt joint.

Square Butt Joint

The square butt joint is a popular weld due to its easy preparation and weldability. A disadvantage is that its edge to edge design is relatively weak. The open and closed variations of the square butt joint are shown in figure 13-1. The open design is generally used for plates between 1/8" to 3/16". The closed design is used for lighter plates and sheet metal. The maximum thickness of the plates for a square butt joint is 3/16 of an inch. Thicker plates must have at least one edge prepared before welding.

PLATE THICKNESS 3/16" MAXIMUM

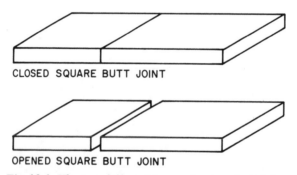

CLOSED SQUARE BUTT JOINT

OPENED SQUARE BUTT JOINT

Fig. 13-1 The two different types of square butt joints

Single-Bevel Groove Joints

Single-bevel groove (one plate beveled) butt joints are stronger than square butt joints. The joint design ensures 100 percent penetration and is more easily welded than the square butt joint. The design makes the cost of the welding job higher because of the time it takes to bevel the plate and fit it together. It also requires more wire deposit. This joint can be welded from one or both sides, either open or closed, and can be single or double grooved. Joints included in this group are the single bevel and the single-J groove, figure 13-2.

PLATE THICKNESS 3/16" MINIMUM

CLOSED SINGLE BEVEL GROOVE JOINT

CLOSED SINGLE J-GROOVE JOINT

OPENED SINGLE BEVEL GROOVE JOINT

OPENED SINGLE J-GROOVE JOINT

CLOSED SINGLE DOUBLE BEVEL GROOVE JOINT

CLOSED SINGLE DOUBLE J-GROOVE JOINT

OPENED SINGLE DOUBLE BEVEL GROOVE JOINT

OPENED SINGLE DOUBLE J-GROOVE JOINT

Fig. 13-2 Single-plate groove joints

Double-Plate Groove Joints

Double-plate groove (both plates beveled) butt joints are used for heavy plates where heavy load conditions are found. They can be welded with 100 percent penetration and are easily welded. This joint design has a higher cost because of the beveling and fit-up involved. It is generally welded from both sides due to the thickness of the material.

It may be a closed or open design and can be welded with a backup bar to increase the quality of the weld. Joints included in the group of double-groove butt welds are the V groove and the U groove, figure 13-3.

PLATE THICKNESS 5/8" MINIMUM

CLOSED DOUBLE BEVEL GROOVE JOINT

CLOSED DOUBLE U GROOVE JOINT

OPENED DOUBLE BEVEL GROOVE JOINT

OPENED DOUBLE U GROOVE JOINT

Fig. 13-3 Double-plate groove joints

JOB 13: BUTT WELD WITH OPEN ROOT, SHORT ARC, FLAT POSITION

Equipment and Materials:

Standard MIG welding equipment
Personal hand tools
Protective clothing and gear

4 pieces, 3/8" x 3" x 6" mild steel plate
E 70S-3 or E 70S-4 wire electrode
Carbon dioxide shielding gas

PROCEDURE	KEY POINTS
1. Bevel the plates at 30 degrees.	1. Grind a 1/8-inch land on each bevel.
2. Turn on the welding equipment and adjust it by running beads on scrap steel.	

PROCEDURE	KEY POINTS
3. Position the plates in the flat position, with a 1/16″ to 3/32″ root opening and tack weld, figure 13-4.	3. Tack weld both samples.

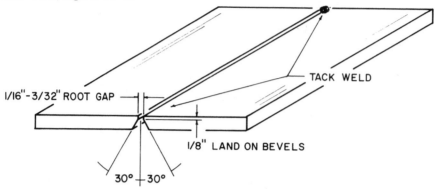

Fig. 13-4 The tack welds should leave a root gap of 3/32 of an inch.

| 4. Turn the tacked samples over into the welding position (groove up) and raise the plates so that the root is above any solid material, figure 13-5. | 4. The root bead will not penetrate if the plate is on a solid surface. Put the sample on scrap steel or in a clamp. |

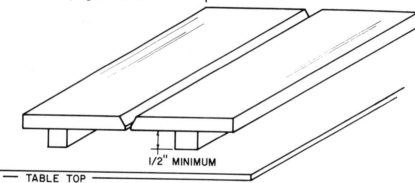

Fig. 13-5 The sample must be raised from a solid surface to allow 100% penetration.

| 5. Begin welding one of the joints, following the bead sequence in figure 13-6. | 5. Use the forehand motion when welding open butt joints. |

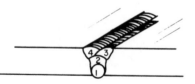

Fig. 13-6 Proper bead sequence for an open single V groove weld

| 6. When running the root bead, make sure the wire is always on the leading edge of the puddle. | 6. Whiskers will appear on the underside of the weld if the wire goes through the groove. |

PROCEDURE	KEY POINTS
7. Examine the completed butt joint for possible weld defects.	7. Make any necessary machine adjustments if any welding flaws are found.
8. Weld the other tacked sample.	8. Prepare more butt weld samples and weld them if more practice is needed.
9. Submit both samples to the instructor for comments and approval before proceeding to the Section Review.	

REVIEW QUESTIONS

Answer the following questions briefly:

1. What is the recommended maximum thickness of metal to be used in a square butt joint?

2. Can the square butt joint be welded from both sides?

3. What is a single-bevel groove butt joint?

4. How is weld cost increased when butt joints are prepared for welding?

5. Are groove joints more likely to have 100 percent penetration than square butt joints?

6. Why is it a good idea to have a butt-joint sample raised slightly off a solid surface while the weld is being made?

SECTION 2: REVIEW A

OBJECTIVE:

- Produce a fillet weld of proper penetration and quality using standard MIG welding equipment and procedures which, when etched with an acid, will display complete penetration and no porosity.

Equipment and Materials:

Standard MIG welding equipment
Personal hand tools
Protective clothing and gear
Grinder
Acid etch solution
1 piece, 3/8" x 5" x 6" mild steel plate
1 piece, 3/8" x 2 1/2" x 6" mild steel plate
Carbon dioxide shielding gas
E 70S-5 wire electrode

PROCEDURE

1. Grind one edge of the smaller piece of plate to a double bevel, figure R2-1.

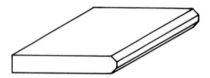

Fig. R2-1 A double-bevel edge

2. Set up and adjust the MIG welding equipment for short-arc welding.

3. Position and tack weld the plates in the proper place, figure R2-2.

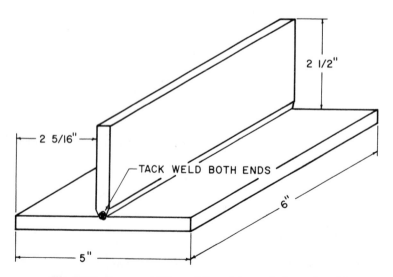

Fig. R2-2 Proper position of the plates and the tack weld

4. Arrange the plates for flat position welding and run a three bead sequence as shown in figure R2-3, using a forehand motion.

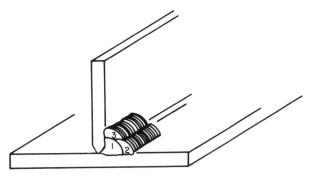

Fig. R2-3 Proper sequence for running beads

5. In the proper sequence, run three more beads on the opposite side of the vertical member using a backhand motion, figure R2-4.

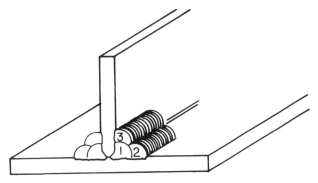

Fig. R2-4 Run beads on both sides of the vertical member.

6. Cut the fillet sample in two, figure R2-5.

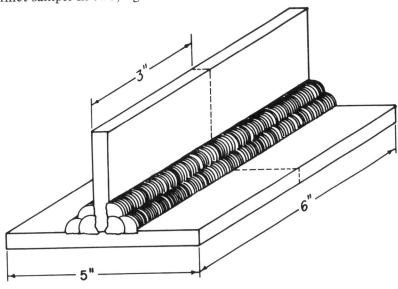

Fig. R2-5 Cutting the fillet sample in two

7. Smooth the surface of the cuts using a grinder and apply an acid etch to each weld surface.

8. Visually check the depth of penetration which appears on each weld.

9. Compare the penetration between the forehand and backhand motion.

10. Examine the sample for porosity or slag inclusion.

11. Submit the sample to the instructor for comments and further instructions.

SECTION 2: REVIEW B

OBJECTIVE:

• Produce a double-plate, double-V groove weld of proper penetration and quality using standard MIG welding equipment and procedures. Subject such a sample to the guided-bend test.

Equipment and Materials:

Standard MIG welding equipment
Personal hand tools
Protective clothing and gear
Grinder
Guided-bend test machine
2 pieces, 3/4" x 4" x 6" mild steel plate
E 70S-5 wire electrode
Argon/oxygen mix shielding gas

PROCEDURE

1. Grind one edge of each plate to a double bevel and set up as shown in figure R2-6.

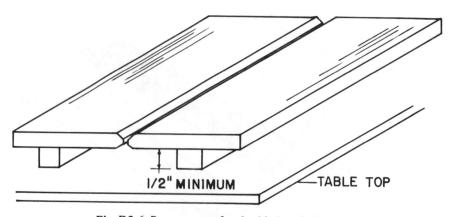

Fig. R2-6 Proper setup for double-bevel plates

2. Tack weld both ends of the joint, figure R2-7.

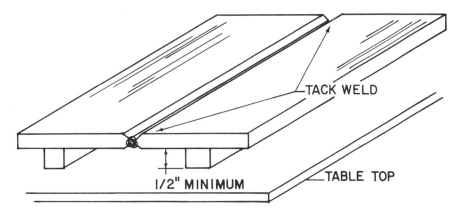

Fig. R2-7 Tack weld both ends of the plates

3. Run three beads in the sequence shown on one side, figure R2-8.

Fig. R2-8 Proper sequence for first side of double V groove weld

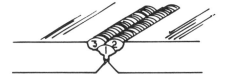

4. Make any adjustments necessary.

5. Run three beads on the opposite side as shown in figure R2-9.

Fig. R2-9 The completed weld with beads run on both sides

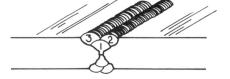

6. Lay out and cut the sample into sections as shown in figure R2-10.

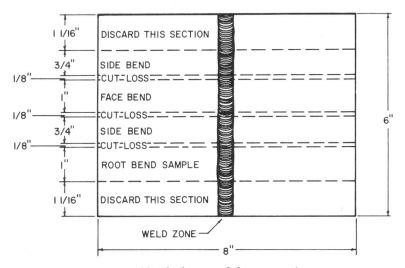

Fig. R2-10 The layout of the test sections

7. Perform a guided-bend test on the face, the root, and the side of the various strips.

8. Submit the bent samples to the instructor for comments and instructions.

SECTION 3:
Welds in the Horizontal Position, Short-Arc Technique

Unit 14 Four Layer Stringer Bead Pad Buildup, Short Arc, Horizontal Position

OBJECTIVES

After completing this unit, the student will be able to:

- define a weld in the horizontal position.
- describe the correct bead motions and gun angles used for horizontal welding.
- describe the methods used to eliminate undercutting and bead sag when making horizontal welds.
- successfully MIG weld a horizontal, four layer pad buildup using the short-arc method of metal transfer.

Horizontal Position

A horizontal weld is made when the plate being welded is in the vertical position, with the bead running across the plate, parallel to the top of the welding bench.

Effective welds can be made in the horizontal position using the MIG process. Horizontal welds are made by the use of certain bead motions, correct angles of the MIG gun, and proper adjustment of the inductance/slope controls. For quality welds, all of these factors must be controlled as closely as possible. Horizontal welding requires a great deal of practice before good welds can be made.

Bead Motions. Bead motions for horizontal welds are of two groups—the stringer bead motion, figure 14-1, and the weave bead motion, figure 14-2. Care must be taken with either motion to prevent undercutting or sagging of the molten welding metal, figure 14-3, page 78. It is not necessary to run the machine at a lower heat output as long as correct gun angle, slope, and motions are followed.

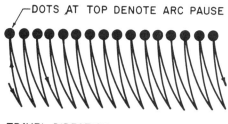

DOTS AT TOP DENOTE ARC PAUSE

TRAVEL DIRECTION
(CAN BE REVERSED)

Fig. 14-1 Proper motion for a stringer bead

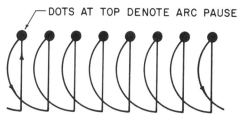

DOTS AT TOP DENOTE ARC PAUSE

TRAVEL DIRECTION
(CAN BE REVERSED)

Fig. 14-2 Proper motion for a weave bead

Undercutting can be eliminated with a correct length of arc pause at the upper part of the weld. The sag of weld beads in the horizontal position is caused by slow travel speed and the speed must be set so that sag is not present. Welding speed should be steady.

Gun Angles. The gun angle used when welding in the horizontal position can vary, depending on the type of joint and the bead sequence. Generally, the gun should be angled down about 5 degrees and the lay-back angle should be held at 20 degrees to 25 degrees. These angles can be used with either the forehand or backhand motion on either stringer beads or weave beads.

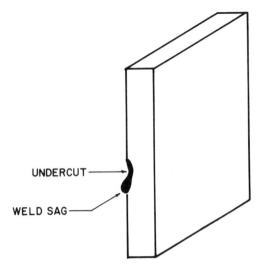

Fig. 14-3 Two common defects of horizontal welding

When welding horizontally, the welder must determine what angle, speed, and bead motion is best for the weld being made.

Inductance Adjustment. The inductance control varies the fluidity or rigidity of the weld puddle. In horizontal welding the inductance should be adjusted to increase the puddle rigidity. A stiffer puddle will help to control both undercutting and bead sag.

JOB 14: FOUR LAYER STRINGER BEAD PAD BUILDUP, SHORT ARC, HORIZONTAL POSITION

Equipment and Materials:

>Standard MIG welding equipment
>Personal hand tools
>Protective clothing and gear
>3/8" x 5" x 6" mild steel plate
>E 70S-3 or E 70S-4 wire electrode
>Carbon dioxide shielding gas

PROCEDURE	KEY POINTS
1. Turn on the machine.	
2. Run a flat position bead on a scrap piece and adjust the inductance for a rigid puddle.	2. It is always a good policy to use scrap material of the metal to be welded to set the machine.
3. With the plate vertical, begin at the bottom and run a horizontal bead across the bottom of the plate, figure 14-4.	

PROCEDURE	KEY POINTS
4. Visually check each bead as it is welded and make motion or gun angle adjustments.	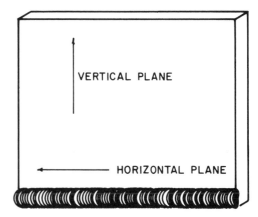 **Fig. 14-4 Proper position of the first bead**
5. Bead #2 should cover the top 1/3 of bead #1, figure 14-5. Follow the bead sequence and completely fill the plate. Rotate the plate 90 degrees so that the second series of welds will cross the first layer.	5. Starting at the bottom of the plate, the bead will form a ledge for bead #2. Bead #3 will flow into bead #2, and so forth.

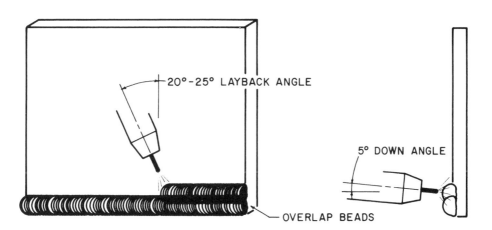

Fig. 14-5 The second bead should overlap 1/3 of the first bead.

6. Repeat the turning of the plate until four layers of weld have crossed each other.	6. Cool the plate as needed between passes.
7. Spatter buildup in the nozzle may be increased from torch angle. The nozzle must be kept clean to maintain a proper gas shield.	
8. Submit the pad to the instructor for comments and instructions.	

REVIEW QUESTIONS

Answer the following questions briefly.

1. What are the two bead motions used for horizontal welds with the MIG process?

2. What are the two main weld defects which can result when welding in the horizontal position?

3. How is undercut eliminated in horizontal welding?

4. What causes weld sag on horizontal beads?

5. What are the recommended gun angles for making horizontal welds?

Unit 15 Butt Weld with Backup Bar, Short Arc, Horizontal Position

OBJECTIVES

After completing this unit the student will be able to:

- define a ferrous metal.
- name the most commonly used alloying agents used to increase toughness, strength, and corrosion resistence in metal.
- name three types of tests used to identify metals.
- successfully complete the MIG welding of a butt weld with a backup bar in the flat position.

Ferrous Metal

Metal materials which contain iron are called ferrous metals. The name comes from the Latin, ferrum, which means iron. The presence of iron produces magnetic properties common to all ferrous metals.

Ferrous steels are alloys of iron and carbon along with other materials, called alloying elements, which give the steel special qualities needed for welding. An alloy is a mixture of two or more materials of which at least one is a metal. The resulting alloy may have totally different characteristics from any of the materials which make it up. The amount of alloying elements should be specified for the kind of job the alloy is to be used for. The alloy steels are made by mixing iron and carbon with any of the following:

- Nickel
- Manganese
- Silicon
- Chromium
- Aluminum
- Cobalt
- Molybdenum
- Vanadium
- Tungsten

Each of these materials adds some toughness, strength, or corrosion resistance to the carbon steels. Figure 15-1 lists a number of steels in accordance with S.A.E. standards.

TYPE OF STEEL	NO.
CARBON STEEL	1XXX
NICKEL STEEL	2XXX
NICKEL CHROMIUM STEEL	3XXX
1.25% NICKEL, .60% CHROMIUM	31XX
1.75% NICKEL, 1.00% CHROMIUM	32XX
3.50% NICKEL, 1.50% CHROMIUM	33XX
3.00% NICKEL, .80% CHROMIUM	34XX
MOLYBDENUM STEEL	4XXX
CHROMIUM	41XX
CHROMIUM NICKEL	43XX

Fig. 15-1 Steels and their S.A.E. numbers

The 3XXX series of stainless steel is the only ferrous metal which is nonmagnetic. This absence of magnetic attraction is due to the high content of nickel in this form of stainless steel.

Metal Identification by Testing

The welding industry has developed tests to help the welder identify materials. The tests have been grouped as the following:

- Spark Test
- Heat Test
- Other Tests

All of the tests can be used alone or together.

Spark tests are made with a power grinder. By watching the sparks given off from the material, a comparison is made which helps identify the metal. Figure 15-2 shows the characteristics of four major types of steel.

While this method is acceptable, there are many different mixtures of metals and the spark test may not always be accurate.

Heat tests are usually made by the use of an oxyacetylene torch. Each metal has a different melting point. This melting point may be checked by the use of a "temp" stick. A temp stick looks like a crayon and, when marked on steel, will melt off when certain temperatures are reached. Figure 15-3 lists the melting point of ten common metals and alloys.

WROUGHT IRON	LOW-CARBON STEEL	HIGH-CARBON STEEL	ALLOY STEEL
COLOR-STRAW YELLOW	COLOR-WHITE	COLOR-WHITE	COLOR-STRAW YELLOW
AVERAGE STREAM LENGTH WITH POWER GRINDER-65 IN.	AVERAGE LENGTH OF STREAM WITH POWER GRINDER-70 IN.	AVERAGE STREAM LENGTH WITH POWER GRINDER-55 IN.	STREAM LENGTH VARIES WITH TYPE AND AMOUNT OF ALLOY CONTENT
VOLUME-LARGE	VOLUME-MODERATELY LARGE	VOLUME-LARGE	COLOR-WHITE
LONG SHAFTS ENDING IN FORKS AND ARROWLIKE APPENDAGES	SHAFTS SHORTER THAN WROUGHT IRON AND IN FORKS AND APPENDAGES FORKS BECOME MORE NUMEROUS AND SPRIGS APPEAR AS CARBON CONTENT INCREASES	NUMEROUS SMALL AND REPEATING SPRIGS	SHAFTS MAY END IN FORKS, BUDS OR ARROW, FREQUENTLY WITH BREAK BETWEEN SHAFT AND ARROWS. FEW, IF ANY, SPRIGS
COLOR-WHITE			

Fig. 15-2 **Different steels give off characteristic sparks.**

Other Tests. Chemical testing is used for more positive identification of metals. This type of test consists of applying the chemical to the material or dipping the material in the solution. The metals are then identified by noting the chemical reaction, if any occurs.

Some common chemicals used are:

- Caustic Soda
- Sodium Hydroxide
- Nitric Acid
- Sulphuric Acid

Sometimes, metal can be identified by sound. An experienced welder can strike metal with a hammer and be able to tell the general carbon range of the metal by the type of tone.

METAL	MELT. TEMP. DEG. F.
ALUMINUM	1215
BRASS (YELLOW)	1640
BRONZE (CAST)	1650
COPPER	1920
IRON, GRAY CAST	2200
LEAD	620
STEEL (.20%) S.A.E. 1020	2800
SOLDER (50-50)	420
TIN	450
ZINC	785

Fig. 15-3 Melting points of common metals and alloys

JOB 15: BUTT WELD WITH BACKUP BAR, SHORT ARC, HORIZONTAL POSITION

Equipment and Materials:

Standard MIG welding equipment
Personal hand tools
Protective clothing and gear
2 pieces, 3/8" x 4" x 6" mild steel plate

1 piece, 3/16" x 2" x 9" mild steel strip
E 70S-3 or E 70S-4 MIG wire electrode
Carbon dioxide shielding gas

PROCEDURE	KEY POINTS
1. Turn on the machine.	
2. The bevels on each plate are 35 degrees (70 degrees included angle). The bevel land is 1/16" to 1/8". The root opening is 3/32".	2. Different welders may prefer different land widths and root openings.
3. Tack weld the beveled plates and backup bar as shown in figure 15-4.	3. Backup bars must fit tightly on the plates.

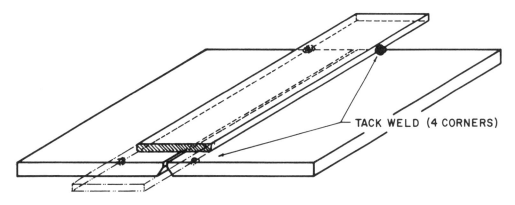

TACK WELD (4 CORNERS)

Fig. 15-4 Proper position for backup bar and tack welds

PROCEDURE   **KEY POINTS**

4. After tacking the backup bar on
    the plates, turn the joint up so the
    groove is in the horizontal plane,
    with the plate in the vertical posi-
    tion, figure 15-5.

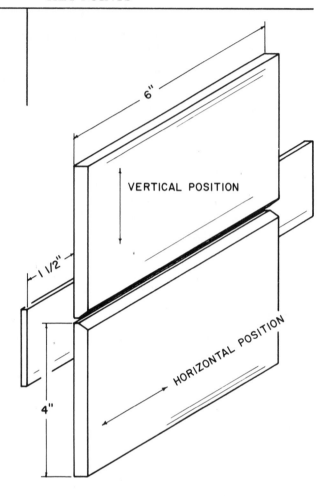

Fig. 15-5 Joint in horizontal position

5. Begin welding on the plates, using the
    bead sequence in figure 15-6.

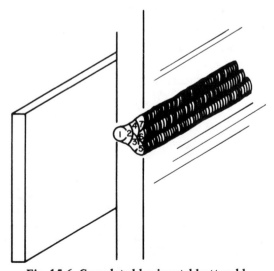

Fig. 15-6 Completed horizontal butt weld

PROCEDURE	KEY POINTS
6. Use the starting and run-off tabs.	
7. Complete the welding of the joint.	7. If more filler is needed, add more layers of weld.
8. Submit the weld to the instructor for comments and further instructions.	

REVIEW QUESTIONS

Answer the following questions briefly.

1. Define the term ferrous metal.

2. Is iron magnetic?

3. Name 6 alloying materials which add toughness, strength, and corrosion resistance to iron.

4. What kind of steel should be used to ensure resistance to acid?

5. Are all stainless steels magnetic?

6. Name three standard methods of testing metal.

7. What is a spark test?

8. Describe one method of finding the melting temperature of a piece of steel.

9. Name four common chemicals used to test steel.

10. How much overlap is required for horizontal beads?

11. Why is it necessary to clean the MIG gun frequently during welding?

SECTION 3: REVIEW

OBJECTIVE:

- Produce a V groove weld in the horizontal position of proper penetration and quality using standard MIG welding equipment and procedures, by subjecting such a sample to the guided-bend test.

Equipment and Materials:

Standard MIG welding equipment
Carbon-arc or oxyacetylene cutting equipment
Personal hand tools
Protective clothing and gear
Guided-bend test machine
2 pieces, 3/8″ x 4″ x 6″ mild steel plate
1 piece, 3/16″ x 2″ x 9″ mild steel strip
E 70S-3 or E 70S-4 wire electrode
Carbon dioxide shielding gas

PROCEDURE

1. Grind the plates to form a V groove joint.

2. Tack weld the plates and the backup bar as shown in figure R3-1.

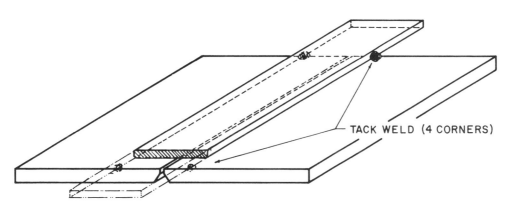

Fig. R3-1 Proper position for backup bar and tack welds

3. Set up the sample so that the plates are vertical and the groove is horizontal, figure R3-2.

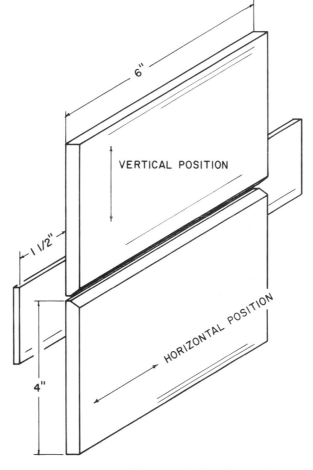

Fig. R3-2 Proper welding position for the sample

4. Run enough beads to fill the groove, figure R3-3.

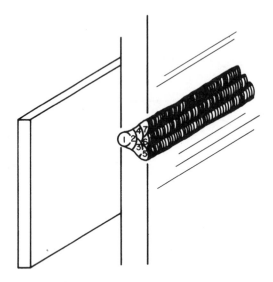

Fig. R3-3 Run enough beads to fill the groove.

5. Remove the backup bar using the carbon-arc or oxyacetylene cutting process.

6. Lay out and cut the sample into sections as shown in figure R3-4.

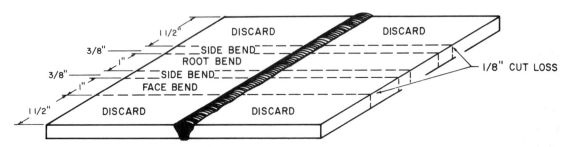

Fig. R3-4 The sample should be cut into sections.

7. Perform a guided-bend test on the face, root, and the sides of the various strips.

8. Examine the bent strips. There should be no separation of weld from the base metal and no porosity.

9. If the welded samples show defects, repeat steps 1-8.

10. When the bent specimens are satisfactory, submit them to the instructor for comments and further instructions.

SECTION 4:
Welds in the Vertical Position, Short-Arc Technique

Unit 16 Four Layer Stringer Bead Pad Buildup, Short Arc, Vertical Position

OBJECTIVES

After completing this unit the student will be able to:

- define a weld in the vertical position.
- name the bead motions used for vertical MIG welding.
- describe a method of correcting undercutting when welding in the vertical position.
- identify the angle of the gun used for vertical welding.
- successfully MIG weld a four layer pad in the vertical position of welding.

Vertical Position

A vertical weld is made when the plate and the joint are both 90 degrees to the horizon. In this position the joint is in the same direction as the force of gravity. The control of the gravitational force pulling on the puddle is a major factor in the production of high quality welds.

Fast and good welds can be made in the vertical position using the MIG process. Vertical welds are made by using certain bead motions, correct angles of the MIG gun, and proper adjustment of the inductance/slope control. For quality welds all of these factors must be properly handled.

Bead Motion. Bead motions for vertical welds are of the same two groups as for horizontal welds—the stringer bead motion, figure 16-1, and the weave bead motion, figure 16-2.

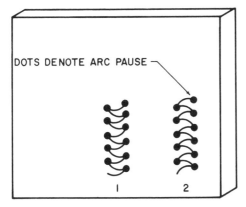

Fig. 16-1 Forehand, uphand stringer motion

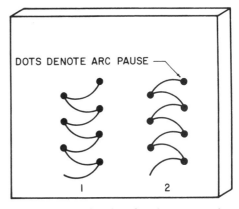

Fig. 16-2 Forehand, uphand weave motion

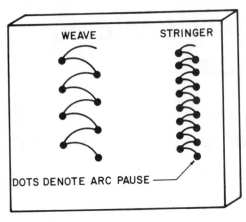

Fig. 16-3 Downhand, downward motion

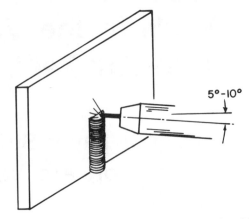

Fig. 16-4 The proper gun angles for vertical welding

The arc swing is an upright "U" or an upside down "U". Figures 16-1 and 16-2 are for uphand welding, from bottom to top, using a forehand technique.

The stringer bead and the weave bead can be run with a down weld, using backhand techniques, by making the "U" motion of the gun in an upside down pattern, figure 16-3.

The welder uses slight variations of arc pause, bead motion, and gun angle to suit his own techniques. Stringer beads and weave beads started at the top of the plate and run in the down welding position have a tendency to undercut the plates. The correct arc pause at the edges of the weld will help overcome undercutting.

Welds run in the upward pattern may sag from the molten puddle. Sagging of the molten puddle can be corrected by adjusting the welding machine so that the current is about 10 percent less for vertical welding than for flat and horizontal welding.

The speed of movement for either up or down welding in the vertical position must be at a smooth, steady rate of travel.

Gun Angles. The gun body should be about parallel with the plate, and pointing upwards about 5 degrees, figure 16-4. The gun must be positioned so that it will help the welder control the molten puddle by using the arc force to hold the metal in position against the pull of gravity. This gun angle can be used for forehand-up motion or backhand-down position.

Inductance. The proper adjustment of the inductance will increase the puddle rigidity. The stiffer puddle will help to control both undercutting and bead sag.

JOB 16: FOUR LAYER STRINGER BEAD PAD BUILDUP, SHORT ARC, VERTICAL POSITION

Equipment and Materials:

    Standard MIG welding equipment
    Personal hand tools
    Protective clothing and gear
    3/8" x 5" x 6" mild steel plate
    E 70S-3 or E 70S-4 wire electrode
    Carbon dioxide shielding gas
    Nozzle cleaning compound

PROCEDURE	KEY POINTS
1. Turn on the machine and apply the cleaning compound to the nozzle.	
2. Adjust the machine on scrap steel.	
3. Position the plate in the vertical position and run a bead from the bottom center to the top center, using a stringer bead motion, figure 16-5.	3. Check each bead and make any machine changes necessary.

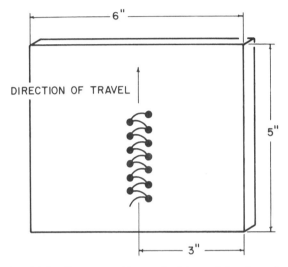

Fig. 16-5 Proper position of plate and bead, and proper bead motion

4. Run a second bead to the right of the first one.	4. Overlap the second bead 1/3 over the first to prevent undercut.
5. Run a third bead to the left of the first.	5. Overlap the third bead 1/3 of the first.
6. Following the bead sequence shown in figure 16-6, fill the plate surface.	6. If the plate overheats, cool it in water.

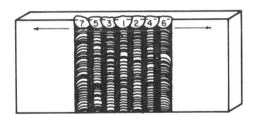

Fig. 16-6 Proper bead sequence for pad buildup

7. Turn the plate 1/4 turn and build another layer across the first layer.	7. Clean the nozzle frequently.
8. Repeat the rotating of the plate until four layers of weld have been made.	
9. Shut down all systems.	
10. Submit the pad to the instructor for comments and further instructions.	

REVIEW QUESTIONS

Answer the following questions briefly.

1. What are the three main factors to observe for vertical welding?

2. What two bead motions are used for vertical welding?

3. How may undercutting be eliminated on the vertical weld?

4. What is the relationship of the gun to the plate which is specified for vertical welding?

5. How is the arc force used to control the puddle in vertical welding?

Unit 17 Fillet Weld, Short Arc, Vertical Position

OBJECTIVES

After completing this unit the student will be able to:

- describe a fillet weld in the vertical position.
- identify the proper bead motion, gun angle, and slope needed to weld a vertical fillet weld.
- successfully MIG weld a fillet weld in the vertical position.

Vertical Fillet Welds

Vertical fillet welds are often encountered in the welding field. After a good deal of practice, the operator will learn to weld high-quality fillet welds in the vertical plane. Once again, using certain bead motions, correct gun angles, and proper adjustment of the inductance controls, will enable the welder to produce good welds.

Bead Motion. Vertical fillet welds require the use of both stringer and weave beads in their construction. The initial bead run in the corner of the joint should be a stringer bead. Any later passes used to strengthen the joint should use a weave motion, figure 17-1. Certain jobs may require the bead to be run uphill or downhill, figure 17-2. In either case the bead motion used remains the same.

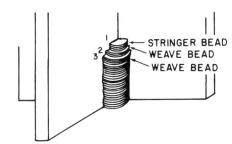

Fig. 17-1 The first bead is of the stringer type. Additional beads require a weave motion.

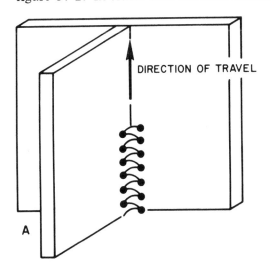

Fig. 17-2 A. Forehand, upward stringer motion

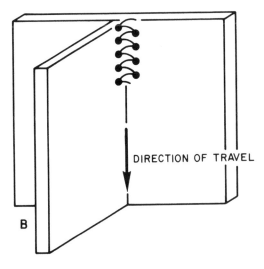

B. Backhand, downward stringer motion

93

Gun Angles. The gun body should split the angle of the two members forming the fillet joint and have a slight upwards direction of five degrees. These gun angles will help to avoid undercutting either plate and keep the puddle from being greatly affected by the force of gravity. This angle is used for both forehand-up or backhand-down welding.

Inductance. All vertical welding needs stiff puddles to resist the force of gravity. Vertical fillet welds must have inductance adjustment that will produce a stiff puddle.

JOB 17: FILLET WELD, SHORT ARC, VERTICAL POSITION

Equipment and Materials:

Standard MIG welding equipment
Personal hand tools
Protective clothing and gear
1 piece, 3/8" x 5" x 6" mild steel plate
1 piece, 3/8" x 2 1/2" x 6" mild steel plate
E 70S-3 or E 70S-4 wire electrode
Carbon dioxide shielding gas

PROCEDURE	KEY POINTS
1. Turn on the machine.	
2. Tack weld the pieces as shown, figure 17-3.	

Fig. 17-3 Tack welded fillet sample

3. Place the sample in the vertical position.	
4. Run a stringer bead all the way up in the corner, welding from bottom to top.	4. Arc pause and gun angle are important.
5. Complete the weld with two weave beads as shown, figure 17-4.	

PROCEDURE

KEY POINTS

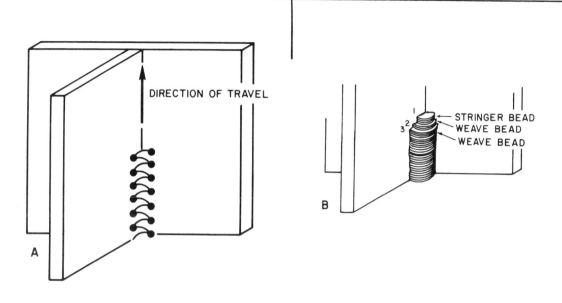

Fig. 17-4 Completing the weld with an upward motion

6. Turn the job around and weld the other side. Weld from top to bottom, figure 17-5.

6. Cool the weld as needed. Check each bead and adjust the machine if welds are not satisfactory.

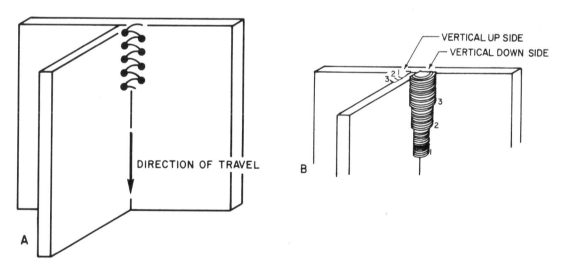

Fig. 17-5 Completing other side using downward motion

7. Shut down all systems.

8. Submit the sample to the instructor for comments and instructions.

REVIEW QUESTIONS

Answer the following questions briefly.

1. What type of bead motion is used for the first bead run in a vertical fillet weld?

2. What type of bead is run after the initial pass on a vertical fillet weld?

3. Why is it important that vertical welds have a stiff puddle?

4. What would be a proper gun angle for forming a vertical fillet weld?

Unit 18 Butt Weld with Backup Bar, Short Arc, Vertical Position

OBJECTIVES

After completing this unit the student will be able to:

- identify two methods of depositing vertical butt welds.
- explain the plate preparation for vertical butt welding.
- describe the difference between up welding and down welding.
- successfully MIG weld a backup bar butt weld in the vertical position.

Vertical Butt Weld Procedures and Motions

Vertical butt welds are easily completed by depositing beads with either an up motion or a down motion. Both methods can pass a destructive or nondestructive (x-ray) test procedure.

Vertical-Up Welding

The vertical-up welding method leaves a relatively large deposit of rod and gives deep penetration. Vertical-up welding is suited to thick metal and single V butt joints. The root bead will require the motion used for stringer beads and each deposit thereafter will use a weave motion. Each deposit of filler metal will be slightly larger than the last until the joint is filled, figure 18-1.

Vertical-Down Welding

Vertical-down welding does not deposit as much filler material as vertical-up welding and the penetration is not as deep. This type of welding is suited to light gauge metal, with square butt joints. The root bead will require stringer bead motion and more passes than up-welded joints to fill the groove. Each deposit after the first uses a weave motion, figure 18-2, page 98.

The backup bar used behind the groove of a vertical butt weld should extend past the edges of the plate, so that the weld can be started and stopped off the base metal.

Bead Sequence

All welds should be made to follow specified bead patterns. Most shops determine

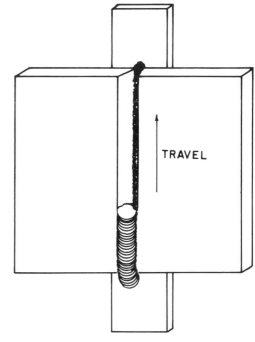

Fig. 18-1 Vertical-up welding

which beads are to be put on first and how each of the other beads should be welded. It is important to follow required bead patterns.

Generally, the first bead is a stringer bead with all successive passes being weave beads. The weave beads are continued until the weld is about flush with the surface.

JOB 18: BUTT WELDS WITH BACKUP BAR, SHORT ARC, VERTICAL POSITION

Equipment and Materials:

> Standard MIG welding equipment
> Personal hand tools
> Protective clothing and gear
> 2 pieces, 3/8" x 4" x 6" mild steel plate
> 1 piece, 1/4" x 2" x 9" mild steel strip
> E 70S-3 or E 70S-4 wire electrode
> Carbon dioxide shielding gas.

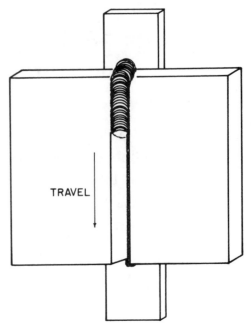

TRAVEL

Fig. 18-2 Vertical-down welding

PROCEDURE	KEY POINTS
1. Turn on the machine.	
2. Tack weld beveled plates and backup bar as shown, figure 18-3.	2. Backup bars must fit tightly on the plate. Any gap increases the chance of defective welds. Bevels are 30 degrees on each plate (60 degrees included angle) and the bevel land is 1/16" to 1/8". The root opening is 3/32".

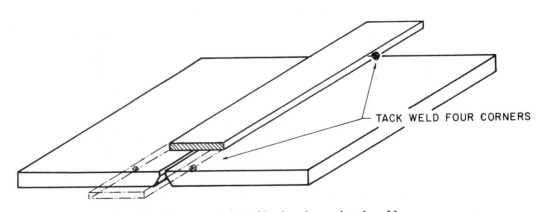

TACK WELD FOUR CORNERS

Fig. 18-3 Proper position of backup bar and tack welds

PROCEDURE	**KEY POINTS**
3. Set the plates in position for a vertical butt weld, figure 18-4.	

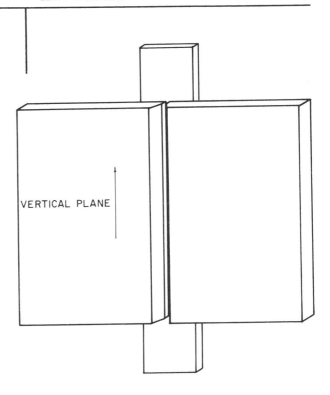

Fig. 18-4 Proper position of sample for welding

4. Begin welding, following the bead sequence shown, figure 18-5. Bead #1 is a stringer bead, beads #2 and #3 are weave beads.	4. Weave bead #3 will be wider than bead #2.

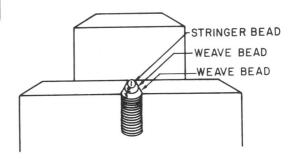

Fig. 18-5 Proper bead sequence for vertical butt weld

5. Use the starting and run-off tabs to begin and end the welds.

6. Submit the completed weld to the instructor for comments and instructions.

REVIEW QUESTIONS

Answer the following questions briefly.

1. What are the two methods of depositing vertical butt weld beads?

2. What is vertical-up welding?

3. What type of butt weld preparation is suited to vertical-up welding?

4. Which method, down welding or up welding, produces the deepest penetration pattern?

5. Which of the two methods is best suited for light gauge sheet metal?

SECTION 4: REVIEW A

OBJECTIVE:

- Produce a fillet weld of proper penetration and quality in the vertical position using standard MIG welding equipment and procedures which, when etched with an acid, will display complete penetration and no porosity.

Equipment and Materials:

Standard MIG welding equipment
Personal hand tools
Protective clothing and gear
Acid etch solution

1 piece, 3/8" x 5" x 6" mild steel plate
1 piece, 3/8" x 2 1/2" x 6" mild steel plate
Carbon dioxide shielding gas
E 70S-3 or E 70S-4 wire electrode

PROCEDURE

1. Tack weld the samples as shown, figure R4-1.

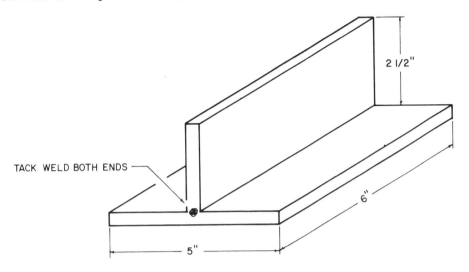

Fig. R4-1 Proper position of plates and tack welds

2. Arrange the plates so the joint is in the vertical position.

3. Run a stringer bead on each side of the joint and then run two weave beads over the stringer bead.

4. Cut the sample in two as shown, figure R4-2.

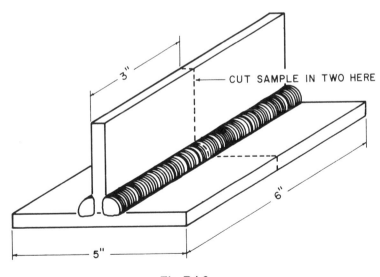

Fig. R4-2

5. Smooth the cut surface and apply an acid etch solution to each surface.

6. Check the depth of the penetration on each side of the joint.

7. Submit the samples to the instructor for comments and instructions.

SECTION 4: REVIEW B

OBJECTIVE:

- Produce a V groove weld of proper penetration and quality in the vertical position using standard MIG welding equipment and procedures, and test the weld by a guided-bend test.

Equipment and Materials:

> Standard MIG welding equipment
> Carbon-arc or oxyacetylene cutting equipment
> Personal hand tools
> Protective clothing and gear
> Guided-bend test machine
> 2 pieces, 3/8" x 4" x 6" mild steel plate
> 1 piece, 3/16" x 2" x 9" mild steel strip
> E 70S-3 or E 70S-4 wire electrode
> Carbon dioxide shielding gas

PROCEDURE

1. Grind the plates to form a V groove joint.

2. Tack weld the plates and the backup bar as shown in figure R4-3.

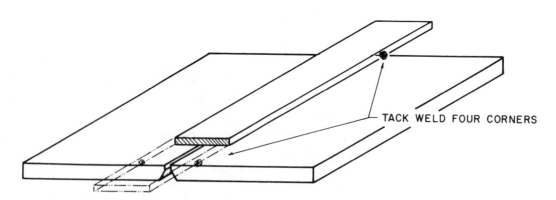

TACK WELD FOUR CORNERS

Fig. R4-3 Proper position of backup bar and tack welds

3. Set up the sample so that the plates and the joint are vertical, figure R4-4.

4. Run enough beads to fill the groove, figure R4-5.

5. Remove the bar using the Carbon-arc or oxyacetylene cutting process.

6. Lay out and cut the sample into sections as shown in figure R4-6.

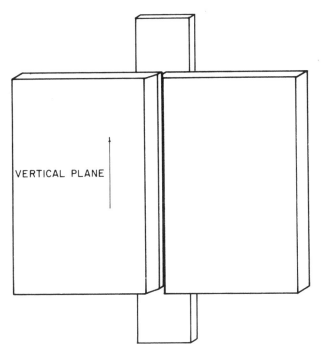

Fig. R4-4 Plates arranged in the vertical plane

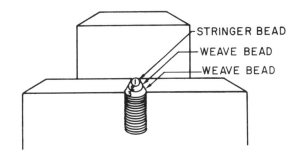

Fig. R4-5 The beads should fill the groove flush to the surface of the plates.

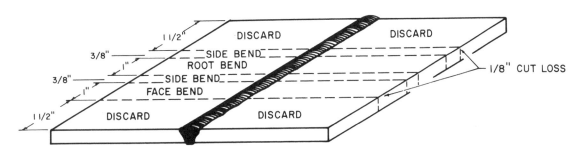

Fig. R4-6 The sample should be cut into sections.

7. Perform a guided-bend test on the face, the root, and the sides of the various strips.

8. Examine the bent strips. There should be no separation of weld from the base metal.

9. If the welded strips show defects, repeat steps 1-8.

10. When the bent specimens are satisfactory, submit them to the instructor for comments and further instructions.

SECTION 5:
Welds in the Overhead Position, Short-Arc Technique

Unit 19 Four Layer Stringer Bead Pad Buildup, ShortArc, Overhead Position

OBJECTIVES

After completing this unit the student will be able to:

- define the overhead position of welding.
- identify the proper bead motion, gun angle, and inductance for overhead welding.
- describe two major types of burns.
- be aware of safety precautions necessary for overhead welding.
- successfully MIG weld a four layer pad in the overhead position.

Overhead Position

An overhead weld is defined as a weld in the horizontal plane, in which the bead is run on the underside of the joint facing the ground. In this position the bead is pulled down away from the joint. Gravity has a greater effect on this welding position than any other. Therefore, due to the effect of gravity and the awkwardness of the welder's position, overhead welding is the most difficult position to master.

Fast, good deposits can be made in the overhead position with the MIG process. Overhead welds are produced by using certain bead motions, correct gun angles, and proper adjustment of the inductance/slope controls.

Bead Motion. Bead motions for overhead welding are in the two groups previously mentioned: the stringer bead motion, figure 19-1, and the weave bead motion, figure 19-2, page 106. Either motion must be used carefully because the molten metal may sag or undercutting may occur. The welding motions must be held to the proper moves. The stringer bead motion is a zigzag pattern and the weave bead is of the "U" type, with a slight backswing.

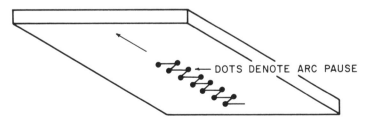

Fig. 19-1 The stringer bead in the overhead position

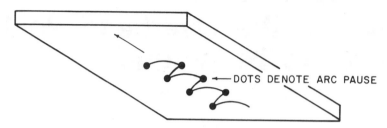

Fig. 19-2 The weave bead in the overhead position

Overhead welds may undercut badly unless an arc pause of the right length is made at the edges of the weld zone. Sagging, caused by the pull of gravity on the molten metal, can be eliminated by using the correct travel speed. All travel movement must be at a smooth, steady rate of speed.

Gun Angles. Generally speaking, the main gun body should be held at about 90 degrees with the plate, nozzle tilted forward in the direction of travel, and a side angle from 5 degrees to 10 degrees, figure 19-3.

Most MIG guns have a curved nozzle so the overhead weld angles do not interfere with the excellence of the weld. The curved nozzle allows the welder to see the weld zone when the weld is made with a forehand motion and a gun side angle is used.

Inductance. Overhead welding requires the stiffest puddle of all the welding positions. Although the welding current is approximately the same as for the flat position, the inductance should be lowered to help control the puddle.

Gun and Cable Handling

Do not drape the cable of the MIG welder over the shoulder when welding. The weight of the gun and cable may seem awkward at first, but the welder soon becomes used to it.

Burn Protection

Out-of-position (all welds except flat welds) procedures are all likely to produce burns more readily than the flat weld. However, overhead welding is the most dangerous. Falling spatter and overhead arcs endanger the operator because of his awkward position. Even minor burns can become serious, due to infection.

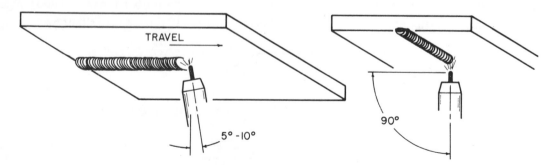

Fig. 19-3 Proper gun angles for overhead pad buildup

Burns fall into two major groups:

- Hot metal burns.

- Electric arc burns.

Available protective clothing and leather goods manufactured for welders, can help to eliminate these dangers. A cap and leather gloves, aprons, and jackets are a must for overhead welding.

JOB 19: FOUR LAYER STRINGER BEAD PAD BUILDUP, SHORT ARC, OVERHEAD POSITION

Equipment and Materials:

Standard MIG welding equipment
Personal hand tools
Protective clothing and gear
Safety glasses

1 piece, 3/8" x 5" x 6" mild steel plate
E 70S-3 or E 70S-4 wire electrode
Carbon dioxide shielding gas
Nozzle cleaning compound

PROCEDURE	KEY POINTS
1. Turn on the machine and apply nozzle cleaning compound.	
2. Position the plate in the overhead welding position and weld a stringer bead in the middle of the plate, the full length. Follow the diagram in figure 19-4 for the bead sequence.	2. After bead #1 is complete, bead #2 should be welded so that it covers up to 1/3 of the width of bead #1. Each bead should be welded in the same manner.

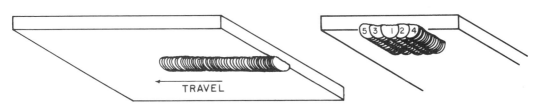

Fig. 19-4 Proper direction of travel and bead sequence for overhead pad buildup

3. Cool the sample between passes.

4. Inspect each bead. Correct any defects by adjusting the machine.

5. When the first layer of weld is completed, rotate the plate 90 degrees and begin another layer of beads across the welded beads.

6. Shut down all systems.

7. Submit the pad to the instructor for comments and instructions.

REVIEW QUESTIONS

Answer the following questions briefly.

1. What are the two bead motions used for overhead welding?

2. Why does the decrease of inductance help control an overhead puddle?

3. Draw a diagram representing the bead motion pattern for a stringer bead in the overhead position.

4. Draw a diagram representing the bead motion pattern for a weave bead in the overhead position.

5. How can undercut be corrected when overhead welding?

6. How can sagging be eliminated when overhead welding?

7. What are the recommended gun angles for overhead welding?

8. What are the two major types of burns?

9. What kind of equipment is provided to prevent burns from overhead welds?

Unit 20 Fillet Weld, Short Arc, Overhead Position

OBJECTIVES

After completing this unit, the student will be able to:

- define a fillet weld in the overhead position.
- discuss the differences between welding in the flat position versus the overhead position.
- successfully MIG weld a T weld in the overhead position.

Overhead Fillet Weld

The overhead fillet weld is one in which the surface in the vertical plane lies beneath the top plate. The weld is then deposited from underneath these plates. The operator must pay strict attention to the factors which influence all out-of-position welds—bead motion, gun angle, and inductance control.

Bead motions for overhead T welds fall into the stringer bead classification. Care must be exercised with the stringer bead motion to prevent undercutting or sagging of the molten weld metal. Welding currents do not need to be sacrificed while welding in the overhead position as long as correct gun angles and motions are observed. The tendency to undercut overhead welds can be eliminated with the correct length of arc pause at the upper portion of the weld zone. The inductance control should be adjusted for a stiffer weld puddle (shorter "on" time of the arc) to allow the operator full control over the liquid weld puddle. The sagging of weld deposits encountered in the overhead position is caused by slow travel speed and can be eliminated by correcting the speed of movement. The bead sequences used in overhead T welds are designed to help the operator control the molten weld. It is a layered sequence and gives the welder a bead to lay the succeeding bead on, figure 20-1.

The gun angles used when welding the overhead T weld will be of three variations, with the layback angle of 20 degrees to 25 degrees in all three cases. Bead #1 and bead #5

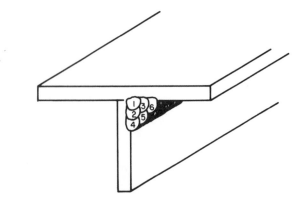

Fig. 20-1 Proper bead sequence for an overhead T weld

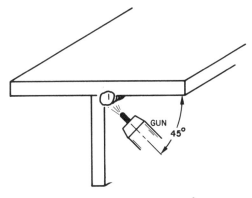

Fig. 20-2 The first bead is run using a 45° gun angle.

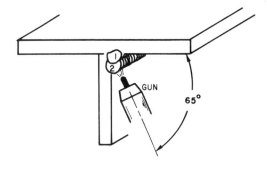

Fig. 20-3 The second bead will use a 65° gun angle into the top plate.

will use a corner gun angle of 45 degrees, figure 20-2. Bead #2 and bead #4 will use a corner gun angle of 65 degrees into the upper plate, figure 20-3. Bead #3 and bead #6 will use a corner gun angle of 75 degrees into the upper plate, figure 20-4.

When welding overhead, slight variations of arc pause, bead motion, and corner gun angles may be made to suit the individual welder.

A great deal of practice and correct machine adjustment are necessary before satisfactory overhead T welds can be produced.

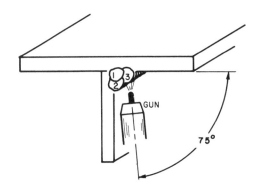

Fig. 20-4 The third bead will use a 75° gun angle into the top plate.

Flat Position Welding Versus Overhead Position Welding

Welds made in the flat position are the best. Speed of application is increased in the flat position, and bead formation and penetration are usually easy to get. Any out-of-position weld is harder to apply because of the pull of gravity on molten metal. Also, the strain of holding the MIG gun and cables affects the quality of the weld. Out-of-position welds must be applied with shallow passes of wire, therefore, more beads must be applied making the welding time greater. The danger to the welder is greater when making out-of-position welds, since hot metal and sparks fall from the weld.

JOB 20: FILLET WELD, SHORT ARC, OVERHEAD POSITION

Equipment and Materials:

Standard MIG welding equipment
Personal hand tools
Protective clothing and gear
2 pieces, 3/8″ x 5″ x 6″ mild steel plate
2 pieces, 3/8″ x 2 1/2″ x 6″ mild steel plate
E 70S-3 or E 70S-4 wire electrode
Carbon dioxide shielding gas

PROCEDURE   **KEY POINTS**

1. Turn on the machine.

2. Position and tack weld the two plates
   as a T weld, figure 20-5.

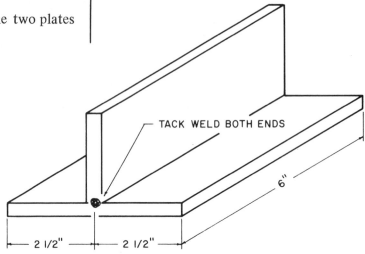

TACK WELD BOTH ENDS

6"

2 1/2"   2 1/2"

Fig. 20-5 Proper position of the plates and the tack welds for a T weld

3. Place the tacked sample in the overhead position and weld on one side.
   Use the proper bead sequence, figure
   20-1, page 110.

3. Check each bead as it is made. Correct gun angle, speed of application,
   or machine settings as needed.

4. Turn the sample and weld the other
   side, following the same bead sequence.

4. Motions and gun angle used in horizontal welding are similar to the overhead weld.

5. Shut down all systems.

6. Submit the sample to the instructor
   for comments and instructions.

REVIEW QUESTIONS

Answer the following questions briefly.

1. Why are flat position welds easier to apply than those made in other
   positions?

2. Draw a diagram showing the bead sequence for a 10-pass overhead T
   weld.

3. Describe an overhead fillet weld.

4. Can out-of-position welds be dangerous to the welder?

SECTION 5: REVIEW

OBJECTIVE:

- Produce a fillet weld of proper penetration and quality in the overhead position using standard MIG welding equipment and procedures. Subject such a sample to a destructive test.

Equipment and Materials:

Standard MIG welding equipment
Personal hand tools
Protective clothing and gear
Hacksaw

1 piece, 3/8″ x 5″ x 6″ mild steel plate
1 piece, 3/8″ x 2 1/2″ x 6″ mild steel plate
E 70S-3 or E 70S-4 filler wire
Carbon dioxide shielding gas

PROCEDURE

1. Turn on and adjust the equipment for an overhead weld.

2. Tack weld the plates as shown in figure R5-1.

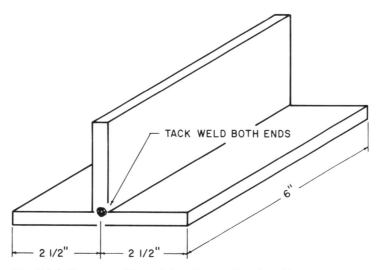

TACK WELD BOTH ENDS

6″

2 1/2″   2 1/2″

Fig. R5-1 Proper position of the plates and tack welds for a fillet weld

3. Place the sample in the overhead position and weld both sides, figure R5-2.

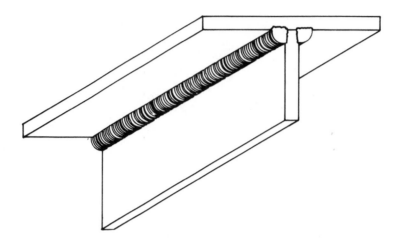

Fig. R5-2 Beads run in the overhead position

4. Saw the welded sample in pieces as shown in figure R5-3.

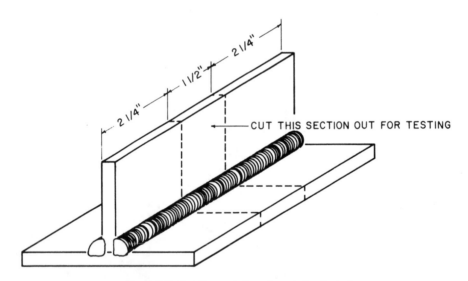

Fig. R5-3 Cutting out the piece to be tested

5. Saw through one side to the joint to the corner as shown in figure R5-4.

6. Perform a destructive test as shown in figure R5-5.

7. Examine the broken weld for defects and proper penetration.

8. Submit the sample to the instructor for comments and instructions.

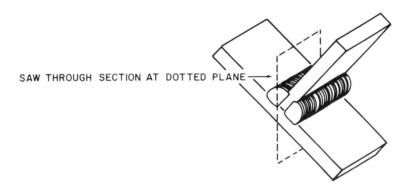

SAW THROUGH SECTION AT DOTTED PLANE→

Fig. R5-4 Preparing the sample piece for testing

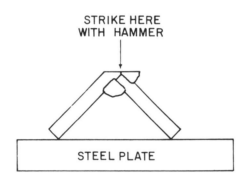

STRIKE HERE
WITH HAMMER

STEEL PLATE

**Fig. R5-5 Performing a destructive test on
a fillet weld**

SECTION 6:
MIG Welding Aluminum

Unit 21 Four Layer Stringer Bead Pad Buildup, Flat Position, Aluminum

OBJECTIVES

After completing this unit, the student will be able to:

- explain the difference between ferrous and nonferrous materials.
- discuss the qualities of aluminum which make it effective for MIG welding.
- explain the characteristics of copper and magnesium.
- identify an aluminum alloy.
- define heat-treatable and nonheat-treatable materials.
- name the reasons why MIG welding is a good process for welding aluminum plates.
- satisfactorily MIG weld a pad buildup on aluminum plate, in the flat position.

Nonferrous Metals

Nonferrous metal is any metal which does not contain iron. These metals oxidize very quickly when subjected to welding heat. This oxidizing characteristic makes MIG welding in a shielded-gas atmosphere a highly successful method of joining nonferrous metals.

Aluminum

Aluminum can be bought in its pure form or in an alloyed form. It is found in many places and is not a scarce metal. It is light in weight and is about 1/3 as dense as steel. Aluminum alloys compare well in strength with carbon steel. Aluminum conducts electricity, transfers heat, and resists corrosion or oxidation. It is easy to shape, non-magnetic, and highly suited for mass production operations. Aluminum is alloyed with other elements such as copper, magnesium, manganese, silicon, and zinc to make it stronger, figure 21-1.

Copper

Copper must have the oxygen removed to make it weldable in the MIG process. Removal of the oxygen is called *deoxidizing*. Copper is over two times as dense as aluminum. It is soft, bends easily, and is tough. The tensile strength (the resistance of a material to

COMMON ALLOY ELEMENTS OF ALUMINUM	
COPPER	HEAT TREATABLE QUALITIES, INCREASES STRENGTH AND HARDNESS.
MAGNESIUM	INCREASES TENSILE STRENGTH, HARDNESS AND WELDABILITY. REDUCES MARINE CORROSION.
MANGANESE	INCREASES STRENGTH AND CORROSION RESISTANCE.
SILICON	LOWERS MELTING POINT, INCREASES FORMABILITY AND CORROSION RESISTANCE.
ZINC	IMPROVES STRENGTH IN HEAT TREATABLE ALLOYS.

Fig. 21-1 Common alloy elements of alluminum

being pulled apart) of copper is approximately 30,000 pounds per square inch. Copper conducts electricity, transfers heat well, and resists corrosion, figure 21-2.

COMMON ALLOY ELEMENTS OF COPPER	
ALUMINUM	INCREASES TENSILE STRENGTH, TOUGHNESS, DUCTILITY AND CORROSION RESISTANCE.
BERYLLIUM	INCREASES FATIGUE RESISTANCE.
NICKEL	INCREASES TOUGHNESS AND DUCTILITY.
SILICON	INCREASES STRENGTH & CORROSION RESISTANCE.
TIN	INCREASES TOUGHNESS, HARDNESS AND FATIGUE RESISTANCE.
ZINC	INCREASES STRENGTH, DUCTILITY AND HARDNESS.

Fig. 21-2 Common alloy elements of copper

Magnesium

Only alloyed magnesium can be used for welding. Pure magnesium is too soft. It is alloyed with aluminum, manganese, zinc, or zirconium for use in the aircraft industries. Magnesium is less dense than aluminum and about three times less dense than steel, figure 21-3.

WEIGHT COMPARISON	
METAL	ATOMIC WEIGHT
ALUMINUM	26.981
COPPER	63.540
MAGNESIUM	24.312
CARBON STEEL	67.858

Fig. 21-3 Comparing metals according to their atomic weight

Magnesium looks like aluminum but will not weld with aluminum wire. One good test to find out if a sample is aluminum or magnesium is to burn a *small* quantity (no more than a 1/4 inch cube) of the material.

CAUTION: Magnesium burns with great intensity. Wear protective clothing and remove any combustibles from the area. Magnesium leaves a white ash residue; aluminum leaves a black ash.

Hot Shortness

Materials which need to be supported when they are hot are said to have hot shortness. Most nonferrous materials possess this characteristic. Ferrous metals, however, will retain their shape when heated. In cases of extreme heat, materials which possess hot shortness will completely fall away.

MIG Welding of Aluminum

Most welders prefer the spray transfer method of welding aluminum. Since aluminum conducts heat very well, more heat is required for welding. Pure aluminum melts at about 1215 degrees Fahrenheit (348 degrees Celsius) and the aluminum alloys vary from 900 degrees to 1220 degrees Fahrenheit (250 degrees to 349 degrees Celsius), depending on the alloy.

Color Change

Aluminum and its alloys show no color change when heated to the welding point. This feature makes it hard for the welder to see when the metal reaches the hot shortness, or fall-away condition. Excessive penetration results.

Aluminum Plate Welding

MIG welding has replaced most other processes for welding heavy aluminum plate.

- Welding of aluminum with the MIG process does away with the use of cleaning agents since the puddle is protected by the inert-gas shield.
- Plate thickness of 3/8" and less can be butt welded with 100 percent penetration without plate preparation.
- Welding speeds are faster when the spray-arc method is used.
- Preheating of thinner aluminum is not necessary since the spray-arc method concentrates more heat.
- No postweld cleaning is necessary.
- Rod stub loss is eliminated.
- MIG welding can be done in any welding position.

Cleaning Action

Only the forehand and vertical-up welds are recommended for aluminum. These positions force the inert-gas shield ahead to clean the metal. When the inert gas and the electric arc combine, the oxide on the aluminum is broken up. Removal of the oxide is necessary for strong fusion in the weld zone.

Striking the Arc

The arc should be struck about one inch in front of where the bead is to begin. The gun is then brought back to the starting point and the weld is made.

Welding Motions

Stringer beads are recommended for MIG welding of aluminum. The gun angle is approximately the same as for carbon-steel welding. Weaving the bead may cause oxidation of the molten metal.

Aluminum Classifications and Identification

Pure aluminum can be mixed (alloyed) with many other elements. The alloys cause it to be grouped into two classifications:

- Wrought Alloys—shaped into structural forms.
- Cast Alloys—molten aluminum poured into molds.

The alloys have been given numbers by the Aluminum Association. The system has four digits. The first digit shows the element which is the major material in the alloy, the second digit shows the alloy's action on impurities. The last two digits show the purity of the aluminum, in hundredths of 1 percent. In the numbering system, number 1 classifies aluminum at 99 percent pure, the number 2 shows aluminum alloyed with copper, and so on. The number 1097 would indicate aluminum 99 percent pure group and .97 percent aluminum, making the aluminum 99.97 percent pure, figure 21-4.

MAJOR ALLOY	INDEX
ALUMINUM 99%	1XXX
COPPER	2XXX
MANGANESE	3XXX
SILICON	4XXX
MAGNESIUM	5XXX
MAGNESIUM & SILICON	6XXX
ZINC	7XXX

Fig. 21-4 Aluminum alloy index

Aluminum Alloys

Two groups of alloys are used:

- Heat-Treatable—materials that can have their molecular structure changed by applying heat.
- Nonheat-Treatable—materials whose physical characteristics are not changed by applying heat.

The welder is concerned with the nonheat-treatable: the 1xxx, 3xxx, 4xxx, 5xxx series. Nonheat-treatable aluminums are easily welded with the MIG process. Some of the heat-treatable aluminums can result in cracking after welding.

Aluminum Welding Recommendations

The following charts list the recommendations for the various thicknesses of material, and types of joints. Figure 21-5, page 120, is for thin plates which do not require plate preparation, while figure 21-6, page 120, is for thicker plates which do require plate preparation. Included are the proper wire diameter, gas flow, amperage, voltage, and wire feed speed.

MIG WELDING—ALUMINUM (SPRAY ARC)						
PLATE THICKNESS (INCHES)	TYPE OF JOINT	WIRE DIAM. (INCHES)	ARGON FLOW (CFH)	AMPERES (DCRP)	VOLTAGE (VOLTS)	APPROXIMATE WIRE FEED SPEED (IPM)
0.040	FILLET OR TIGHT BUTT	0.030	30	40	15	240
0.050	FILLET OR TIGHT BUTT	0.030	15	50	15	290
0.063	FILLET OR TIGHT BUTT	0.030	15	60	15	340
0.093	FILLET OR TIGHT BUTT	0.030	15	90	15	410

(Linde Co.)

Fig. 21-5 Recommendations for thin plate aluminum welding

MIG WELDING—ALUMINUM (SPRAY ARC)					
PLATE THICKNESS	PREPARATION	WIRE DIAM. (INCHES)	ARGON FLOW (CFH)	AMPERES (DCRP)	VOLTAGE
.250	SINGLE VEE BUTT (60° INCLUDED ANGLE) SHARP NOSE BACKUP STRIP USED	3/64	35	180	24
	SQUARE BUTT WITH BACKUP STRIP	3/64	40	250	26
	SQUARE BUTT WITH NO BACKUP STRIP	3/64	35	220	24
.375	SINGLE VEE BUTT (60° INCLUDED ANGLE) SHOP NOSE, BACKUP STRIP USED	1/16	40	280	27
	DOUBLE VEE BUTT (75° INCLUDED ANGLE, 1/16-IN. NOSE). NO BACKUP. BACK CHIP AFTER ROOT PASS	1/16	40	260	26
	SQUARE BUTT WITH NO BACKUP STRIP	1/16	50	270	26
.500	SINGLE VEE BUTT (60° INCLUDED ANGLE) SHARP NOSE. BACKUP STRIP USED	1/16	50	310	27
	DOUBLE VEE BUTT (75° INCLUDED ANGLE, 1/16-IN. NOSE). NO BACKUP BACK CHIP AFTER ROOT PASS	1/16	50	300	27

(Linde Co.)

Fig. 21-6 Recommendations for thick plate aluminum welding

JOB 21: FOUR LAYER STRINGER BEAD PAD BUILDUP, FLAT POSITION, ALUMINUM

Equipment and Materials:

Personal hand tools
Protective clothing and gear
Stainless steel wire brush
1 piece, 3/8" x 5" x 6" aluminum plate
1/16" 4043 or 5356 wire electrode
Argon shielding gas

PROCEDURE	KEY POINTS
1. Set up the machine for spray-arc transfer and change the flowmeter for argon gas.	
2. Set the voltage at 22 and make later adjustments as necessary.	
3. Set the current at 250 amperes (30 inches per minute wire feed) and make later adjustments as necessary.	
4. Set the flowmeter at 40 cubic feet per hour.	4. Purge the gas system to eliminate any left over carbon dioxide.
5. Strike an arc on scrap aluminum to adjust the machine.	
6. Position the plate in the flat position and run a bead all the way across the center, figure 21-7.	6. The melting temperature of aluminum is low, the heat input of the spray arc is high so the heat buildup will be heavy. Cool the plate as needed.

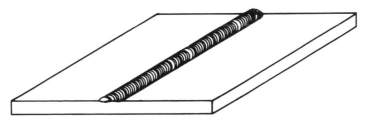

Fig. 21-7 Run the first bead across the center of the plate.

7. Continue to run stringer beads until the plate is covered. Follow the bead sequence shown, figure 21-8.

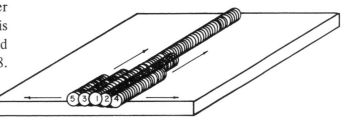

Fig. 21-8 Proper bead sequence for running the beads

PROCEDURE	KEY POINTS
8. Turn the sample 90 degrees so the beads will cross in the opposite direction.	8. Spatter buildup may be heavy. Clean the nozzle and apply compound often.
9. Repeat the process, turning the plate until four layers have been welded, figure 21-9.	

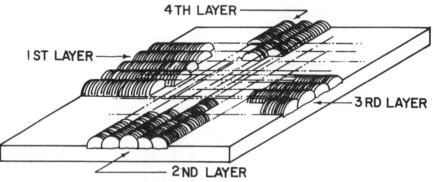

Fig. 21-9 Procedure for completing pad

10. Shut down all systems.	
11. Submit the pad to the instructor for comments and instructions.	

REVIEW QUESTIONS

Answer the following questions briefly.

1. What is the difference between a wrought alloy and a cast alloy?

2. What is a heat-treatable material?

3. What is a nonferrous metal?

4. What is the weight ratio between aluminum and steel?

5. Is aluminum magnetic?

6. Is copper heavier than aluminum?

7. What is the tensile strength of copper?

8. What is "hot shortness"?

9. Is magnesium lighter than aluminum?

10. What quick test can the welder use to determine if the metal about to be welded is aluminum or magnesium?

11. Name five common alloying elements used with aluminum.

12. Why is it difficult to determine when aluminum and aluminum alloys have reached the melting point?

13. Name five reasons why MIG welding is preferred for aluminum plate welding.

14. Why is aluminum plate generally welded in the vertical-up position?

15. Define the cleaning action which breaks up aluminum oxide.

16. Where is the initial arc struck on the aluminum plate?

17. What type of welding motion is preferred for aluminum MIG welding?

Unit 22 Fillet Weld, Flat Position, Aluminum

OBJECTIVES

After completing this unit, the student will be able to:

- explain the use of argon gas for aluminum welding.
- describe the cleaning processes used on aluminum.
- explain the three main differences in aluminum and steel joint preparation.
- successfully weld an aluminum fillet with the MIG process, using the spray-arc method of metal transfer in the flat position.

Shielding Gas

Pure argon is the shielding gas most commonly used for aluminum MIG welding. Argon makes a smooth, stable arc when spray transfer is being used. Penetration is deep and narrow and best for most joint designs where only a small amount of plate preparation is used, figure 22-1. The bead outline of argon gas shows a narrow, deep penetration which other gases do not make. Better rod transfer is made with the argon shield and there is less spatter in the weld zone. Argon is used for plates up to 1″ thick. Plates larger than this are generally welded with an argon and helium mix.

Argon Cylinders

Argon cylinders are high pressure cylinders, like oxygen, and need to be fastened upright when in use. The pressure in the cylinder is high, and care should be taken that when the cylinder is used it is fastened so it cannot fall. When the cylinder is moved, the protective valve cap should be screwed on securely to prevent damage if it is dropped. It should not be stored in extremely warm areas.

Material Cleaning Procedure

To make good welds, nonferrous metals must be cleaned to remove the oxide film. Aluminum oxide has a higher melting temperature (3800 degrees Fahrenheit, 1145 degrees Celsius) than the aluminum (1215 degrees Fahrenheit, 348 degrees Celsius) itself. The oxide must be removed before welding. Aluminum can be cleaned by chemical or mechanical

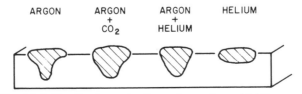

Fig. 22-1 Depth of penetration using various shielding gases

means. The usual method of cleaning is with a stainless steel wire brush. If circular wire brushes are used with power tools, protective eye devices must be worn.

Fillet Welds

Fillet welds are used in many aluminum weldments. They are a major joint design. The fillet does not require plate preparation on metal 3/8″ thick or less. Grooving of the joint on heavier aluminum is used to ensure penetration.

There are few differences in the design of either steel or aluminum. The main differences in welding of aluminum joints are:

- Narrower joint spacing.
- Less sharp bevels.
- Thicker bevel lands.

JOB 22: FILLET WELD, FLAT POSITION, ALUMINUM

Equipment and Materials:

Standard MIG welding equipment	1 piece, 3/8″ x 5″ x 6″ aluminum plate
Personal hand tools	1 piece, 3/8″ x 2 1/2″ x 6″ aluminum plate
Protective clothing and gear	4043 or 5356 wire electrode
Stainless steel wire brush	Argon shielding gas

PROCEDURE	KEY POINTS
1. Turn on the welding machine.	
2. Set the controls for spray-arc transfer.	
3. Position and tack weld the plates as shown in figure 22-2.	3. Wire brush the joints to be welded with a stainless steel brush.

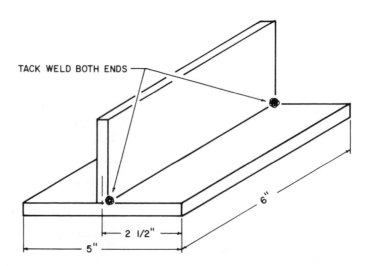

Fig. 22-2 Proper position of the plates and the tack welds

PROCEDURE	KEY POINTS
4. Position the sample and begin welding, following the bead sequence shown, figure 22-3.	4. Gun angles are the same as those used for steel fillets. Be sure the inert-gas coverage is 100 percent.

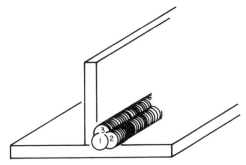

Fig. 22-3 Proper bead sequence for completing the joint

PROCEDURE	KEY POINTS
5. Complete the joint. Rotate it and weld the other side.	5. Check each bead as it is made and adjust the machine if the weld is not satisfactory.
6. Shut down all systems.	
7. Submit the weld to the instructor for comments and instructions.	

REVIEW QUESTIONS

Answer the following questions briefly.

1. Why is argon usually used for MIG welding of aluminum?

2. What is the difference in the penetration appearance in metal when argon is used for shielding gas?

3. How is aluminum customarily cleaned before MIG welding operations?

4. Does aluminum oxide melt at a higher or lower temperature than aluminum?

5. What are the three main differences in joint preparation between aluminum and steel?

Unit 23 Butt Weld, Flat Position, Aluminum

OBJECTIVES

After completing this unit, the student will be able to:

- explain the three most commonly found problems related to aluminum welding.

- recognize the reasons for porosity in aluminum welds.

- recognize and correct poor penetration in aluminum welding.

- successfully MIG weld an aluminum butt weld in the flat position, using the spray-arc method of metal transfer.

Weld Defects

The most common weld defects in aluminum welding are:

- Cracks

- Porosity

- Poor penetration

Aluminum Weld Cracks

Cracking of aluminum welds may be found in the weld, or the area around the weld. They generally run the length of the weld or appear in craters. Long cracks are generally not weld defects, but are flaws which happen because the weld deposit is overloaded. Crater cracks are only as long as the crater itself.

To decrease the chance of cracks forming in the heat zones, jigs or holding fixtures can be used. Sometimes preheat or postheat may help control cracking.

Crater cracks can be eliminated by backing the weld into the end of a bead to increase the amount of metal. Run-off tabs to begin and end the weld are a great help.

Porosity

Porosity is occasionally found in aluminum welds. A small amount of porosity does not have much effect, but large areas of holes may make the joint weak. Porosity is generally the result of gas trapped in the molten metal of the puddle. The puddle must be worked to allow the gas to escape. The trapped gas is usually hydrogen.

Other reasons porosity occurs:

- Wrong machine settings
- Worn contact tips
- Dirty filler wire

- Dirty base metal
- Poor gas shield
- Improper gun angle, motion, and distance

Poor Penetration

Poor penetration is the failure to get correct fusion of the weld to the base metal.

Where a prepared groove joint is used, poor penetration is an incomplete mixture of filler and base metal at the bottom of the joint.

The major cause of poor penetration is poor cleaning of the metal before welding, or not enough cleaning between passes. If the bead has a gray, dirty appearance, it is not clean enough for the next pass. A well cleaned bead is bright and shiny.

JOB 23: BUTT WELD, FLAT POSITION, ALUMINUM

Equipment and Materials:

Standard MIG welding equipment	2 pieces, 3/8" x 4" x 6" aluminum plate
Personal hand tools	1 piece, 3/8" x 2" x 9" aluminum plate
Protective clothing and gear	1/16" 4043 or 5356 wire electrode
Stainless steel wire brush	Argon shielding gas

PROCEDURE	KEY POINTS
1. Turn on the welding machine.	
2. Position and tack weld the sample as shown in figure 23-1.	2. Bevel each plate 30 degrees. The root opening is 3/32" and the land 1/16". Use a silicon carbide grinding wheel designated for soft metals.

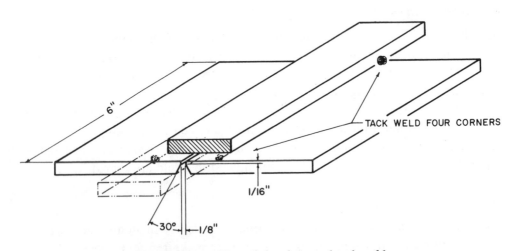

Fig. 23-1 Proper position of the plates and tack welds

3. After tacking, turn the piece over so the groove is up.	
4. Weld the sample, following the bead sequence shown, figure 23-2.	4. Use the forehand welding technique with a stringer bead motion. Brush the joint with a stainless steel brush before welding.

PROCEDURE   **KEY POINTS**

BEGIN AND END WELD ON RUN-OUT TAB

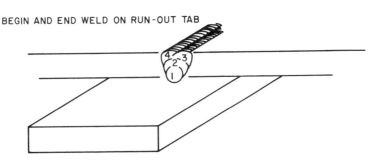

Fig. 23-2 Proper sequence for running beads

5. Complete the welding.

6. If the completed welds do not appear satisfactory, repeat the entire procedure.

7. Submit the weld to the instructor for comments and instructions.

REVIEW QUESTIONS

Answer the following questions briefly.

1. What are the three most common weld defects in aluminum welding?

2. What causes longitudinal cracks?

3. What is a crater crack?

4. How may a crater crack be avoided?

5. Name five sources of porosity in aluminum MIG welding.

6. What is poor penetration?

Unit 24 Four Layer Stringer Bead Pad Buildup, Horizontal Position, Aluminum

OBJECTIVES

After completing this unit, the student will be able to:

- describe the gun motion for horizontal welding of aluminum.
- name the best filler wires for five different aluminum base metals and describe the reasons for their use.
- successfully weld a four layer, stringer bead aluminum pad buildup in the horizontal position of welding, using the spray-arc method of metal transfer.

Horizontal Aluminum Welding

The welding of aluminum in the horizontal position requires attention to the same factors as in horizontal mild steel welding. The bead motion, gun angle, and slope control are all approximately the same in both cases.

Constant observation of the puddle is necessary as aluminum tends to form a fluid puddle. A lowering of the inductance will help to maintain a rigid puddle.

The bead motion should be restricted to the stringer motion to keep the heat concentrated in one area. The use of a weave bead will allow the heat to disperse throughout the base metal and possibly cause the piece to lose its original shape.

Filler Wire Selection

Four things are considered before a correct aluminum alloy can be chosen for a welded part:

- Joint design
- Metal thickness
- Welding position
- Service requirements

The aluminum alloy to be used makes the choice of filler wire important. Different filler wires add special features to the welded part. It is important to know which alloy has been chosen for the base metal. After both the alloy and the weld requirements of the part are known, the manufacturer's wire charts will help to make a correct choice for the filler wire, figure 24-1.

Some aluminum wires are used more often than others. All wires have basic core makeup and these are classified and numbered. A chart may be obtained from any welding supplier. The chief alloys used in wire electrodes are:

- Copper
- Magnesium
- Manganese
- Silicon
- Zinc

RECOMMENDED FILLER FOR SOME ALUMINUM ALLOYS			
WELDED BASE METAL OR ALLOY	BEST MATCH UP FOR		
	STRENGTH	DUCTILITY	SALT WATER CORROSION RESISTANCE
1100	4043	1100	1100
6061	5356	5356	4043
6063	5356	5356	4043
6003	5356	1100	1100
6052	5356	5356	4043

Fig. 24-1 Proper filler wires for various materials and conditions

By adding these elements to the wire, strength, corrosion resistance, and weldability are increased.

Recommended Wires

The wire best suited to add the strength, ductility (ability to be hammered or drawn out), crack resistance, and corrosion resistance are those numbered 4043 (silicon-containing) and 5356 (magnesium containing) alloys. It is necessary to match the wire to the service requirements of the base metal and the type of joint to be welded.

Weld Cracking

The most common defect of aluminum welding is the weld crack. The wire selection must be made so that it will offer a completed weld that will decrease the chances of cracks.

Weld cracks can be limited to a minimum by using a filler wire of higher alloy content than the base metal.

JOB 24: FOUR LAYER STRINGER BEAD PAD BUILDUP, HORIZONTAL POSITION, ALUMINUM

Equipment and Materials:

Standard MIG welding equipment
Personal hand tools
Protective clothing and gear
Stainless steel wire brush
3/8" x 5" x 6" aluminum plate
1/16" 4043 or 5356 wire electrode
Argon shielding gas

PROCEDURE	KEY POINTS
1. Turn on the welding equipment.	
2. Adjust the machine for horizontal welding by practicing on scrap aluminum.	2. Use only the forehand and stringer bead motions on aluminum.

PROCEDURE	**KEY POINTS**

3. Place the plate vertical to the welding table. Begin at the bottom of the plate and weld across, using forehand motion, figure 24-2.

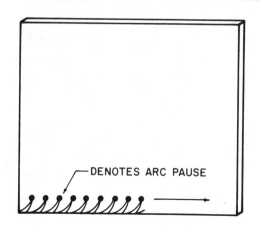

Fig. 24-2 Using a stringer bead motion across the bottom of the plate

4. Check the beads as they are completed and make corrections in machine settings as needed.

4. The gun motion and angles used for carbon steel welding are satisfactory for aluminum.

5. Bead #2 should cover the top third of bead #1, figure 24-3.

5. Follow the bead sequence indicated and completely fill the plate.

6. Rotate the plate 90 degrees so that the next layer of beads crosses the first.

7. Keep turning the plate until four layers of beads have been welded.

8. Shut down all systems.

9. Submit the pad to the instructor for comments and instructions.

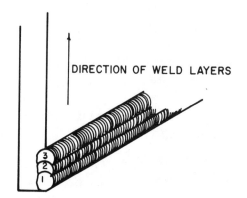

Fig. 24-3 Proper bead sequence for horizontal pad buildup

REVIEW QUESTIONS

Answer the following questions briefly.

1. In the following statements, match up the filler wire recommended in the text with the metal or alloy noted, and the quality needed:

   a. If the base metal is 1100 aluminum and salt water corrosion will be a problem, _____ filler wire is best.

   b. If 3003 base metal is used, and strength is needed, _____ filler wire is best.

   c. If 1100 base metal is used, and ductility is needed, _____ filler
      wire is best.

   d. If 6063 base metal is used, and strength is needed, _____ filler
      wire is best.

2. What four things must be considered before a base metal is chosen for a
   welded joint?

3. Name 5 alloys generally used to mix with the aluminum for MIG wire.

4. What is considered the number one weld defect in MIG welding?

Unit 25 Butt Weld, Horizontal Position, Aluminum

OBJECTIVES

After completing this unit the student will be able to:

- describe the methods of preparing aluminum plates for butt welding.
- understand the necessity for cleaning aluminum before MIG welding takes place.
- successfully MIG weld an aluminum butt in the horizontal position, using the spray-arc method of metal transfer.

Aluminum Butt Weld Joint Design

The edges of aluminum plate may be made ready for welding by sawing, planing, machining, chipping, or carbon-arc cutting. Oxyacetylene cutting has not been noted here because the melting temperature of the aluminum oxide which forms on the base metal has a higher melting temperature than the plate itself. Aluminum which has been cut with the oxyacetylene torch is burned in appearance and not usable for welding without considerable cleaning of the cut. When cutting is done on aluminum plates with the carbon-arc process, all carbon particles must be removed before any welding is done.

Aluminum joint designs are the same as those used for steel, except for the methods of plate preparation. Butt joints are a major design in modern aluminum welding. Square butt joints (without any preparation except for cleaning) can be MIG welded in metal up to 1/4" in thickness, and the resulting weld has full penetration and strength.

On thicker materials, the edges of the plates must be beveled so that each plate has a 60 degree angle on its edge. Since the grinding of an angle on a plate will leave a very thin edge at the bottom of the weld, it is necessary to grind a land on the bottom of each bevel. Always use a silicon carbide grinding wheel designated for soft metals.

The extra thickness created by grinding the land keeps the arc from burning through too fast and creating excessive penetration. Also, when thick metal is used, a small gap, called the root opening, is left between the lands of the plates, to ensure penetration, figure 25-1.

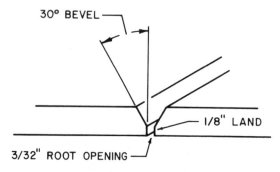

Fig. 25-1 Proper bevel angle, land, and root opening for aluminum single V butt weld

Root Pass

The first bead put into the butt weld is called the root pass. It is necessary that this first pass melt the base metal together completely at the bottom of the groove. Penetration must be 100 percent through the root opening and join the lands of the two plates.

JOB 25: BUTT WELD, HORIZONTAL POSITION, ALUMINUM

Equipment and Materials:

Standard MIG welding equipment
Personal hand tools
Protective clothing and gear
Stainless steel wire brush
2 pieces, 3/8" x 4" x 6" aluminum plate
1 piece, 3/8" x 2" x 9" aluminum plate
1/16" 4043 or 5356 wire electrode
Argon shielding gas

PROCEDURE	KEY POINTS
1. Turn on the welding machine.	
2. Bevel the plates to be butt welded.	2. Bevels are 60 degrees on each plate. Use a silicon carbide grinding wheel.
3. Grind a 1/8-inch land on each bevel.	
4. Tack weld the plates together with 3/32-inch root opening. Tack the plates to a backup bar, figure 25-2.	4. The backup bar should be tight and should extend past the edges of the butt weld 1 1/2 inches at each end.

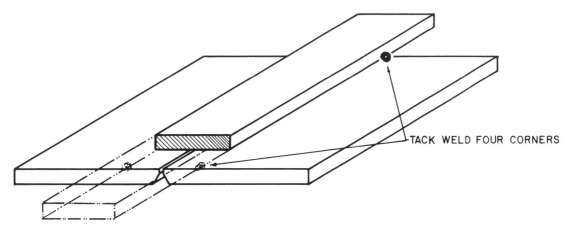

TACK WELD FOUR CORNERS

Fig. 25-2 Proper position of the plates and the tack welds

5. Thoroughly brush the entire joint with a stainless steel brush.	
6. Position the plates for a horizontal weld.	

PROCEDURE	KEY POINTS
7. Weld the plates, using the sequence shown, figure 25-3.	7. Inspect each bead as it is welded. Clean the weld area between passes.

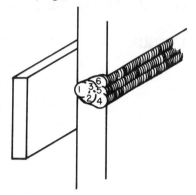

Fig. 25-3 Proper bead sequence for the horizontal, aluminum, butt weld

8. Examine the weld, repeat the entire procedure if more practice is needed.

9. Submit the sample to the instructor for comments and instructions.

REVIEW QUESTIONS

Answer the following questions briefly.

1. What methods are generally used to prepare aluminum plates for butt welding?

2. Butt welds can be made on some aluminum plates without preparing the edges of the plates. When should a plate be ground?

3. What kind of abrasive wheel is used to grind soft metals?

Unit 26 Four Layer Stringer Bead Pad Buildup, Vertical Position, Aluminum

OBJECTIVES

After completing this unit, the student will be able to:

- explain the care and storage of aluminum wire electrode material.
- satisfactorily MIG weld a four layer stringer bead pad buildup in the vertical position.

Vertical Aluminum Welding

The welding of aluminum in the vertical position not only requires the proper bead motion, gun angle, and inductance, but also the correct amperage. Beads in the vertical position should be of the stringer variety and slightly larger than those used for vertical mild steel welding. The amperage, as well as the inductance, should be lowered to increase the rigidity of the weld.

Care and Storage of Aluminum Wire Electrodes

High-quality welds can only be made from wire which has been properly stored and taken care of. Even though the wire and base metal has been matched up, the weld may still be poor. Inferior welds may be the result of carelessly handled filler wire.

Successful welds in aluminum must begin with high-quality wire and they can only be made with wire which is the correct size and alloy. The diameter of the wire must be constant so that it can pass through the contact tip of the MIG gun without interference. The wire should be free of moisture, lubricants, and any other substances.

Care of the wire while the machine is not being used is extremely important. To avoid contamination, the filler wire should be covered and stored in a dry place at a uniform temperature. Wire spools left on idle machines should be covered with cloth or plastic. If the wire is not to be used for some time, it should be removed from the welding machine and stored in the original carton and sealed.

If rod ovens are in the shop, cartons of wire should be stored in them to prevent moisture from accumulating on the wire.

Backup Bars and Beads

When aluminum joints are welded from one side only, some form of penetration control is needed. Backup material may be left in place, or be a removable type.

The removable type of backup bar is generally used when welding aluminum. The bars are grooved directly under the root penetration zone to allow room for the penetration to fall through. Sometimes the groove in the backup bar is used to let argon gas flow under the welding zone to keep it from sucking atmospheric gases into the weld, figure 26-1, page 140.

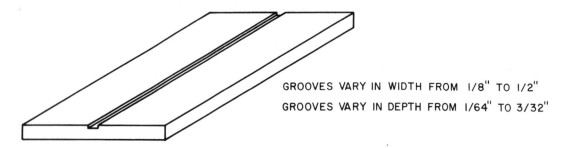

GROOVES VARY IN WIDTH FROM 1/8" TO 1/2"

GROOVES VARY IN DEPTH FROM 1/64" TO 3/32"

Fig. 26-1 Removable grooved backup bar

Removable backup bars are made from steel, stainless steel, or copper.

Backup beads may be used where a weld can be made on both sides of a joint. A weld is made on one side, the back of the weld is chipped down into the deposit and then welded, figure 26-2.

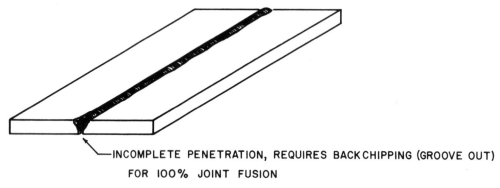

INCOMPLETE PENETRATION, REQUIRES BACK CHIPPING (GROOVE OUT) FOR 100% JOINT FUSION

Fig. 26-2

JOB 26: FOUR LAYER STRINGER BEAD PAD BUILDUP, VERTICAL POSITION, ALUMINUM

Equipment and Materials:

Standard MIG equipment
Personal hand tools
Protective clothing and gear
Stainless steel wire brush

1 piece, 3/8" x 5" x 6" aluminum plate
1/16" 4043 or 5356 wire electrode
Argon shielding gas

PROCEDURE	KEY POINTS
1. Turn on the welding equipment.	
2. Wire brush a piece of scrap material and set it in the vertical position. Run practice beads to adjust the machine.	2. Amperage for vertical welds must be reduced for puddle control.
3. Use the forehand, vertical-up motion. Run a stringer bead up the center of the plate all the way to the top, figure 26-3.	3. Vertical stringer beads must be a little wider than beads used for other positions.

PROCEDURE	KEY POINTS

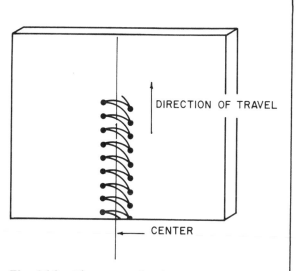

Fig. 26-3 The proper bead motion for vertical aluminum welding

4. Run bead #2 to the right of bead #1, overlapping bead #1. Run bead #3 to the left of bead #1, overlapping bead #1, figure 26-4.

4. Brush each completed weld and check for flaws.

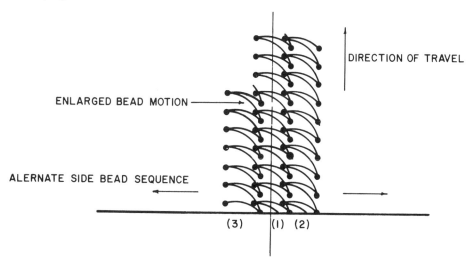

Fig. 26-4 The proper bead sequence for vertical aluminum pad buildup

5. When the plate is filled, rotate it 90 degrees so the following welds will cross the first welds. Continue this procedure until four layers are completed.

6. Submit the pad to the instructor for comments and instructions.

REVIEW QUESTIONS

Answer the following questions briefly.

1. Why is it necessary that the diameter of filler wire be of uniform size?

2. If the wire spool is to be left on a machine for long periods of time, how can it be protected?

3. How can rod ovens be used to store filler wire?

4. How should the amperage be adjusted for vertical aluminum welding?

5. What type of bead motion should be used for vertical aluminum welding?

Unit 27 Fillet Weld, Vertical Position, Aluminum

OBJECTIVES

After completing this unit, the student will be able to:

- compare and contrast fillet welds and T welds.
- successfully MIG weld T weld on aluminum plates, in the vertical position.

Fillet Welds and T Welds

A fillet weld may be defined as any weld made between two plates where they meet at an angle. A T weld is a fillet weld which has two plates meeting at a 90 degree angle. The word "fillet" means reinforcement.

T Welds in Industry

The T weld is the most common fillet weld used in industrial construction projects, and it is necessary that a welder be able to make strong, acceptable welds on this type of joint. Penetration into both plates must be maximum, whether welded from one side or both sides. There is a tendency to overweld this joint. In general, too much welding does not increase the strength and will add to the cost of welding, besides creating warpage which may make a bad fit for other parts to be welded to it.

Vertical T Welds

T welds on aluminum plate, made in the vertical position, are difficult and much practice must be devoted to mastering the job. The gun has to be directed into the corner so that the weld is laid equally on each plate. Travel up the joint has to be at a slow, even rate so that the penetration can soak into the corner where the two plates meet. Rapid welding up the joint will not give enough penetration.

Fitup and Tack Welds on Aluminum

Many times the success or failure of aluminum welds depends on the fitup and tacking procedures followed.

Correct joint fitup saves time and materials and has a direct bearing on the production of high-quality welds.

Tack welds are used to position material in place before the welding operations and to attach the start and run-off tabs when needed. Tack welds should be small, and of a sufficient number to hold the material in line. Tack welds used for aluminum should be slightly longer than those used for steel. All tack weld craters should be backfilled and any cracks in the tacks should be removed before welding the joint.

JOB 27: FILLET WELD, VERTICAL POSITION. ALUMINUM

Equipment and Materials:

Standard MIG welding equipment
Personal hand tools
Protective clothing and gear
Stainless steel wire brush

2 pieces, 3/8" x 5" x 6" aluminum plate
2 pieces, 3/8" x 2 1/2" x 6" aluminum plate
1/16" 4043 or 5356 wire
Argon shielding gas

PROCEDURE	KEY POINTS
1. Turn on the equipment.	
2. Tack weld both pieces, figure 27-1.	

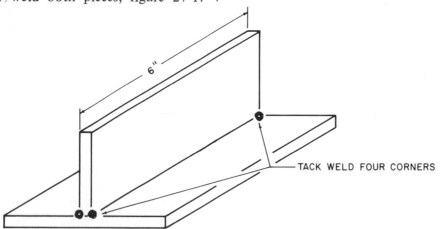

Fig. 27-1 Proper position of the plates and the tack welds

3. Wire brush each joint before welding.

4. Position the joint in the vertical position. Begin the weld at the bottom of the joint and run a stringer bead from the bottom to the top, figure 27-2.

4. Use the same gun angles and motions as used in the vertical pad buildup.

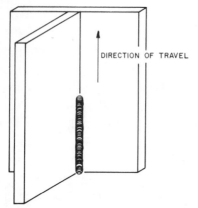

Fig. 27-2 Run a stringer bead using the vertical-up welding method.

PROCEDURE	KEY POINTS
5. Brush each bead after it is finished.	
6. Rotate the joint and weld the other side.	
7. Submit the sample to the instructor for comments and instructions.	

REVIEW QUESTIONS

Answer the following questions briefly.

1. What two poor results may come from overwelding the T joint?

2. Where must the gun be directed to make a good T weld?

3. Why should the welder travel up the joint at a slow, even rate?

4. When are tack welds used?

SECTION 6: REVIEW A

OBJECTIVE:

- Produce an aluminum fillet weld of proper penetration and quality in the flat position using standard MIG welding equipment and procedures. Subject such a sample to a destructive test.

Equipment and Materials:

Standard MIG welding equipment
Personal hand tools
Protective clothing and gear
Stainless steel wire brush

1 piece, 3/8" x 5" x 6" aluminum plate
1 piece, 3/8" x 2 1/2" x 6" aluminum plate
4043 or 5356 wire electrode
Argon shielding gas

PROCEDURE

1. Turn on and adjust the equipment for spray-arc transfer.

2. Position the plates and tack weld as shown in figure R6-1.

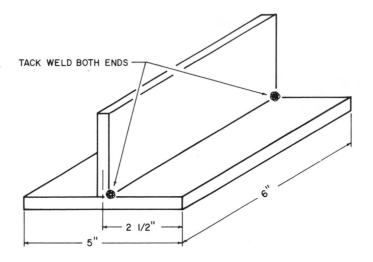

Fig. R6-1 Proper position of the plates and tack welds

3. Arrange the plates so the joint is in the flat position and begin welding.

4. Follow the bead sequence shown in figure R6-2, and complete the weld as shown.

5. Shut down all systems.

6. Perform a destructive test as shown in figure R6-3.

7. Examine the weld for defects such as incomplete penetration, porosity, and cracks.

8. Submit a satisfactory sample to instructor for comments and instructions.

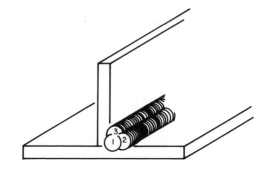

Fig. R6-2 Proper bead sequence for aluminum fillet weld

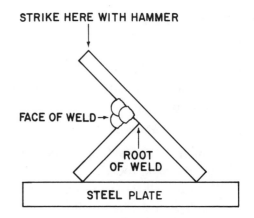

Fig. R6-3 Destructive test of a fillet weld

SECTION 6: REVIEW B

OBJECTIVE:

- Produce an aluminum single V butt weld of proper penetration and quality in the flat position using standard MIG welding equipment and procedures. Subject such a sample to the guided-bend test.

Equipment and Materials:

Standard MIG welding equipment
Personal hand tools
Protective clothing and gear
Stainless steel wire brush
Carbon-arc cutting equipment or hacksaw
2 pieces, 3/8" x 4" x 6" aluminum plate
1/16" 4043 or 5356 wire electrode
Argon shielding gas

PROCEDURE

1. Turn on and adjust the equipment for spray-arc welding.

2. Bevel the plates and grind a land on each bevel.

3. Tack weld the plates in place, figure R6-4.

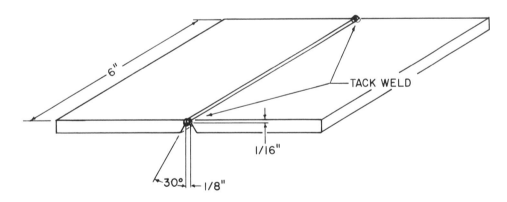

Fig. R6-4 Proper position of the tack welds

4. Position the plates in the flat position.

5. Run the beads as shown in figure R6-5.

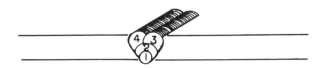

Fig. R6-5 Proper bead sequence aluminum single V butt weld

6. Cut the sample into sections as shown in figure R6-6.

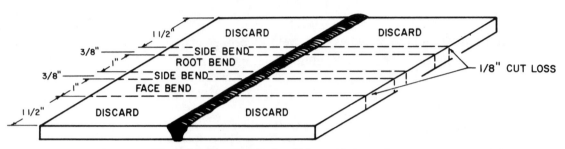

Fig. R6-6 The sample should be cut into sections

7. Perform a guided-bend test on the face, the root, and the sides of the various strips.

8. Examine the bent strips. There should be no separation of weld from the base metal and no porosity.

9. If the welded samples show defects, repeat steps 1-8.

10. When the bent specimens are satisfactory, submit them to the instructor for comments and further instructions.

SECTION 7:
MIG Welding Stainless Steel

Unit 28 Stringer Beads, Flat Position, Stainless Steel

OBJECTIVES

After completing this unit the student will be able to:

- explain the composition of stainless steel.

- explain the difference in stainless steel alloys.

- describe the types of shielding gases used for stainless steel welding.

- successfully MIG weld stringer beads in the flat position using the short arc method of metal transfer, and the spray-arc method.

Stainless Steel

Stainless steels are alloys of iron, chromium, and other materials which do not oxidize easily, even at high temperatures.

Basically, stainless steels are composed of iron and chromium. Iron is present in all stainless steels. Iron oxidizes (rusts) easily, so chromium is added at 11.5 percent or more. Corrosion resistance gets higher as more chromium is added to iron. The corrosion resistance is formed when a thin layer of chromium oxide is formed on the surface. This film stops oxidation from taking place. Some other elements can be added to stainless steels to produce more worthwhile features.

Nickel increases corrosion resistance when added to stainless steel alloys. It also helps improve ductility, and increases the strength of the metal. The element molybdenum is also sometimes added to help keep the steel from expanding when the temperature of the metal is raised by heat.

Straight Chromium Stainless Steels

Straight chromium stainless steels are classified into the 400 series. They are called "straight chromes." The 410, 430, and 446 types are widely used in industry. The difference between these three, and other alloys, is the amount of chromium alloyed with the iron. The percentage range is from 11.5 percent to 29 percent chromium. All of the straight chromes are magnetic. The welding of these alloys is difficult since they must be preheated and postheated. They require special welding procedures.

Chromium-Nickel Stainless Steels

The chromium-nickel steels are the ones most generally welded in industry. They are classified into the 300 series. The one element in the chromium-nickel steels which is not a welding factor in the straight chromium steels is carbon. In order to keep from having trouble with the carbon in the weld, a stabilizer (columbium or titanium) is added. If the carbon content is limited to a maximum of .03 percent the stainless steel is called "extra low carbon," (ELC). The chromium content of this steel is 16 percent to 26 percent with the nickel content at 3.5 percent to 22 percent. The chromium-nickel steels are not magnetic due to the nickel alloy. The steels are not difficult to weld after training.

Welding Stainless Steels

When welding stainless steel, the wire electrode must match up with the base metal and the shielding gas must cover both the front and back of the joint. If the shielding gas does not cover the back of the joint a weld defect known as "honeycombing" (a form of porosity) will appear.

Shielding Gas

The recommended shielding gas for stainless steel welding of thicknesses of less than 3/16 inch is argon, with the addition of 1 percent oxygen. This mix improves metal transfer, arc stability and wetting action in the welding zone, figure 28-1. Thicker plates should use a mix of argon and 2 percent oxygen.

SHIELDING GAS PREFERENCE – STAINLESS STEEL CONSUMABLE WIRE ELECTRODES	
PLATE THICKNESS	GAS MIXTURES
0 – 3/16"	ARGON + 1 OR 2% OXYGEN
3/16" – 1"	ARGON + 2% OXYGEN

Fig. 28-1 The thickness of the plate determines which shielding gas is to be used.

Welding Procedures

The welding procedures used for stainless steel with the MIG process are similar to those used for carbon steels. Either the short-arc or spray-arc transfer can be used, and the same bead motions will work. Forehand motions give a better cleaning action from the shielding gas. The short arc is generally used for thin materials, poor fitups, and all out-of-position welds. The spray arc produces high speed, quality welds on heavy metal in the flat position. Heat does not spread through the stainless steels as fast as it does through the carbon steels. The heat stays in the weld zone longer. This tends to create warpage and shrinkage, especially in thin gauge metal. These faults can be helped by the use of jigs and holding fixtures.

Recommendations for Welding Stainless Steel

The welder should follow all recommendations made by the material manufacturers and the wire manufacturers. While the stick-arc process is still used, MIG welding is replacing the process because of the following advantages:

- There is no flux or slag to clean off after welding.

- The weld zone is visible since no flux is present.

- Continuous wire is fed into the weld zone.

- There is no rod stub loss.

- There is reduced welding time.

The following chart lists the recommendations for the various thicknesses of material, types of joints, and welding positions, figure 28-2. Included are the proper wire diameter, amperage, gas, and gas flow.

MIG WELDING – STAINLESS STEEL					
SHEET OR PLATE THICKNESS (IN.)*	FILLER WIRE DIAM. (IN.)	CURRENT AMPS.	WIRE FEED (IN./MIN.)	GAS AND FLOW CFH	WELDING POSITION**
1/16***	0.035	110/140	230/260	He 20/30	F, H, VD
1/8-3/16	0.035	110/140	230/260	He 20/30	F, H, OH, VD
1/4-1	0.035	110/140	230/260	A + 1% O_2 20/30	V, OH
1/4-1	0.035	170/190	330/360	A + 1% O_2 20/30	F, H, OH‡, VD
1/2-1	0.045	140/180	160/200	A + 1% O_2 20/30	V, OH
3/16-3/8	0.045	190/310	210/340	A + 1% O_2 30/40	F, H
7/16 & UP	0.045	190/310	210/340	A + 1% O_2 20/40	OH‡
1/4 & UP	1/16″	280/350	240/330	A + 1% O_2 30/40	F, H

*ALL JOINT TYPES – BUTT, LAP & FILLET   (Linde Co.)
**F–FLAT; H–HORIZONTAL; OH–OVERHEAD; VD–VERTICAL DOWN; V–VERTICAL UP
  WEAVE BEAD
***GOOD FIT-UP REQUIRED
‡ STRINGER BEAD

Fig. 28-2 Recommendations for stainless steel welding

JOB 28: STRINGER BEADS, FLAT POSITION, STAINLESS STEEL

Equipment and Materials:

Standard MIG welding equipment
Personal hand tools
Protective clothing and gear
Side cutting pliers
Stainless steel wire brush
1 piece, 1/4″ x 5″ x 6″ stainless steel plate
.045″, 308L wire
Argon/oxygen 1 percent shielding gas

PROCEDURE	KEY POINTS
1. Set up and adjust the wire electrode and shielding gas system to those required for stainless steel MIG welding.	

PROCEDURE	KEY POINTS
2. Turn on the machine.	
3. Set the flowmeter for 25-30 cubic feet per hour (CFH).	3. Adjust the gas flow system for ample coverage on the plate.
4. Purge the gas system.	
5. Set the wire feed unit for short-arc transfer.	
6. Set the voltage for 24 volts.	
7. Remove the contact tip and move the wire through the gun, then replace the contact tip.	7. Always remove the contact tip when different wires are placed on the machine.
8. Use the forehand motion and begin welding across the plate along one edge of the material, figure 28-3.	

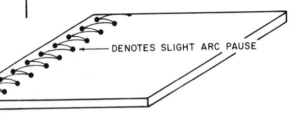

Fig. 28-3 **Use the forehand motion and run a stringer bead across the plate.**

9. Space the stringer beads about 1/2" apart, figure 28-4 Inspect each bead and make machine. adjustments if necessary.	9. Cool the plate in water if it becomes too hot. Brush each bead with a stainless steel wire brush.

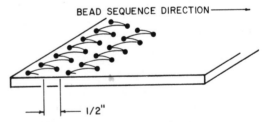

Fig. 28-4 **Run a bead every 1/2 inch.**

10. Continue welding stringer beads until full control is felt.	10. Change to the backhand motion and make a comparison of the different results. Change back to forehand motion.
11. Set the wire feed unit for spray-arc transfer.	
12. Turn the plate over and run spray-arc stringer beads on the clean side.	

PROCEDURE	KEY POINTS
13. Space the stringer beads 1/2″ apart. Run beads until full control is felt.	
14. Submit the beads to the instructor for comments and instructions.	

REVIEW QUESTIONS

Answer the following questions briefly.

1. What is stainless steel?

2. What element is added to stainless steel to make it resist oxidation?

3. What does the addition of nickel do to stainless steel alloys?

4. What is the percentage range of chromium in a "straight chrome"?

5. What is extra low carbon stainless steel?

6. How is the shielding gas used for stainless steel welding?

7. What kind of shielding gas is recommended for stainless steel welding?

8. Which welding motion gives the most satisfactory results for stainless steel welding?

9. Why does thin gauge stainless steel shrink or warp easily?

10. What is the flowmeter setting recommended for stainless steel welding? The voltage setting?

Unit 29 Fillet Weld, Flat Position, Stainless Steel

OBJECTIVES

After completing this unit the student will be able to:

- list the important factors for stainless steel welding.

- explain stainless steel classifications.

- successfully MIG weld a fillet joint on stainless steel in the flat position using the short-arc and the spray-arc method of metal transfer.

Stainless Steel Fillet Weld

Fillet welds of stainless steel are fabricated in the same manner as mild steel fillet welds. Care must be taken to keep the MIG gun pointed equally at both plates, and into the corner, so that penetration remains equal in both pieces.

Important Factors for Stainless Steel Welding

- Iron and chromium are the two basic factors present in stainless steel.

- To make a true stainless steel, chromium must be added in amounts of 11.5 percent or more.

- Other alloy elements are nickel, columbium, titanium, and molybdenum.

- The 400 series is often referred to as the "straight chromes."

- The 300 series is often referred to as the "chrome nickels."

- The principal alloying element in the 300 series is nickel.

- 18-8 stainless is so called because it contains 18 percent chromium and 8 percent nickel.

- The addition of nickel to stainless steel improves weldability, increases corrosion resistance and electrical resistance, and adds impact strength, ductility, and durability.

- The addition of nickel also decreases the heat spread in the metal.

- The addition of molybdenum increases the stretch strength at higher temperatures and reduces the pitting action.

- The addition of columbium and titanium stabilizes the carbon in stainless steel.

- The letters ELC means extra low carbon.

A.I.S.I.

The American Iron and Steel Institute has set up numbers to identify various kinds of stainless steel alloys. The numbers contain three digits. The first digit is of concern to the welder because it shows the alloys that may have to be welded, figure 29-1. Of these groups, the 300 series is the one most used for welding.

INDEX	GROUP
2XX	CHROMIUM-NICKEL-MANGANESE
3XX	CHROMIUM-NICKEL
4XX	CHROMIUM

Fig. 29-1 Stainless steel alloy index

JOB 29: FILLET WELD, FLAT POSITION, STAINLESS STEEL

Equipment and Materials:

Standard MIG welding equipment
Personal hand tools
Protective clothing and gear
Stainless steel wire brush
2 pieces, 3/16" x 5" x 6" stainless steel plate
2 pieces, 3/16" x 2 1/2" x 6" stainless steel plate
308L wire electrode
Argon/oxygen 1 percent shielding gas

PROCEDURE	KEY POINTS

1. Turn on the machine.

2. Adjust the machine for short arc.

3. Position the plates for flat welding and tack weld both samples, figure 29-2.

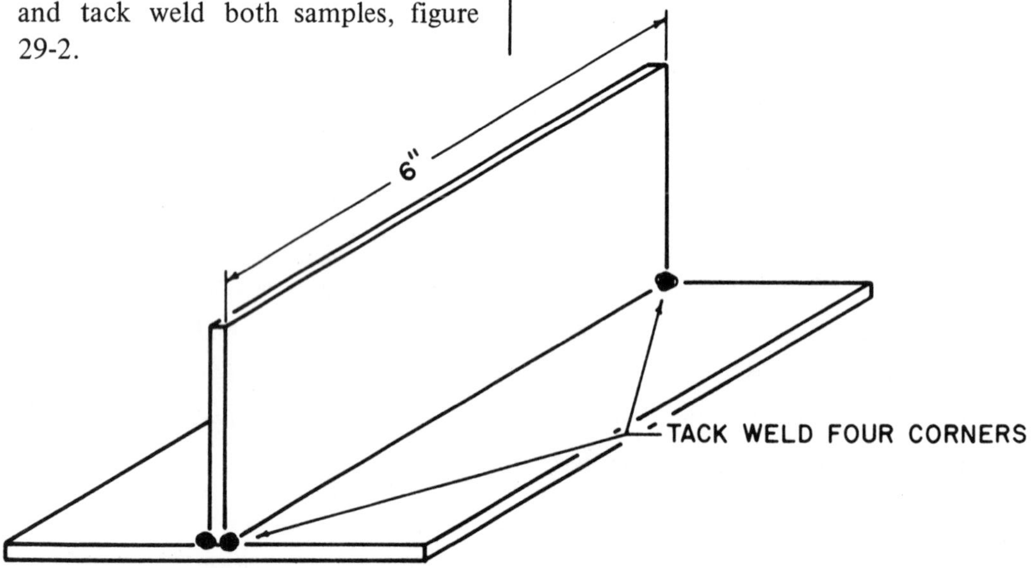

6"

TACK WELD FOUR CORNERS

Fig. 29-2 The proper position of the plates and tack welds for a stainless steel fillet weld

PROCEDURE	KEY POINTS
4. Weld one side of the first sample, following the bead sequence shown, using the forehand motion, figure 29-3.	4. Brush metal before and during welding operations with a stainless steel wire brush.

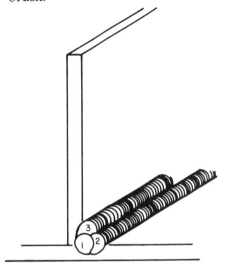

Fig. 29-3 The proper bead sequence to complete the joint

PROCEDURE	KEY POINTS
5. Weld the other side of the sample using the same procedure.	5. Cool the sample often.
6. Adjust the machine for spray arc.	
7. Weld one side of the second sample, following the same bead sequence and using the forehand motion.	
8. Weld the other side of the sample.	
9. Shut down all systems.	
10. Submit both samples to the instructor for comments and instructions.	

REVIEW QUESTIONS

Answer the following questions briefly.

1. What are the two basic materials present in stainless steel?

2. How much chromium must be added to iron to make a true stainless steel?

3. Name four reasons that nickel may be added to stainless steel.

4. What is the meaning of ELC in the welding field?

Unit 30 Butt Weld Flat Position, Stainless Steel

OBJECTIVES

After completing this unit the student will be able to:

- identify weld defects found in stainless steel butt welds.

- name some methods of avoiding inclusions, undercutting, and weld cracking in stainless steel MIG welding.

- successfully MIG weld a stainless steel butt weld in the flat position using the short-arc method of metal transfer.

Butt Welds

When two pieces of metal lying in the same plane are joined together with a weld it is called a butt weld. Stainless steel under 3/16" in thickness, should have the edges ground at an angle and the two angles placed together to form a V. On metal over 1/4" in thickness it is generally advisable to leave a small gap (called a root gap) between the plates so that 100 percent penetration is assured in the butt weld. Since the bottom of the ground angle (bevel) is very thin, it may burn away and too much penetration may take place. It is therefore necessary to square the bottom of the bevel by grinding (called a land) before welding, figure 30-1. Butt weld plates should be securely tack welded together at the ends to keep the gap between the pieces from spreading or growing smaller.

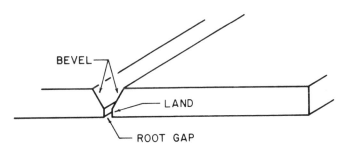

Fig. 30-1 Proper fit up of a single V butt joint

Defects in Stainless Steel Butt Welds

The three most common defects in this type of weld are: inclusion, undercutting, and cracking.

Inclusions are particles which are embedded in the molten metal during welding. The inclusion will weaken the weld because of a tendency for cracks to start at that point. The force of the arc drives the particles into the puddle where they are trapped as the puddle cools. Correct cleaning before welding will eliminate this defect. Stainless steel wire brushes should be used to clean the metal as a plain steel wire brush could leave particles of carbon or iron on the surface of the metal to be welded.

Undercutting of stainless steel joints is the result of any of the following:

- Incorrect travel speed.
- Arc pause.
- Wrong arc voltage settings.
- Wrong wire feed settings.
- Wrong wire electrodes.
- Wrong gas shielding.

To correct these faults, the welder can decrease the travel and wire feed speeds, adjust the voltage setting, and select the right filler wire and gas shield needed for the material being welded. Manufacturer's charts should be checked for correct filler wire and gas shield.

Cracks are defects in the form of tears or splits in metal. They are the result of shrinkage or strains inside the weld, built up by heat input into the plates. The defect can be controlled by using holding fixtures or copper backup bars (called chill bars). The copper backup bar takes the heat away from the molten puddle.

Another major reason for cracks is the incorrect selection of filler wire. Use manufacturers' charts to choose the right wire.

JOB 30: BUTT WELD, FLAT POSITION, STAINLESS STEEL

Equipment and Materials:

Standard MIG welding equipment
Personal hand tools
Protective clothing and gear
Stainless steel wire brush
2 pieces, 3/16" x 4" x 6" 308 stainless steel plate
1/8" x 3" x 7" copper backup bar
.045", 308 wire electrode
Argon/oxygen 1 percent shielding gas

PROCEDURE	KEY POINTS
1. Turn on the machine.	
2. Adjust the machine for short-arc welding.	2. Use scrap stainless steel to adjust the welding.
3. Tack weld the samples, figure 30-2.	3. Put the tacks on the edge of the plates so the copper backup bar will fit tightly on the back of the joint.

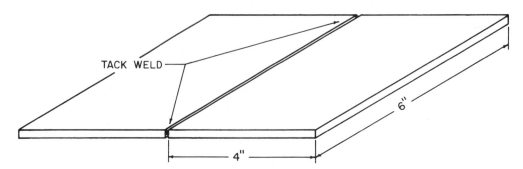

TACK WELD

6"

4"

Fig. 30-2 Proper position of the plates and the tack welds

PROCEDURE	KEY POINTS
4. Position the copper bar, figure 30-3. Clamp the bar in place.	4. Brush the back of the stainless steel before the copper bar is put on.

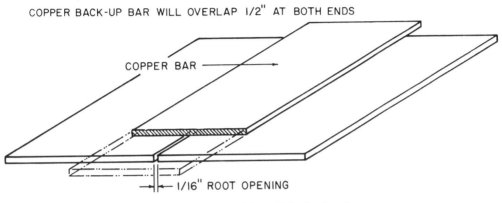

COPPER BACK-UP BAR WILL OVERLAP 1/2" AT BOTH ENDS

COPPER BAR

1/16" ROOT OPENING

Fig. 30-3 Proper position of the backup bar

5. Turn the tacked sample over and lay it on the work table.	5. Brush the front side of the plate surface.
6. Weld the sample, following the bead sequence shown, figure 30-4.	6. One pass should be sufficient.

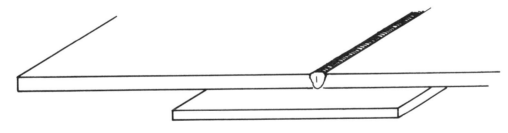

Fig. 30-4 Running the bead on the joint

7. Shut down all systems.	
8. Inspect the completed weld.	
9. Unclamp and remove the copper backup bar.	9. The copper will not fuse with the stainless steel.
10. Inspect the root penetration.	
11. If more practice is needed, repeat all procedures.	
12. Submit the weld to the instructor for comments and instructions.	

REVIEW QUESTIONS

Answer the following questions briefly.

1. What is a weld inclusion?

2. How may inclusions be eliminated?

3. Name four reasons which may cause undercutting to occur.

4. How may undercutting be corrected?

5. What causes a weld crack?

6. What is the reason for using a copper backup bar on a butt weld?

Unit 31 Four Layer Stringer Bead Pad Buildup, Horizontal Position, Stainless Steel

OBJECTIVES

After completing this unit the student will be able to:

- discuss the different phases of stainless steel cleaning operations.
- explain the negative results that can occur if cleaning operations are performed incorrectly.
- successfully weld a four layer pad on stainless steel in the horizontal position using standard MIG welding procedures.

Horizontal Stainless Steel Welding

Welding of stainless steel in the horizontal position is very similar to that of mild steel in the horizontal position. Attention must be paid to the gun angle, bead motion, and inductance.

Stainless Steel Welding Preparation

Stainless steels are high alloy steels and need special cleaning operations to make high-quality welds.

The cleaning phases for stainless steel are:

- preweld cleaning,
- interpass cleaning,
- finish weld cleaning.

Preweld Cleaning. Stainless steel is like aluminum in that the outside oxide formed on the metal melts at a higher temperature than the metal. It is therefore necessary to remove the oxide before welding begins. Preweld cleaning can be done by machinery or by hand cleaning. In both cases, the brush must be of stainless steel. The use of regular wire brushes can result in particles of carbon or steel being left on the stainless steel. These particles can be trapped in the molten welding puddle and result in inclusions in the weld. Cleaning must be done carefully and completely to make sure that no oil or grease which may be on the metal is left on the edges and surface.

Interpass Cleaning. Multipass welding (more than one pass) is generally necessary on stainless steel over 1/4″ thick. The number of passes required depends on the thickness of the material and on the diameter of the filler wire used.

Clean each bead with a stainless steel wire brush. Cleaning the passes helps reduce weld defects and the weld zone has a chance to cool while the cleaning is taking place, which prevents overheating of the weld zone.

Finish Weld Cleaning. In arc welding, the finish weld cleaning is more costly than the actual welding time. The MIG process reduces the finish weld cleaning to almost zero. If the filler wire has been matched with the base metal, a final brushing of the completed weld with a stainless steel wire brush will result in a finished weld. There is no flux or slag to clean from the weld.

JOB 31: FOUR LAYER STRINGER BEAD PAD BUILDUP, HORIZONTAL POSITION, STAINLESS STEEL

Equipment and Materials:

Standard MIG welding equipment
Personal hand tools
Protective clothing and gear
Stainless steel wire brush
1 piece, 3/16" x 5" x 6" 308 stainless steel plate
.045 308L stainless steel wire
Argon/oxygen 1 percent shielding gas

PROCEDURE	KEY POINTS
1. Turn on the machine.	
2. Adjust the machine for short arc on scrap stainless steel plate.	
3. Place the plate in the vertical plane.	
4. Run the first stringer bead across the bottom of the plate, figure 31-1.	4. Spatter buildup in the nozzle of the MIG gun may be heavy because of the welding position. The nozzle must be kept clean so that the gas coverage is steady. Use spatter compound.

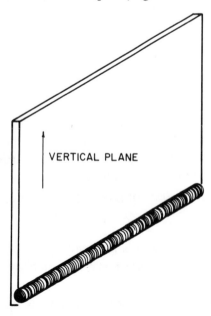

VERTICAL PLANE

Fig. 31-1 Run the first bead across the bottom of the plate.

PROCEDURE

KEY POINTS

5. Check each completed bead to correct any welding faults.

6. Bead #2 should cover the top 1/3 of bead #1. Follow the bead sequence as shown in figure 31-2.

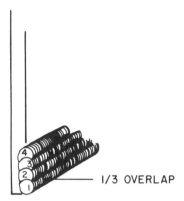

1/3 OVERLAP

Fig. 31-2 Proper bead sequence to complete one layer

7. Rotate the plate so that the next series of beads runs across the first beads and fills the plate with welds.

8. Repeat the rotation of the plate until four layers of weld have been built up, figure 31-3.

7. Different arc pauses, rates of travel and motions should be tried by the welder to establish a personal method.

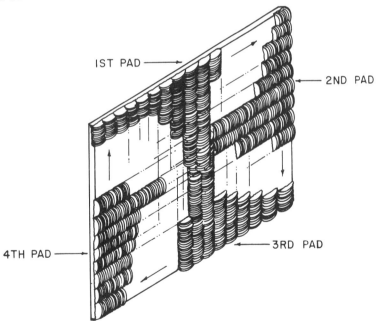

IST PAD

2ND PAD

3RD PAD

4TH PAD

Fig. 31-3 After a layer is complete, rotate the plate 90° until four complete layers have been made.

9. Shut down all systems.

10. Submit the pad to the instructor for comments and instructions.

REVIEW QUESTIONS

Answer the following questions briefly.

1. What are the three cleaning phases listed for stainless steel?

2. Why is it necessary to remove the oxide from stainless steel before welding?

3. Why is a stainless steel wire brush recommended for cleaning stainless steel?

4. Give two reasons for cleaning each bead of a multipass weld before proceeding with the next pass.

5. Why is there little finish-weld cleaning required on stainless steel?

Unit 32 Butt Weld, Horizontal Position, Stainless Steel

OBJECTIVES

After completing this unit the student will be able to:

- list the procedures used in industry to weld stainless steel.
- successfully MIG weld a stainless steel butt joint in the horizontal position, using the short-arc method of metal transfer.

Stainless Steel Welding Procedures from Industry

- Remove all moisture from the joint by blowing with dry air or heating with a torch.
- Remove all dirt from the joint by brushing with a stainless steel wire brush.
- Do not use stainless steel brushes that have been used on other materials.
- For best corrosion resistance, weld deposits should be run at a minimum practical amperage and in stringer bead patterns, rather than weave beads.
- Avoid high amperages and long arcs to prevent spatter, cracks, undercuts, and poor bead appearance.
- Always backweld end craters to avoid crater cracks.
- Brush each completed bead before welding the next pass.
- Keep penetration at a minimum practical level. Excessive penetration leads to cracked welds, loss of corrosion resistance and honeycombing at the backside of the joint.
- Use jigs and fixtures to hold down distortion.
- Use backup bars for materials of 14 gauge and thinner.
- Use a planned bead sequence to control stresses.

JOB 32: BUTT WELD, HORIZONTAL POSITION, STAINLESS STEEL

Equipment and Materials:

Standard MIG welding equipment
Personal hand tools
Protective clothing and gear
2 pieces, 3/16" x 4" x 6" 308 stainless steel plate
1/8" x 3" x 7" copper backup bar
.045" wire
Argon/oxygen 1 percent shielding gas

PROCEDURE	KEY POINTS
1. Turn on the machine.	
2. Adjust the MIG machine for short arc.	2. Use scrap stainless steel to adjust the machine.
3. Tack weld the samples as shown in figure 32-1.	3. Put the tack welds on the edges of the plates so the copper backup bar will fit tightly on the back of the joint. The root opening is 1/16″.

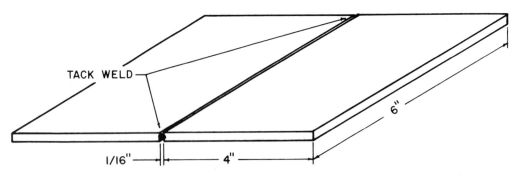

Fig. 32-1 Proper position of the tack welds for butt joint

4. Position the copper bar and clamp it in place, figure 32-2.	4. Brush the back of the plates before the bar is placed.

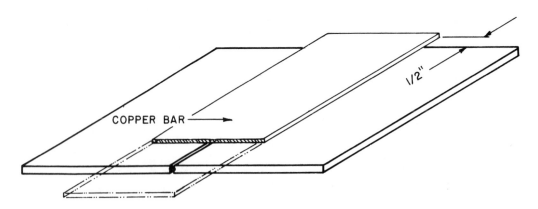

Fig. 32-2 Proper position for copper backup bar

5. Turn the plate over and position it for horizontal welding.	5. Brush the front side of the plate surface.
6. Weld the sample, following the bead sequence shown, figure 32-3.	

PROCEDURE   **KEY POINTS**

7. Shut down all systems.

8. Inspect the completed weld.

9. Unclamp and remove the copper backup bar.

10. Inspect the root penetration.

11. Examine the weld. If any defects appear, repeat steps 1-10.

12. Submit a satisfactory weld to the instructor for comments and instructions.

Fig. 32-3   Proper bead sequence to complete butt weld

9. The copper backup bar will not fuse with the stainless steel.

REVIEW QUESTIONS

Answer the following question briefly.

1. List ten different industrial stainless steel welding procedures.

Unit 33 Four Layer Stringer Bead Pad Buildup, Vertical Position, Stainless Steel

OBJECTIVES

After completing this unit the student will be able to:

- review the advantages of the MIG welding process as they relate to the welding of stainless steel.

- understand the defects which appear during MIG welding of stainless steel and explain the methods of correcting such defects.

- successfully complete the MIG welding of a four layer stringer bead pad buildup in the vertical position, using the short-arc method of metal transfer.

Advantages of MIG Welding of Stainless Steel

Although the stick-arc welding of stainless steel is still very much in demand, MIG welding of stainless steel is on the increase. Thin stainless materials are welded successfully with the TIG (Tungsten Inert Gas) process and the thicker sections are completed with the MIG process.

MIG welding of stainless steel offers the following advantages:

- Produces sound, high-quality welds at high production speeds.

- The absence of fluxes reduces postweld cleaning operations and the weld zone is more exposed for viewing by the welder.

- A high percentage of consumable wire electrode is deposited in the joint.

- Rod stub losses are eliminated.

- Weld defects caused by starts and stops in the joint are eliminated.

- MIG welding saves on labor costs.

- MIG welding offers easily trained operators.

- Excellent finish matchup is provided with MIG welding.

Inspection of Completed Welds

After a weld is done, look it over carefully for defects. If the weld is uniform in height and width, and no surface holes or cracks show, it is generally a good weld.

The defects listed in the chart, figure 33-1, are most often observed in stainless steel welding.

DEFECT	REASON	CORRECTION
Weld Cracking	a. Improper joint preparation.	Practice correct joint preparation procedures.
	b. Incorrect filler material.	Use manufacturers' charts for selection.
	c. Improper weld procedures.	Use correct welding technique.
	d. Excessive welding amperage.	Correct the machine output.
	e. Improper machine settings.	Review machine settings.
Warping & Distortion	a. Excessive heat at joint.	Use correct amperage and chill bars where necessary.
	b. Improper joint preparation.	Review joint preparation.
	c. Improper bead sequence.	Review bead sequences.
	d. Improper holding jigs or fixtures.	Clamp the metal correctly.
Poor Weld Appearance	a. Wrong filler wire.	Use manufacturers' charts.
	b. Improper joint preparation.	Follow correct preparation procedures.
	c. Improper machine settings.	Adjust machine.
	d. Improper welding procedures.	Correct bead sequence.
Undercut	a. Excessive amperage.	Adjust machine.
	b. Improper gun manipulation.	Review procedures.
	c. Wrong gun angle.	Review gun angles.
Porous Welds	a. Welding speed too fast.	Slow travel rate.
	b. Wrong gun angle.	Correct angle for proper gas coverage.
	c. Welding operations in drafty conditions.	Use wind shields to protect welding zone.
	d. Wrong inert gas.	Use recommended gas for shield.
	e. Dirty base metal.	Use proper preweld and interpass cleaning.

Fig. 33-1

JOB 33: FOUR LAYER STRINGER BEAD PAD BUILDUP, VERTICAL POSITION, STAINLESS STEEL

Equipment and Materials:

Standard MIG welding equipment
Personal hand tools
Protective clothing
Stainless steel wire brush

1 piece, 1/4″ x 5″ x 6″ stainless steel plate
.045″, 308L wire
Argon/oxygen 1 percent shielding gas

PROCEDURE	KEY POINTS
1. Turn on the machine.	
2. Adjust the machine for short arc.	2. Adjust for vertical welding on scrap steel.
3. Position the plate for vertical welding.	

PROCEDURE	KEY POINTS
4. Run the first bead up the middle of the plate.	4. The vertical beads should be slightly wider than in other positions.
5. Follow the outlined bead sequence, figure 33-2.	5. Keep the nozzle clean and add spatter compound often.

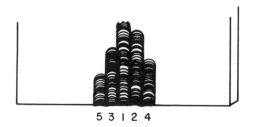

5 3 1 2 4

Fig. 33-2 Proper bead sequence for vertical-up pad buildup

6. Check each completed bead to correct any defects.	6. Wire brush and inspect each bead.
7. Bead #2 should cover to the crown of #1, bead #3 to the crown of #1, and so on.	7. Experiment with different gun manipulations until welds are acceptable.
8. When first layer is complete, rotate the plate 90 degrees so the next layer will cover the first layer, figure 33-3.	8. Four layers are required.

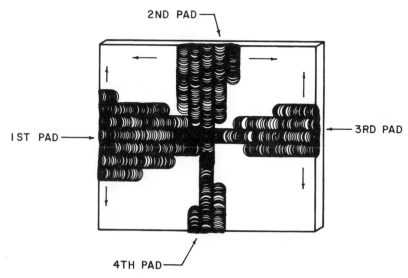

2ND PAD

1ST PAD

3RD PAD

4TH PAD

Fig. 33-3 Rotate the plate 90° after each complete layer. Four layers are required.

9. Shut down all systems.

10. Submit the pad to the instructor for comments and instructions.

REVIEW QUESTIONS

Answer the following questions briefly.

1. Name 8 advantages of the MIG process over other welding processes.

2. Name 4 reasons for weld cracking, and list proper procedures for correcting the problems.

3. Name 3 reasons for warping and distortion and list proper procedures for correcting the problems.

4. What causes poor weld appearances?

Unit 34 Fillet Weld, Vertical Position, Stainless Steel

OBJECTIVES

After completing this unit the student will be able to:

- discuss the industrial uses of the fillet design and the advantages and disadvantages created by using the joint design.
- successfully MIG weld a T weld on stainless steel, in the vertical position, using the short-arc method of metal transfer.

Fillet Welds and T Welds

Penetration on fillet joints must be maximum, whether the joint is welded from one side or both sides. There is a tendency to overweld this kind of joint and when stainless steel base metal is used, the damage in the form of warpage and distortion can be great. Stainless steel has a lower rate of heat spread than carbon steels and, therefore, the heat remains in the weld zone longer. This fact, along with the high heat expansion rate, increases shrinkage, stresses, and warpage. Correct heat input and weld deposits along with the use of holding fixtures will overcome these difficulties.

Vertical T Welds

T welds on stainless steel plate, made in the vertical position, are difficult and much practice is needed to master them. The MIG gun has to be directed into the corner so that the weld is laid equally on each side of the joint. The travel speed up the joint has to be at a slow, even rate of speed so that the penetration can soak into the corner where the two plates meet.

Rapid welding of the joint will not give maximum penetration.

Protective clothing must be worn while practicing this weld, since molten spatter from the weld zone may cause burns.

All T joints require the addition of filler metal to provide the buildup for strength in the joint. The number of passes on each side of the joint (if both sides are welded) depends on the thickness of the metal and the size of the weld needed for strength. When 100 percent penetration is called for, the welder should make sure that the current setting of the MIG machine is high enough for the thickness of the material.

JOB 34: FILLET WELD, VERTICAL POSITION, STAINLESS STEEL

Equipment and Materials:

Standard MIG welding equipment	2 pieces, 1/4" x 2 1/2" x 6" stainless steel plate
Personal hand tools	2 pieces, 1/4" x 5" x 6" stainless steel plate
Protective clothing and gear	308L stainless steel wire
Stainless steel wire brush	Argon/oxygen 1 percent shielding gas

PROCEDURE

KEY POINTS

1. Turn on the machine.

2. Adjust the machine for short arc.

2. Adjust the machine on scrap steel.

3. Position the plate in the flat position and tack weld both samples, figure 34-1.

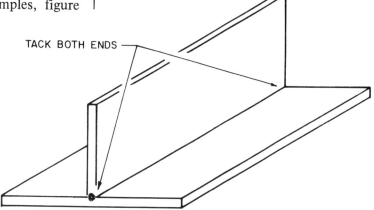

Fig. 34-1 Proper position of the plates and the tack welds for a T joint

4. Position for vertical welding.

5. Weld one side of the first sample following the bead sequence as shown in figure 34-2. Use the forehand method with stringer bead.

5. Brush the metal before welding with a stainless steel wire brush.

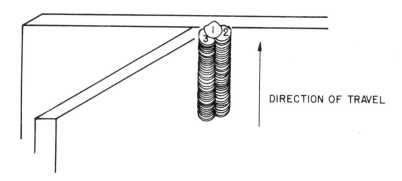

Fig. 34-2 Proper bead sequence and direction of travel for vertical-up welding

6. Use single bead passes on each side of the joint.

7. Inspect the welded beads.

8. Repeat all procedures on the other sample.

9. Shut down all systems.

10. Submit the samples to the instructor for comments and instructions.

REVIEW QUESTIONS

Answer the following questions briefly.

1. Why is the danger of warpage in a stainless steel joint greater than in carbon steel joints?

2. In what direction is the MIG gun directed for a successful weld on a T joint?

3. Should the T weld in the vertical position be made rapidly?

SECTION 7: REVIEW A

OBJECTIVE:

• Produce a stainless steel T weld of proper penetration and quality in the flat position using standard MIG welding equipment and procedures. Subject such a sample to a destructive test.

Equipment and Materials:

Standard MIG welding equipment
Personal hand tools
Protective clothing and gear
Stainless steel wire brush
1 piece, 1/4" x 5" x 6" stainless steel plate
1 piece, 1/4" x 2 1/2" x 6" stainless steel plate
308L stainless steel wire electrode
Argon/oxygen 1 percent shielding gas

PROCEDURE

1. Turn on and adjust the equipment for spray-arc transfer.

2. Position the plates and tack weld as shown in figure R7-1.

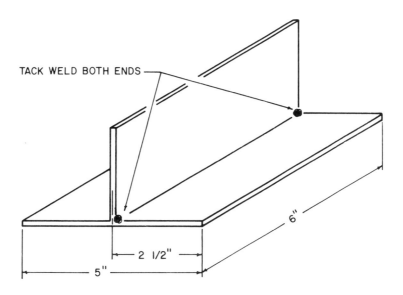

TACK WELD BOTH ENDS

2 1/2"

5"

6"

Fig. R7-1 Proper position of the plates and tack welds

3. Arrange the plates so the joint is in the flat position of welding.

4. Brush the joint.

5. Run the beads following the bead sequence shown in figure R7-2.

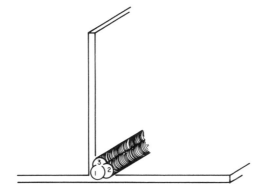

Fig. R7-2 Proper bead sequence for stainless steel fillet weld

6. Shut down all systems.

7. Perform a destructive test as shown in figure R7-3.

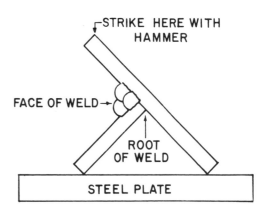

STRIKE HERE WITH HAMMER

FACE OF WELD→

ROOT OF WELD

STEEL PLATE

Fig. R7-3 Destructive test of a fillet weld

8. Examine the weld for defects such as incomplete penetration, porosity, and cracks.

9. Submit a satisfactory sample to instructor for comments and instructions.

SECTION 7: REVIEW B

OBJECTIVE:

- Produce a stainless steel single V butt weld of proper penetration and quality in the flat position using standard MIG welding equipment and procedures. Subject such a sample to the guided-bend test.

Equipment and Materials:

Standard MIG welding equipment
Personal hand tools
Protective clothing and gear
Stainless steel wire brush
Power cutting equipment or hacksaw

2 pieces, 3/16" x 4" x 6" stainless steel plate
1 piece, 1/8" x 3" x 7" copper backup bar
.045" 308 wire electrode
Argon/oxygen 1 percent shielding gas

PROCEDURE

1. Turn on and adjust equipment for short-arc welding.

2. Bevel the plates and grind a land on each bevel.

3. Tack weld the plates as shown in figure R7-4.

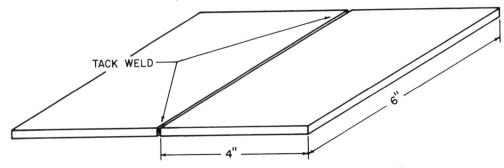

TACK WELD

6"

4"

Fig. R7-4 Proper position of the plates and the tack welds

4. Position and clamp the backup bar in place, figure R7-5.

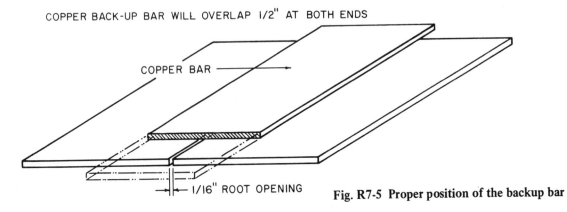

COPPER BACK-UP BAR WILL OVERLAP 1/2" AT BOTH ENDS

COPPER BAR

1/16" ROOT OPENING

Fig. R7-5 Proper position of the backup bar

5. Position the plates in the flat position.

6. Run the beads as shown in figure R7-6.

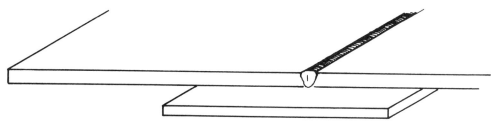

Fig. R7-6 Running the bead on the joint

7. Lay out and cut the sample into sections as shown in figure R7-7.

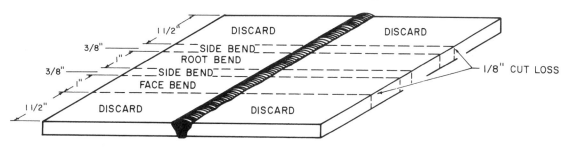

Fig. R7-7 The sample should be cut into sections.

8. Perform a guided-bend test on the face, the root, and the sides of the various strips.

9. Examine the bent strips. There should be no separation of base metal from the weld and no porosity.

10. If the welded samples show defects, repeat steps 1-9.

11. When the bent specimens are satisfactory, submit them to the instructor for comments and further instructions.

SECTION 8:
Fundamentals of TIG Welding

Unit 35 Setting Up TIG Welding Equipment

OBJECTIVES

After completing this unit the student will be able to:

- briefly describe the TIG welding process.
- explain the advantages of the TIG process over other methods of welding.
- name and explain the use of the equipment used for TIG welding.
- describe the different electric currents used for the process.
- successfully set up TIG welding equipment in preparation for welding.

The TIG Process

TIG (tungsten inert-gas) welding is a completely manual welding process, developed in the 1940 s for fusion of materials that are difficult to weld. According to the American Welding Society, the proper name for this process is *gas tungsten-arc welding* (GTAW), although TIG welding is the term most commonly used.

It is similar to MIG welding in that an electric arc is maintained between the work and the electrode in an inert-gas atmosphere. As in MIG welding, the elimination of the earth's atmosphere from the weld zone prevents the metal from oxidizing.

The major difference between TIG and MIG welding centers on the electrode. The TIG welding electrode is *nonconsumable,* requiring filler material to be added through the use of a hand held rod. The electrode is mounted in the TIG torch which directs the flow of the inert-gas to the weld zone, figure 35-1.

Advantages of the TIG Process

Tungsten inert-gas (TIG) welding produces sound, high-quality welds without the use of fluxes. Since cleaning agents and fluxes are not used, postweld cleaning is not always needed.

The TIG process can be closely controlled, so it is in great demand by industry for the welding of thin gauge metal.

The welding takes place without sparks, fumes, or spatter and it can be done on nearly all commercial metals, including the hard-to-weld materials; aluminum, stainless steel, magnesium, nickel, copper, copper nickel, brass, silver, phosphor bronze, and silicon copper. It is also very effective on carbon steels, low-alloy steels, and cast iron.

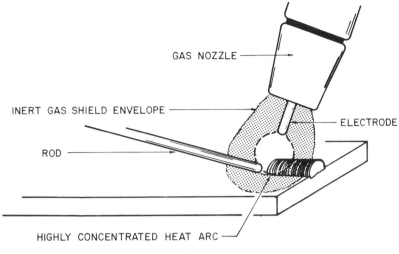

GAS NOZZLE

INERT GAS SHIELD ENVELOPE

ELECTRODE

ROD

HIGHLY CONCENTRATED HEAT ARC

Fig. 35-1

TIG Welding Equipment

The TIG welding equipment is highly specialized and expensive. Setup procedures should be followed exactly to avoid damage to the parts. The welder must set up and operate the machine and attachments in accordance with rules and regulations established by the welding industry.

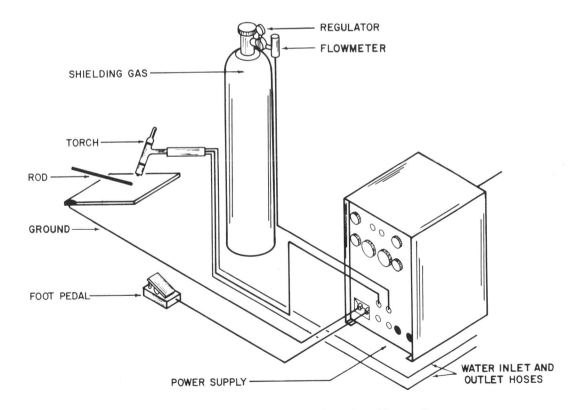

REGULATOR

FLOWMETER

SHIELDING GAS

TORCH

ROD

GROUND

FOOT PEDAL

POWER SUPPLY

WATER INLET AND OUTLET HOSES

Fig. 35-2 The basic components of a TIG welding outfit

TIG welding equipment consists of four primary elements, figure 35-2, p. 181.

- The power supply with a high-frequency unit
- The welding torch with attachments
- Shielding gas attachments
- A water-cooling system for water-cooled torches

The Power Source. TIG Welding operations are made with constant current AC (Alternating Current) DC (Direct Current) machines, equipped with a superimposed, high-frequency current, figure 35-3. The high-frequency current jumps the air gap between the tungsten electrode and the workpiece. This prevents the contamination of the electrode which would occur if the tungsten were touched to the metal to start the arc. Contamination (base metal or filler wire) on the tungsten causes the arc to waver or jump to the side, so the high frequency is used to complete the arc without touching the metal of the workpiece.

The high-frequency control switch consists of three positions:

- Off
- Start Only (used for stainless steel welding)
- Continuous (used for aluminum and its alloys).

The power source for TIG welding may be either alternating current or direct current depending on the electrical polarity recommended for the material to be welded.

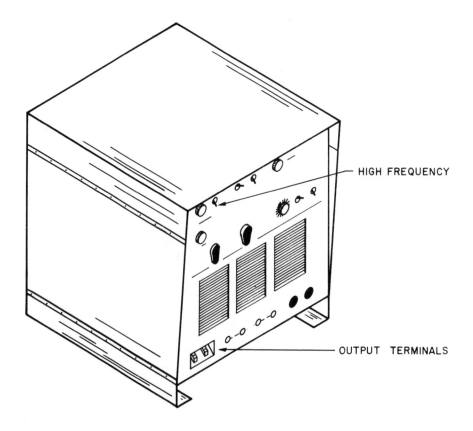

HIGH FREQUENCY

OUTPUT TERMINALS

Fig. 35-3

Direct Current. Direct current welding operations may be either DC+ (reverse polarity) or DC– (straight polarity). DC– is recommended over the DC+ because 2/3 of the heat created by the arc of the reverse polarity current is contained in the tungsten electrode. For any given welding amperage used on a job, a larger tungsten must be used to contain the heat, figure 35-4.

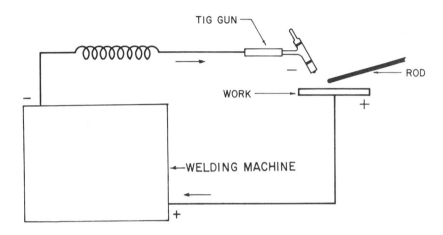

Fig. 35-4 Direct current, straight polarity

In straight polarity (DC–) welding 2/3 of the heat is in the workpiece, which makes cooler operating heat on the electrode and a deeper penetration into the material. The heat input creates better welding action, while the larger electrodes which must be used for DC+ polarity produce a wide, shallow penetration bead contour, figure 35-5.

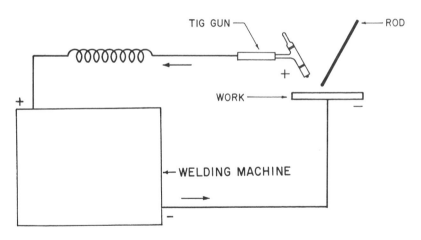

Fig. 35-5

Alternating Current. Alternating current (AC) is a combination of both DC– and DC+ with each direct current polarity sharing equal time of the current cycle, figure 35-6, page 184.

The conditions on the surface of the base metal (water, mill scale, and oxides) will retard the DC+ cycle of the AC polarity. The superimposed high frequency is used to correct the current.

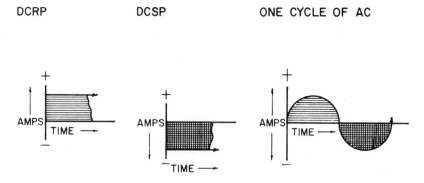

Fig. 35-6

The high-frequency current adds the following advantages to AC current polarity:

- The arc jumps the air gap between the electrode and the workpiece.
- Arc stability is increased.
- Longer arc lengths are possible.
- The life of the tungsten electrode is increased.
- Increased current ranges are possible for each diameter electrode.
- There is a decrease of the input amperage.

Figure 35-7 shows the penetration and shape of the beads made with the use of DC+, DC–, and AC high frequency (ACHF).

WELD BEAD COMPARISON

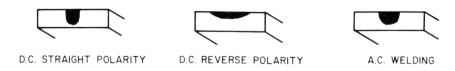

D.C. STRAIGHT POLARITY   D.C. REVERSE POLARITY   A.C. WELDING

Fig. 35-7

Power-Source Panel. The power source panel generally contains the following switches and controls:

- Current setting
- Remote or panel switch
- Soft-start switch
- Power switch
- Process-selection switch
- High-frequency control, phase shift, and intensity
- Current range switch
- Postpurge timer
- Polarity switch

Current Setting.   The current setting is used to adjust the amperage output to the TIG torch.   It is adjusted along with the current-range selection switch which has settings for special-low amperage, low amperage, medium amperage, and high amperage.

Remote or Panel Switch.   When in the panel position, this switch controls the amperage output from the panel of the power source.   When the switch is in the remote position the amperage output is controlled with the foot control equipment, figure 35-8.   The remote foot control allows the welder to adjust the amperage as the job conditions change.

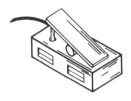

Fig. 35-8

Soft-Start Switch.   This switch will allow the arc to be made in a soft manner (reduced current) rather than the harsh, digging, full current manner that will occur when the switch is not used.

Power Switch.   The power switch turns the power source on and off.   When the welder leaves the machine, it should be turned off.

Process-Selection Switch.   This switch determines the type of welding that will be done.   The switch has two settings—tungsten inert-gas welding (TIG) or stick-arc welding (SMAW).   This will only hold true when the power source is a combination stick-arc and TIG welding machine.

High-Frequency Control Switch, Phase Shift, and Intensity Control.   The high-frequency control switch has three positions: off, start only, and continuous.   The switch is off when stick-arc welding.   It is placed on "start only" when TIG welding carbon steel and stainless steel, and on "continuous" when TIG welding aluminum and magnesium alloys.   The phase shift and intensity controls are used to start TIG welding operations, to control the amount of high-frequency current that is needed at the TIG torch, for the type of base metals to be welded.

Current-Range Switch.   The current range switch may consist of four settings:
- Special Low (2 to 16 amperes)
- Low (5 to 30 amperes)
- Medium (25 to 130 amperes)
- High (80 to 250 amperes)

These ranges will change according to manufacturers' specifications.

Postpurge Timer.   This control allows the inert-gas flow to continue for several seconds after the arc has been stopped.   The inert-gas coming out after welding cools the hot tungsten electrode and protects the molten puddle from being contaminated by the air around the weld.   The timer can be adjusted for different postflow lengths of time.

Polarity Switch.   Combination shielded stick-arc and TIG machines will have three polarity switch selections—AC, DC+ and DC–.   The polarity switch must be shifted to the polarity needed for the weld to be made.

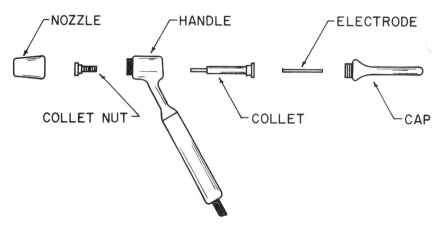

Fig. 35-9

TIG Welding Torch. The TIG welding torch consists of six primary parts, figure 35-9.

- Torch cap
- Tungsten electrode
- Electrode holder (collet)

- Torch body
- Electrode-holder body (collet body)
- Inert-gas cup

The TIG torch is delicate and easily broken. The torch transfers the amperage and inert-gas to the weld zone. If the torch is of a water-cooled design, it contains water inlets and outlets.

Air-cooled torches will weld steadily up to approximately 100 amperes. When welding currents are higher than 100 amperes, water must be run through the torch to remove the heat. A power cable is used to take the current of electricity to the torch, from the power source. The inert gas is carried to the torch in a separate hose, inside the power cable to the weld, from the inert-gas supply. Water-transfer hoses for water-cooled TIG torches have hoses attached to the torch body. All the hoses are made of plastic and care must be taken to avoid burning holes in them. All power cables and transfer hoses should have leather shields around them.

The welding current is transferred to the weld zone from the power cable to the collet and then on to the tungsten electrode. The high frequency then jumps the air gap and the welding arc is established.

The shielding gas is transferred to the weld zone from the gas supply to the passages in the torch body, through the gas cap to the weld zone. Some TIG torches have a control valve which must be opened before the gas flow starts—other torches have automatic inert-gas switches which start the gas flow when the remote control is pressed.

Shielding Gas for TIG Welding. There are two major inert-gases used with the TIG process: argon and helium.

Argon. Argon is one of the many gases that are combined to make up the air around us. It is an inert-gas, will not have any effect on the weld, and is the most popular inert-gas used for the TIG process. It is heavier than helium and, therefore, will be cheaper to use since it stays over the weld longer.

Argon is put into high-pressure cylinders similar to those which hold oxygen (but of a different color) and all the safety requirements for handling high-pressure cylinders must be followed.

Helium. Helium is the gas originally used in inert-gas welding. TIG welding is sometimes called "heliarc welding," which resulted from the use of helium as the shielding gas. Since helium is not popular for shielding any longer, the words Tungsten Inert Gas (or TIG) have replaced the name heliarc.

Helium is found in large quantities over oil and gas field deposits. Because it is about 9 times lighter than argon, it requires more amperage for welding and more of the gas is used on a weld job.

Helium is also bottled in high-pressure cylinders of a different color than oxygen or argon and all safety requirements for handling high-pressure cylinders must be followed.

Shielding-Gas Attachments. Shielding gas must be stored in a safe manner, and the high pressure in the cylinder must be brought down to a working pressure. The gas must be released to the torch in a pure, dry condition.

Shielding-gas equipment consists of:

- Cylinder of inert gas
- Flowmeter (cylinder pressure regulator)
- Transfer hoses (to move the gas from the flowmeter to the power source, through the torch to the cup, and into the weld zone).

As the gas shield leaves the torch cup, its flow pressure and density push atmospheric elements out and away from the weld zone, keeping the molten puddle clean.

Water-Cooling System. When welding amperage is above 100 amperes, the TIG power cable and TIG torch will overheat and water must flow through the system to prevent heat damage. The water-cooling system uses a water-cooled torch, inlet hose, outlet hose, water-storage tank, and a pump to push the water through the system, figure 35-10.

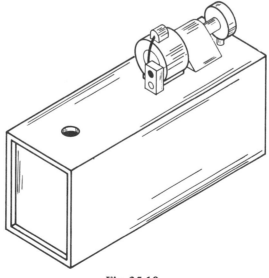

Fig. 35-10

Ground-Clamp Leads

Large diameter flexible leads (cables) are used to transfer the electric currents. The cables are covered with rubber insulation and a woven reinforcement. They are subjected to much wear and should be checked often for flaws which might be in the insulation, figure 35-11.

Fig. 35-11

Ground Clamps

The ground cable (lead) from the machine to the work is connected to a spring-loaded clamp to attach to the work. Clamps should be heavy enough to carry the electric current required for the welding job, figure 35-12.

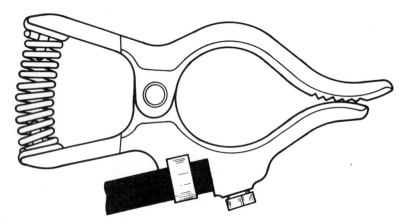

Fig. 35-12

JOB 35: SETTING UP TIG WELDING EQUIPMENT

Equipment and Materials:

> TIG welding machine, AC/DC constant current
> TIG Welding torch complete with cup
> 3/32″ − 2 percent tungsten electrode
> Collet
> Collet body
> Cap
> Power Cable
> Inert-gas hose and water inlet and outlet hoses
> Flowmeter
> Inert-gas hose and fittings
> Argon cylinder
> Personal hand tools

PROCEDURE	KEY POINTS
1. Attach the ground clamp to the ground lead.	1. Use the adjustable wrench from the personal hand tools as needed for all tightening.
2. Attach the ground cable to the terminal on the welding maching.	
3. Attach the ground clamp to the workpiece or the bench.	
4. Connect the power cable, inert-gas hose and water inlet and outlet to the welding machine and torch body.	4. Use the correct size wrench on all hose fittings. Do not use pliers as this will round off the fitting nuts.
5. Install the leather shield around the cables and hoses from the welding machine to the TIG torch.	5. Leather shields generally have snaps. Make sure all snaps are fastened.
6. Insert the collet and collet body into the TIG torch. Insert the 3/32″ tungsten electrode into the collet and collet body. Tighten the electrode cap to hold the tungsten in place. Fasten the inert-gas cup to the collet body. Loosen the electrode cap and extend the tungsten past the end of the cup 1/8″. Tighten the cap to hold the tungsten in place.	6. The collet goes inside the collet body and the collet body screws into the torch body.
7. Make certain the power source is connected to the electrical supply. Make certain the power source is properly grounded.	
8. Attach the power cable to the "electrode" terminal of the welding machine.	
9. Obtain a cylinder of argon inert gas and fasten it next to the welding machine in a vertical position.	9. All high-pressure gas cylinders must be fastened tightly in an upright position. When moving bottles always put the valve-protector cap in place on the bottle.

PROCEDURE	KEY POINTS
10. Crack the cylinder valve.	10. "Cracking" the valve means to open it quickly. This action allows a small amount of gas to escape and blow out any dirt which may be in the valve. Do not stand in front of the valve when cracking the cylinder. Do not point the valve towards a wall since the force of the gas may cause particles to bounce back into the eyes.
11. Attach the inert-gas regulator flowmeter to the cylinder.	
12. Connect the inert-gas hose to the regulator and the welding machine.	12. Inert gas used for TIG welding is very expensive. Check and correct leaks in connections and hoses. A mixture of soap and water may be used to find the leaks. Escaping gas will make bubbles if leaks are occurring.
13. Have the instructor check all steps in the procedure.	

REVIEW QUESTIONS

Answer the following questions briefly.

1. What are the advantages listed for the TIG process over other welding processes?

2. List the four primary units of the TIG welding process.

3. What advantages to TIG welding result from the use of a high-frequency current?

4. List the functions of the following:

(a) Current setting

(b) Remote or panel switch

(c) Soft-start switch

(d) Power switch

(e) Process selection switch

(f) High-frequency control switch, phase shift, and intensity control

(g) Current-range switch

(h) Post-purge timer

(i) Polarity switch

5. Name the major parts of a TIG torch.

6. What is an "inert" gas?

Unit 36 Striking the Arc and Adjusting the TIG Welding Machine

OBJECTIVES

After completing this unit the student will be able to:

- describe the arc action and the function of the high-frequency arc in TIG welding.
- understand the use of the torch nozzle and the purging action of the shielding gas.
- successfully strike an arc and adjust the machine in preparation for TIG welding.

The Arc

When the ground and the tungsten electrode carrying an electrical charge are brought close enough together to close an electrical circuit, an arc occurs. The superimposed, high-frequency arc jumps the air gap between the tungsten and the ground, making ultraviolet and infrared rays and large amounts of heat. The rays are very harmful to the eyes and must be filtered out with a filter lens. The rays made by the TIG welding process are stronger than those of conventional stick-arc welding because the arc is more open, and most of the welding is done on bright metal like stainless steel or aluminum.

AC TIG Welding

The AC TIG welding arc is struck by holding the torch above the work and pressing the remote control. As the remote control is pressed, two things happen:

- The inert-gas flow starts.
- The electric current starts.

The tungsten does not contact the surface of the material being welded. The superimposed, high-frequency current jumps from the tungsten to the work and the arc is established.

If the end of the tungsten does come in contact with the metal being welded, small pieces of the molten puddle will contaminate the tungsten. This will cause the arc to waver, resulting in poor welds and possibly forcing the shielding gas to one side or the other.

After the arc is struck, the welder must adjust the arc length for good welds. The electrode should be held about 1/8" above the metal during the welding operation, figure 36-1.

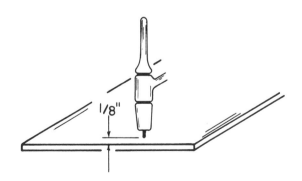

Fig. 36-1 The proper arc length is 1/8 inch.

DC TIG Welding

The TIG torch is held in a horizontal position approximately 2″ above the workpiece, then rapidly lowered until the electrode is about 1/8″ above the work, figure 36-2. The tungsten electrode must touch the metal to produce the arc. When the arc is struck, move the electrode back up 1/8″ above

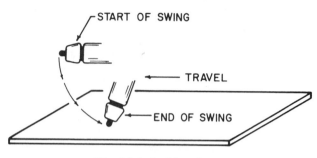

Fig. 36-2 Striking the arc

the metal to avoid contaminating the tungsten. If the welding machine is equipped with a high frequency for "start only," it can be used to start the arc without touching the metal with the electrode. In the start only position, the high frequency automatically turns off when the arc ignites.

Since the shielding gas will start flowing out of the torch when the remote foot control is pressed, the torch should be moved down to strike the arc very quickly. This ensures that the shielding gas will cover the weld zone when the molten puddle forms.

TIG Nozzle Construction

The inert gas is released to the weld zone through a gas nozzle that fastens onto the collet body. The material used for the nozzles has to have high heat resistance and it also has to conduct heat away from the lower edge of the nozzle. Gas nozzles are generally made from a ceramic material which reflects the heat. Glass and metal are also used. Ceramic and glass nozzles at welding heat must not be allowed to contact anything at room temperature. The sudden change in temperature frequently causes the nozzle to break. The torch must either be hand held or returned to a specially designed rack after use. Metal caps are used when the welding currents go above 200 amperes and the torches are water cooled.

Tungsten Electrode Stick Out

The tungsten electrode should be adjusted so that it sticks out past the end of the gas cup from 1 1/2 to 2 times the diameter of the tungsten being used, figure 36-3.

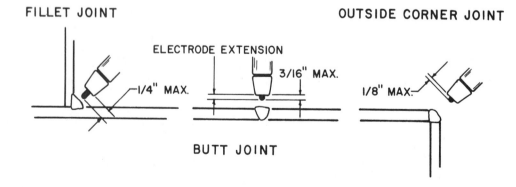

Fig. 36-3 Electrode extension for various types of weld joints

TIG Welding Amperage

TIG welding amperage (heat) is set by the welder on the front panel of the welding machine. When the amperage for welding is properly adjusted, a bead should be run across the plate in a steady, forward motion and the electrode must be held above the workpiece at the same height at all times. If the operation is done too fast, the weld bead will not look good and the weld will not penetrate deep enough into the plate. Fast progress may also leave holes inside the weld because the fast welding does not let the shielding gas have a chance to cover the molten puddle.

Direction of Travel

TIG welding is generally done with the forehand motion. When the torch is held so that the nozzle is about straight up-and-down and pointing in the direction of travel, it is called a "forehand" motion, figure 36-4. The forehand motion is used so that the weld puddle can be watched and the bead buildup can be kept equal. In TIG welding, the bead is clean inside the gas shield and there is no time lost cleaning the weld after it is made.

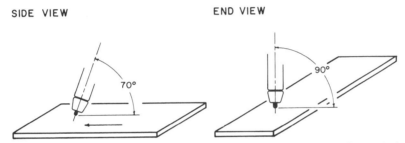

Fig. 36-4 The forehand motion progresses in the direction the torch is pointing.

Penetration

When the arc is struck on the plate, a crater or melted spot is made, figure 36-5. This crater must be filled with filler rod which must be of the same material as the base metal. The process of making the crater and filling it is called penetration. The filler rod and the plate must be well mixed for a good weld.

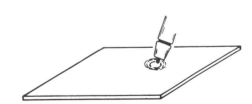

Fig. 36-5 Form the puddle using a circular motion.

Purging

When the gas is turned on to start a weld, it is a good idea to let some of the gas go through the lines before the weld starts. This action makes sure that only pure inert gas is in the lines. This action is called purging.

Adding Filler Rod to the Puddle

Four steps are necessary to run a bead, figure 36-6, page 196.

- Bring the TIG electrode down to the metal and strike the arc. Move the electrode in a small circle to heat the puddle, figure 36-6A, page 196.

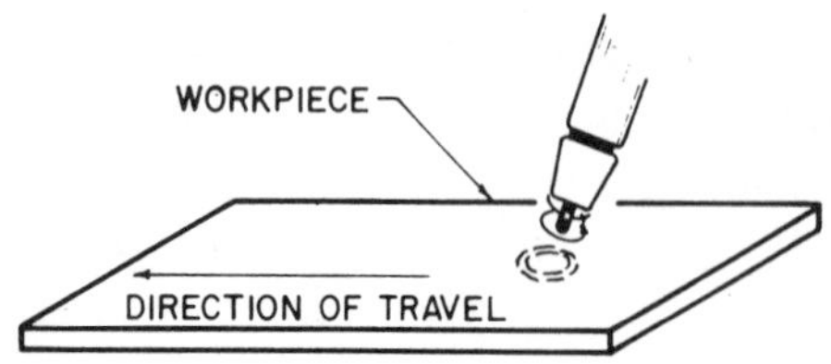

DEVELOP THE PUDDLE

MOVE TORCH BACK

Fig. 36-6A

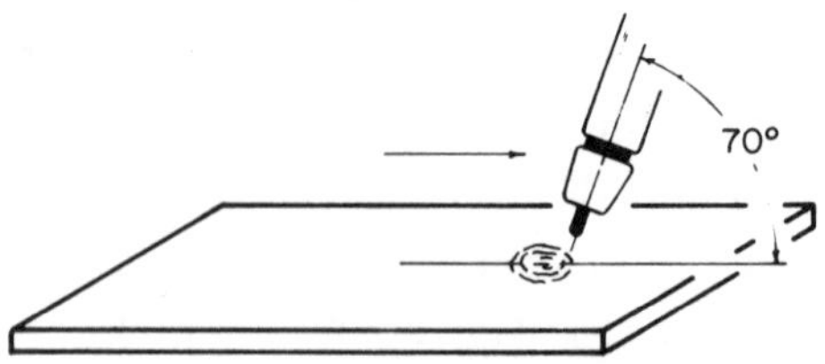

Fig. 36-6B

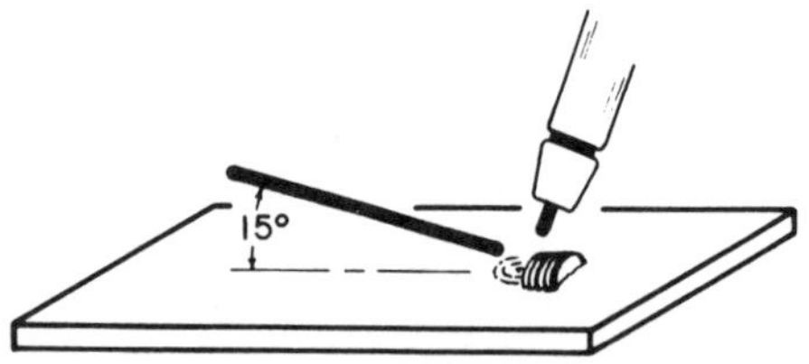

ADD FILLER METAL

MOVE TORCH TO LEADING
EDGE OF PUDDLE

Fig. 36-6C

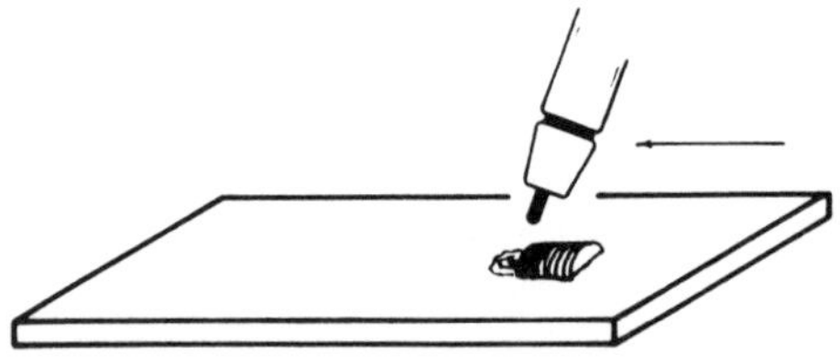

Fig. 36-6D

Fig. 36-6 The four steps to running a bead

- Move the torch back to the back edge of the puddle, figure 36-6B.
- Add the filler rod at the front of the puddle, figure 36-6C.
- Move the rod out of the puddle and move the torch to the front edge of the puddle to melt another part of the plate, figure 36-6D.

These motions are all made continuously as the welder moves the bead across the plate.

Practice will enable the welder to make all of these motions in a regular pattern so that the bead presents the appearance of overlapping shingles, or fish scales.

JOB 36: STRIKING THE ARC AND ADJUSTING THE TIG WELDING MACHINE

Equipment and Materials:

Standard TIG welding equipment
Personal hand tools
Protective clothing and gear

1/8″ 4043 aluminum filler rod
1/8″ x 6″ x 6″ aluminum plate
Argon shielding gas

PROCEDURE	KEY POINTS
1. Check all inert gas and electrical fittings.	
2. Connect the ground clamp to the work.	

PROCEDURE	KEY POINTS
3. Open the inert-gas cylinder slowly and let the pressure build up, then open it fully.	3. Always stand to one side when opening valves on high pressure cylinders.
4. Turn on the welding machine.	
5. Adjust the flowmeter to set the gas flow at 20 cubic feet per hour (CFH).	5. Inert gases are expensive, do not waste them.
6. Shift the remote and panel switch to "remote."	
7. Shift the polarity switch to "AC" polarity.	
8. Purge the gas system by pressing and releasing the remote control.	
9. Set the soft-start switch to "on."	
10. Set the process-selection switch to "inert-gas welding."	
11. Set the high-frequency switch to "continuous."	
12. Set the phase shift to "half scale."	
13. Set the intensity to half scale.	
14. Set the current control to half scale.	
15. Set the current range on medium heat.	
16. Set the postflow timer on 4 seconds.	
17. Turn on the water-cooling system (if the machine is so equipped).	17. Always turn on the water-cooling system or the power cable or TIG torch will overheat.
18. Put on helmet and gloves.	
19. Lower the helmet over the face and strike the arc. Begin welding and add filler material as needed for proper bead width and buildup.	
20. Run a bead 3″ long.	20. Adjust the current control switch for more or less heat, increase or decrease the phase shift and intensity as required.
21. Examine the bead and submit it to the instructor for comments.	

PROCEDURE	KEY POINTS
22. Run more short beads on the plate until the entire plate is used.	
23. Shut off the machine and the water system.	
24. Turn off the inert-gas cylinder.	
25. Turn the flowmeter dial to zero.	
26. Hang up TIG torch and disconnect the ground clamp.	
27. Clean up the work area.	

REVIEW QUESTIONS

Answer the following questions briefly.

1. How is the arc struck for the AC TIG process; the DC TIG process?

2. How is the high-frequency spark used in AC welding?

3. How much gap should be held between the electrode and the metal being welded?

4. How far out of the holder should the tungsten protrude for TIG welding?

5. What two things happen when the remote control is pressed?

6. Which motion of welding is generally used for TIG welding?

7. Define the word "purging" as it applies to TIG welding.

Unit 37 Four Layer Stringer Bead Pad Buildup, Flat Position, Aluminum

OBJECTIVES

After completing this unit the student will be able to:

- explain the procedure for selection of filler wire.
- discuss the various types of tungsten electrodes.
- describe the preparation of tungsten electrodes for DC and AC welding.
- produce a satisfactory four layer pad buildup on aluminum, using standard TIG welding procedures.

TIG Welding Wire Selection

The welder must know the type of base metal to be used for a weld before the right filler wire can be chosen. Other job conditions also have an effect on wire choice.

- Joint design
- Base metal thickness
- Welding position
- Type of shielding gas
- Service requirements

After the base metal is known, a manufacturer's chart can be used to aid in the filler wire selection. The aluminum wire used most often in TIG welding is number 4043. All filler wires have a basic core wire makeup and these are classified and numbered. A filler material chart may be secured from any welding supply distributor. The aluminum filler wires best suited to add strength, ductility, crack resistance, or corrosion resistance are those numbered 4043 (containing silicon) and 5356 (containing magnesium).

Tungsten Electrodes

The electrodes used for TIG welding are called "nonconsumable" because they are not melted into the puddle of weld. In reality, though, they are consumed, since each time they become dirty from contact with the base metal they must be broken off and reshaped for the welding being done.

Tungsten is an element with a melting temperature around 6100 degrees Fahrenheit (3371 degrees Celsius) and the electrodes for TIG welding are available in three classifications.

- Pure tungsten
- 1 percent or 2 percent thoriated (added)
- 1 percent zirconium (added)

Tungsten used on AC current should have a rounded end. Too much electric current used with the tungsten may cause the end of the electrode to enlarge, resulting in poor

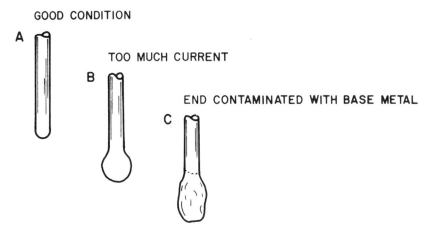

Fig. 37-1 Various shapes of AC tungsten electrode

welds. If the electrode is touched to the base metal, some of the metal can stick to the electrode and cause a large glob of metal on the end of the electrode, spoiling the weld. When this condition occurs, the end of the tungsten is cut off with pliers, and the tungsten must be reground to shape, figure 37-1.

Tungsten electrodes used with DC current should have a sharp pointed end, figure 37-2. The end of the tungsten can be sharpened by using a pedestal type grinder which should not be used for any other purpose. If the grinder is used for other grinding, it may contaminate the end of the tungsten electrode. The tungsten must be ground so that the point is in the center of the electrode. If the point is to one side, it will cause the gas flow to wander.

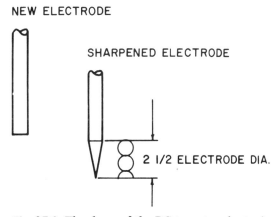

Fig. 37-2 The shape of the DC tungsten electrode

Pure Tungsten. Pure tungsten is the least expensive electrode and is used with AC welding. It should have a rounded end, with a diameter slightly larger than the original electrode.

Thoriated Tungsten Electrodes. These electrodes are used on DC– polarity settings. They are more expensive than pure tungsten, but they are not consumed as fast. Thoriated tungstens run cooler and their tips do not become overheated from arcing. They do not pick up metal from the base metal as fast as pure tungsten.

Zirconium Tungsten Electrodes. Used successfully with AC polarities, the costs of these electrodes are higher than pure tungsten. They have the same characteristics as the thoriated tungsten.

Aluminum Welding Recommendations

The following chart lists the recommendations for the various thicknesses of material, types of joints, and welding positions, figure 37-3. Included are the proper amperage, electrode diameter, gas flow, and filler rod for each possibility.

Stock Thickness (inch)	Type of Joint	Amperes, AC Current			Electrode Diameter (inch)	Argon Flow 20 psi		Filler Rod Diameter (inch)
		Flat	Horizontal & Vertical	Overhead		lpm	cfh	
1/16	Butt	60-80	60-80	60-80	1/16	7	15	1/16
	Lap	70-90	55-75	60-80	1/16	7	15	1/16
	Corner	60-80	60-80	60-80	1/16	7	15	1/16
	Fillet	70-90	70-90	70-90	1/16	7	15	1/16
1/8	Butt	125-145	115-135	120-140	3/32	8	17	1/8
	Lap	140-160	125-145	130-160	3/32	8	17	1/8
	Corner	125-145	115-135	130-150	3/32	8	17	1/8
	Fillet	140-160	115-135	140-160	3/32	8	17	1/8
3/16	Butt	190-220	190-220	180-210	1/8	10	21	5/32
	Lap	210-240	190-220	180-210	1/8	10	21	5/32
	Corner	190-220	180-210	180-210	1/8	10	21	5/32
	Fillet	210-240	190-220	180-210	1/8	10	21	5/32
1/4	Butt	260-300	220-260	210-250	3/16	12	25	3/16
	Lap	290-340	220-260	210-250	3/16	12	25	3/16
	Corner	280-320	220-260	210-250	3/16	12	25	3/16
	Fillet	280-320	220-260	210-250	3/16	12	25	3/16

TIG WELDING–ALUMINUM

psi – pounds per square inch
lpm – liters per minute
cfh – cubic feet per hour

(Linde Co.)

Fig. 37-3 Recommendations for aluminum TIG welding

JOB 37: FOUR LAYER STRINGER BEAD PAD BUILDUP, FLAT POSITION, ALUMINUM

Equipment and Materials:

Standard TIG welding equipment
Personal hand tools
Protective clothing and gear
Stainless steel wire brush
1/8" x 6" x 6" aluminum plate
1/8" #4043 filler wire
Argon shielding gas

PROCEDURE	KEY POINTS
1. Set up the TIG machine for ACHF.	1. Position the plate for the flat position of welding.
2. Turn on the machine.	

PROCEDURE	KEY POINTS
3. Turn on the cooling system.	
4. Turn on the inert-gas system.	
5. Run a stringer bead across the center of the plate.	
6. Continue to run beads until the plate is covered, as shown in figure 37-4.	6. Overlap the beads about 1/3. Cool the weld specimen as needed.

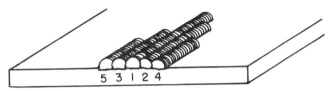

Fig. 37-4 Proper bead sequence for four layer pad buildup

PROCEDURE	KEY POINTS
7. Brush the beads with a stainless steel wire brush.	7. Aluminum plate forms an outside oxide that must be removed by brushing.
8. Turn the sample 90 degrees and weld across the completed beads.	8. Complete four layers. Brush each pass, and each layer.
9. Submit the pad to the instructor for comments and instructions.	

REVIEW QUESTIONS

Answer the following questions briefly.

1. How is the correct filler wire selected for TIG welding?

2. What is a "nonconsumable" electrode?

3. What shape should the tungsten have on the end when used for AC welding?

4. What shape should the tungsten have on the end when used for DC welding?

5. Why does aluminum require brushing between welds?

Unit 38 Lap Weld, Flat Position, Aluminum

OBJECTIVES

After completing this unit the student will be able to:

- discuss the factors which effect the construction of a lap weld in the flat position.

- list the lens shades used for various TIG welding amperages.

- discuss the general safety practices of TIG welding.

- successfully TIG weld a lap joint on aluminum in the flat position, using standard TIG welding procedures.

Lap Weld

Lap welds of thin materials require no edge preparation. The only requirements for making a satisfactory lap weld are that the edges be clean and positioned tightly together. Lap joints may be fused with or without filler material. Proper support of the work should also be provided to avoid distortion and warping, figure 38-1.

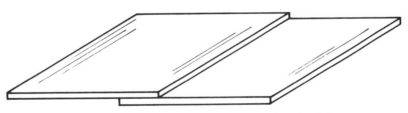

Fig. 38-1 **Proper position and support for a lap joint**

Solid tack welds are necessary to keep the metal plates together to eliminate movement from the heat of the arc, figure 38-2.

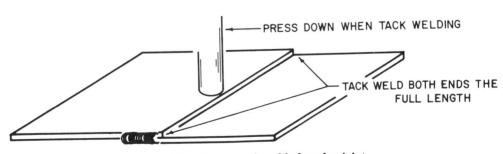

Fig. 38-2 **Proper tack welds for a lap joint**

The travel speed should be regulated so the bead width is approximately 3 times the diameter of the electrode being used. If no filler rod is being used, the electrode should be centered on the edge of the top of the plate with an arc length equal to the diameter of the

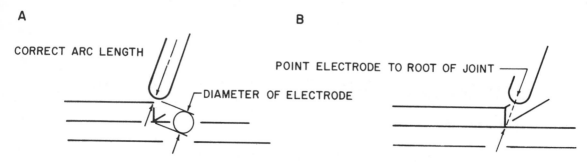

Fig. 38-3 Correct electrode position for a lap weld

electrode. If filler rod is used, the electrode should point at the root of the joint, figure 38-3.

Safe Practices for TIG Welding

The TIG welding process is not any more dangerous than other types of welding if safety precautions are observed. As with any form of electric welding the potentials of electric shock, skin burns, or fire hazard are always present.

In addition, welding heat on certain metals or alloys may produce toxic gases. The operator should always be protected from the rays of the arc which are more intense in TIG welding, because of the exposed arc and the reflections from the material welded.

Good protection requires protective clothing to cover all the exposed skin and a welding helmet which has a darker filter lens than required for regular stick-arc welding. The shade of the filter lens depends on the intensity of the arc, figure 38-4.

SHADE NUMBER	WELDING AMPERAGE
8	0 TO 70 AMPERES
9	50 TO 100 AMPERES
10	80 TO 200 AMPERES
12	180 TO 400 AMPERES
14	ABOVE 400 AMPERES

Fig. 38-4 Shade of lens for various amperage settings

Practice all of the following safety rules while doing TIG welding:

- Always use a welding helmet equipped with a filter lens of a shade recommended for the amperage set on the machine.

- Always wear a welding cap to protect the hair and scalp from possible burns.

- Cover all skin with protective clothing to protect the body surface from burns resulting from sparks, ultraviolet or infrared rays, and welding spatter.

- Recommended protective equipment includes the following:

  A. Welding helmet   E. Gauntlet gloves

  B. Welding cap   F. Leather apron

  C. Safety glasses   G. Heavy flame-resistant shirt and pants (no cuffs)

  D. Leather welding jacket H. High-top safety shoes

- Do not weld near flammable gases, powders, chemicals, or liquids.

- All containers which have held flammable gases, powders, chemicals, or liquids must be cleaned with steam or other retardant actions before any welding is done on them.

- Remove all combustible materials from the welding area.

- Know where all available fire extinguishers are, and learn how to use them.

- The welding area should be dry and clean to avoid electric shock and falls.

- Do not weld in areas where adequate ventilating systems are not available. Use special precautions where toxic fumes are suspected.

- Use a non-reflective shield around the welding when other workers are present.

- Visually inspect the TIG torch, power source, service leads, ground leads, inert-gas system, and all protective equipment. Never use faulty equipment for any reason.

- Never change any machine setting while the machine is under a load.

- Do not drape the torch power cable over the shoulder while welding is in progress.

- Know and practice all safety instructions.

JOB 38: LAP WELD, FLAT POSITION, ALUMINUM

Equipment and Materials:

    Standard TIG welding equipment
    Personal hand tools
    Protective clothing and gear
    Stainless steel wire brush
    4 pieces, 1/8" x 3" x 6" aluminum sheet
    1/8" 4043 filler rod
    Argon shielding gas

PROCEDURE	KEY POINTS
1. Turn on the machine, the gas system and the water-cooling system (if the welder is so equipped).	1. Check the water-return hose to make sure the cooling system is working.
2. Set the machine for TIG aluminum welding.	
3. Position the base metal and tack weld both samples, as shown in figure 38-2.	
4. Thoroughly wire brush each joint.	4. Brush each joint just before welding.

PROCEDURE	**KEY POINTS**
5. Weld one side of one joint using a forehand motion and filler rod, figure 38-5.	5. The lap weld requires a full, single-pass, fillet weld.

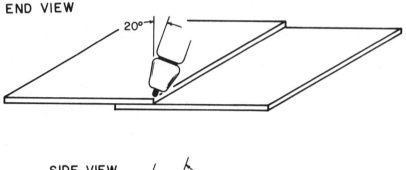

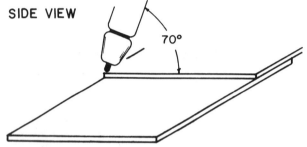

Fig. 38-5 Proper torch angles for a lap joint

6. Weld the other side of the joint using the forehand motion, without the use of filler rod.	6. Compare the filler wire joint and the fusion melt joint.
7. Position the other sample and weld both sides of the joint forehand, with filler rod.	
8. Shut down all systems.	
9. Submit both samples to the instructor for comments and instructions.	

REVIEW QUESTIONS

Answer the following questions briefly.

1. What are the two requirements listed for making a good lap joint?

2. What is the bead width recommended for a lap joint?

3. What two reasons are given for the arc light being brighter from the TIG process than from other processes?

4. What shade number lens is specified for welding in the 80 to 200 ampere range?

5. List the 8 items recommended as protective equipment in the text.

6. What is the cleaning method recommended for the lap joint to be welded in Job 38?

Unit 39 T Weld, Flat Position, Aluminum

OBJECTIVES

After completing this unit the student will be able to:

- describe the setup procedures for a flat position T weld.
- list the steps of machine setting for alternating current, high-frequency welding of aluminum.
- list the color code markings for tungsten electrodes.
- satisfactorily TIG weld a T joint on aluminum plate, in the flat position using standard TIG welding procedures.

T Weld

The T weld requires a bead very similar to that used for a lap weld. Extra attention must be paid to the position of the electrode so as not to undercut one of the plates. The torch should be held at a 70 degree angle up from the flat workpiece and point into the joint at approximately 40 degrees, figure 39-1.

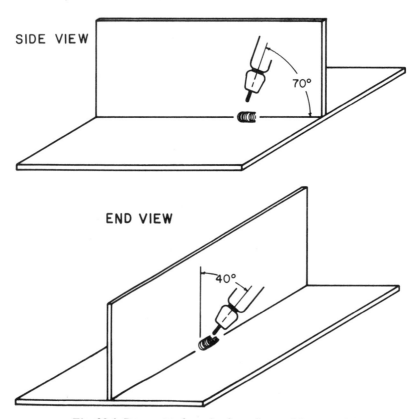

Fig. 39-1 Proper torch angles for a flat position T weld

The bead width of a T weld should equal 3 times the diameter of the electrode used to make the weld, figure 39-2.

The plates of the T weld should be firmly tack welded at both ends, figure 39-3.

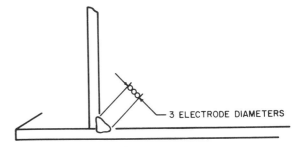

Fig. 39-2 Proper bead width for a flat position T weld

Review of Machine Settings for ACHF Aluminum Welding

1. Set the current setting to half scale. After striking the arc readjust as necessary to make the weld.

2. Shift the remote or panel switch to "remote," so the welder can control the heat with either a foot or hand remote control.

3. Soft-start switch should be placed in the "on" position, for reduced current at the start of the arc.

4. Turn the power switch to the "on" position to receive electricity from the power source.

5. Shift the process-selection switch to "Inert-Gas Welding."

6. Set the phase shift and intensity control dials to half scale, and after striking the arc, readjust as needed.

7. Shift the current-range switch to the medium setting.

8. Set the postpurge timer on 3 seconds afterflow. Watch the end of the tungsten after the arc is stopped. If it turns a dark color, increase the postflow time length.

9. Shift the polarity switch to AC polarity.

10. Slowly open the valve on the high-pressure argon cylinder. Always stand to one side of the regulator when opening the cylinder. When the tank pressure is shown, open the valve all the way.

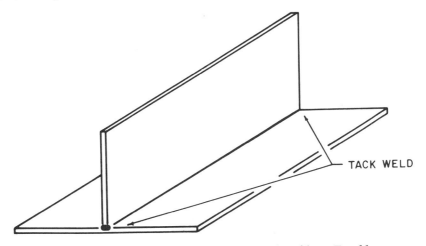

Fig. 39-3 Proper position for plates and tack welds on T weld

11. Purge the gas system and set it at 15-20 CFH (cubic feet per hour).

12. Turn on the water-cooling system (if applicable) and check the return flow to ensure proper operation.

13. Strike the arc and readjust machine settings as necessary (see steps #1, #6, #7, #8).

14. Begin the welding operations.

Tungsten Electrode Color Code

Tungsten electrodes are either pure tungsten (AC polarity) color code green, tungsten 1 percent thoriated (AC or DC polarity) color code yellow, tungsten 2 percent thoriated (AC or DC polarity) color code red, or tungsten-zirconium (AC or DC polarity) color code brown. All four types are available in lengths of 3 to 24 inches. The 7 inch tungsten is the one most commonly used.

JOB 39: T WELD, FLAT POSITION, ALUMINUM

Equipment and Materials:

Standard TIG welding equipment
Personal hand tools
Protective clothing and gear
Stainless steel wire brush

2 pieces, 1/8" x 5" x 6" aluminum sheet
2 pieces, 1/8" x 2 1/2" x 6" aluminum sheet
1/8" 4043 filler rod
Argon shielding gas

PROCEDURE	KEY POINTS
1. Turn on the machine, the gas system, and the cooling system.	1. Check the water return for operation.
2. Set the machine for ACHF aluminum welding.	2. Try out the machine on scrap aluminum.
3. Position and tack weld both samples, figure 39-3, page 211.	3. Thoroughly brush the joint just before welding.
4. Position one sample and complete the welding of one side using the forehand motion and single-pass sequence, figure 39-4.	

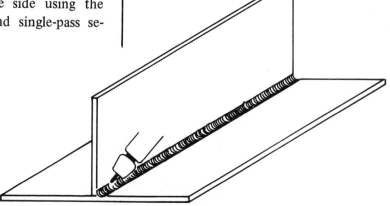

Fig. 39-4 Run a single pass on one side of the T.

PROCEDURE	KEY POINTS
5. Rotate the joint and weld the other side, figure 39-5.	5. Cool the sample in water as needed to control the heat.

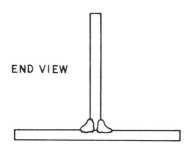

END VIEW

Fig. 39-5 Completed T weld

6. Position the other sample and weld both sides.
7. Shut down all systems.
8. Submit the samples to the instructor for comments and instructions.

REVIEW QUESTIONS

Answer the following questions briefly.

1. What bead width should be used for a T weld?

2. What torch angle should be used for a flat T weld?

3. How is a T weld tack welded?

4. What kind of tungstens are represented by the colors listed?
   (a) Green
   (b) Yellow
   (c) Red
   (d) Brown

SECTION 8: REVIEW A

OBJECTIVE:

- Produce an aluminum butt weld of proper penetration and quality in the flat position using standard TIG welding procedures. Subject such a sample to the guided-bend test.

Equipment and Materials:

Standard TIG welding equipment
Personal hand tools
Protective clothing and gear
Stainless steel wire brush
Carbon-arc cutting equipment or hacksaw
2 pieces, 1/8" x 4" x 6" aluminum sheet
1/8" 4043 filler rod
Argon shielding gas

PROCEDURE

1. Turn on the equipment and adjust for ACHF aluminum welding.

2. Position and tack weld the joint as shown in figure R8-1.

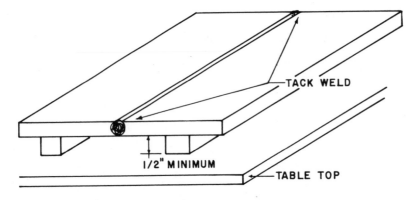

Fig. R8-1 Proper position of the work and the tack welds

3. Weld the sample using a forehand technique and a stringer bead motion, figure R8-2.

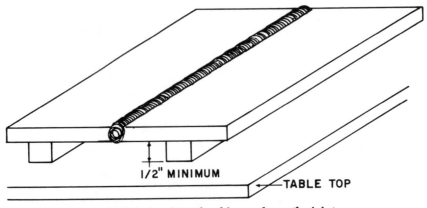

Fig. R8-2 A stringer bead is run down the joint.

4. Lay out and cut the sample into sections as shown in figure R8-3.

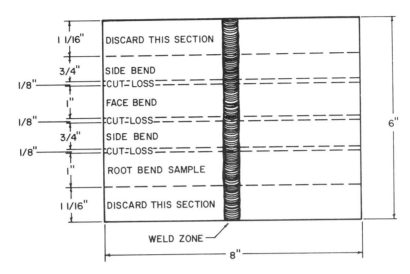

Fig. R8-3 Proper layout of sample for specimens to be used in guided bend test

5. Perform a guided-bend test on the face and root of the various strips.

6. Examine the bent strips. There should be no separation of base metal from the weld and no porosity.

7. If the welded samples show defects, repeat steps 1-6.

8. When the specimens are satisfactory, submit them to the instructor for comments and further instructions.

SECTION 8: REVIEW B

OBJECTIVE:

- Produce an aluminum outside corner weld of proper penetration and quality in the flat position using standard TIG welding procedures. Subject such a sample to a destructive test.

Equipment and Materials:

    Standard TIG welding equipment
    Personal hand tools
    Protective clothing and gear
    Stainless steel wire brush
    2 pieces, 1/8" x 2 1/2" x 6" aluminum sheet
    1/8" 4043 filler rod
    Argon shielding gas

PROCEDURE

1. Turn on the equipment and adjust for ACHF aluminum welding.

2. Position and tack weld the corner joint as shown in figure R8-4.

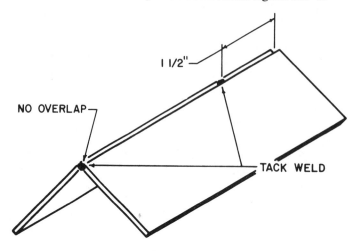

Fig. R8-4 Proper position of the workpieces and tack welds for an outside corner joint

3. Weld the sample using a forehand technique, a stringer bead motion, and the angles shown, figure R8-5.

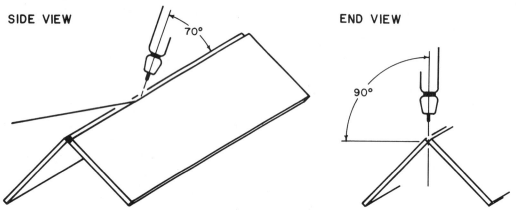

Fig. R8-5 Proper torch angles for an outside corner joint

4. Perform the destructive test as shown in figure R8-6.

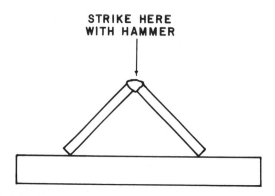

Fig. R8-6 Performing a destructive test on an outside corner weld

5. Submit the sample to the instructor for comments and further instructions.

SECTION 9:
Horizontal and Vertical Welds of Aluminum, TIG Process

Unit 40 Four Layer Pad Buildup, Horizontal Position, Aluminum

OBJECTIVES

After completing this unit the student will be able to:

- describe the horizontal position of welding and the difficulties presented by the position.

- discuss the inert-gas flow rate used for different conditions of welding and joints.

- successfully TIG weld a four layer pad buildup in the horizontal position using standard TIG welding procedures.

TIG Welding Aluminum in the Horizontal Position

Horizontal welds are much more difficult to weld than flat welds, since the molten puddle is pulled down by the force of gravity. Because of aluminum's relatively low melting point, this factor becomes extremely important. The operator must make the proper equipment adjustments and use the proper welding motions and torch angles. Figure 40-1 shows a bead being run in the horizontal position.

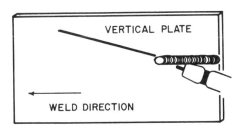

Fig. 40-1 TIG welding a horizontal bead on a vertical plate.

Inert-Gas Shield Flow Rates

The main function of the shielding gas is to keep the free oxygen in the atmosphere away from the weld zone and the tungsten electrode. The flow of the gas is directed down around the electrode and out of the nozzle to the weld zone, figure 40-2, page 218.

The flow rate of the inert gas must be large enough to push the air from the weld zone and to keep from being pushed away by air movement around the weldment. Too much

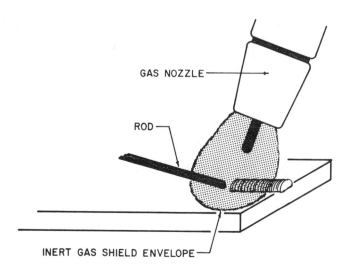

Fig. 40-2 Closeup of the weld zone showing the inert-gas shield

gas flow will raise the costs and interfere with welding. When TIG welding is in progress many different things may affect the flow of the gas shield. Some of these things are:

1. The type of base metal being welded will control the flow of the shield. Aluminum, which readily combines with the oxygen in the atmosphere, requires higher flow rates than does stainless steel.

2. Joint design has to be considered in the correct rate of flow. Inside corner joints require less flow than outside corner joints and flat joints (butt welds, lap welds) need a setting somewhere in between the outside and inside corner joints, figure 40-3.

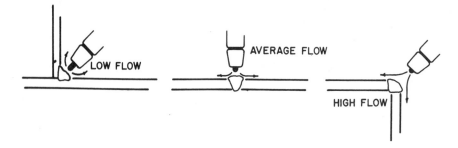

Fig. 40-3 Gas flows vary depending on the joint design.

3. As welding amperages are increased, the flow of shielding gas is also increased. More current requires larger electrodes and gas nozzles, which in turn cause larger weld puddles that need more gas coverage.

4. Gas shielding nozzles are numbered (4, 5, 6, 7, etc.). These numbers mean the nozzle size in 1/16ths of an inch. For example, a nozzle numbered "4" would equal 4/16" = 1/4", which is the inside cup diameter. The larger the nozzle size, the more gas flow needed.

5. The type of shielding gas used will determine the rate of flow. Helium is much lighter than argon and, therefore, requires higher flow rates for the same coverage.

6. The position of the joint to be welded will also determine the rate of flow.

7. A check of the end of the tungsten electrode will indicate if proper gas flow rates are being used. The right gas flow will leave the end of the electrode smooth and rounded on the tip. Incorrect gas flow increases burning, scale, and deformation of the electrode.

JOB 40: FOUR LAYER PAD BUILDUP, HORIZONTAL POSITION, ALUMINUM

Equipment and Materials:

Standard TIG welding equipment
Personal hand tools
Protective clothing and gear
Stainless steel wire brush
1 piece, 1/8″ x 5″ x 5″ aluminum sheet
1/8″ 4043 filler rod
Argon shielding gas

PROCEDURE	KEY POINTS
1. Turn on the machine, the gas system (15-20 CFH) and the cooling system.	1. Check the water return system for operation.
2. Adjust the machine by welding on scrap aluminum.	2. The TIG torch should slant towards the direction of travel about 5 degrees.
3. Place the plate vertical to the table. Begin at the bottom of the plate and weld a bead across, using forehand motion, figure 40-4.	3. Use only the forehand and stringer bead motions for horizontal welding.

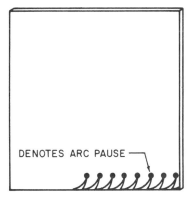

DENOTES ARC PAUSE

Fig. 40-4 Proper bead motion for a horizontal stringer bead

4. Check the beads as they are completed and make necessary corrections to the machine.

PROCEDURE	KEY POINTS
5. Follow the bead sequence indicated. Bead #2 should cover the top third of bead #1, figure 40-5.	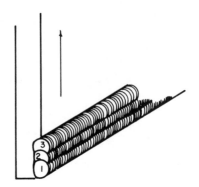
6. Rotate the plate 90 degrees so that the next layer of beads crosses the first.	
7. Keep turning the plate each time a layer is completed until four layers have been welded on the plate.	
8. Shut down all systems.	**Fig. 40-5 Proper bead sequence and position for a pad buildup**
9. Submit the pad to the instructor for comments and instructions.	

REVIEW QUESTIONS

Answer the following questions briefly.

1. What is the main function of the shielding gas?

2. What would the number "5" signify on a shielding gas nozzle?

3. Is helium gas lighter than argon?

4. How can the welder determine if the gas flow is correct when checking the tungsten electrode after completing a weld?

Unit 41 Lap Weld, Horizontal Position, Aluminum

OBJECTIVES

After completing this unit the student will be able to:

- draw a lap weld in the horizontal position and identify the side preferred for welding.

- describe the types of filler metals used for manual TIG operations.

- understand and correct common weld defects which occur in TIG welding.

- successfully TIG weld a horizontal lap weld on aluminum using standard TIG welding procedures.

Lap Welds in the Horizontal Position

In the welding of a lap weld in the horizontal position, there is a definite preference as to which side the welder should work. One side of the lap joint will have an edge that will help support the molten puddle. Obviously, this is the side which the welder should work on, figure 41-1. Although it is possible to weld the other side, it is much more difficult to produce a satisfactory weld without a great deal of practice.

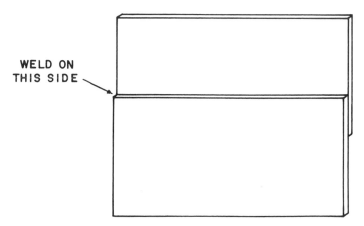

WELD ON
THIS SIDE

Fig. 41-1 Proper position for a lap weld in the horizontal position

Filler Materials

The type of filler wire to be used is determined by the job to be accomplished. In manual TIG operations, it is recommended that the length of the filler wire used be 36". The filler wire is available in several diameters: 1/16", 3/32", 1/8", 5/32", and 1/4".

Filler wire is made either flat or round. To receive the maximum benefit from filler material it should have the following:

- Cleanliness

- Controlled chemical makeup

- Straightness and surface smoothness

- Melting temperatures equaling the base metal

Filler materials of the same makeup as the base metal will deposit welds of high quality and strength. The completed weld will closely match the color of the base material. Being the same composition, the wire will not harden quicker than the base metal and cracks which might occur due to shrinkage stresses are not common.

TROUBLE SHOOTING TIG WELDING		
WELD DEFECT	**REASON**	**CORRECTION**
Excessive Tungsten Usage	1. Incorrect inert-gas flow resulting in oxidation of the electrode. 2. Welding on DC+ polarity. 3. Excessive amperage for electrode diameter. 4. Contaminated electrode—from touching work piece or filler material. 5. Oxidation of electrode during cooling period.	1. Increase the gas flow to the correct CFH. 2. Increase the diameter of the tungsten electrode. 3. Increase diameter of electrode. 4. Break off tungsten and reshape for the polarity used. Increase high-frequency current. 5. Increase after-flow time of inert gas.
Arc Wandering	1. Base metal dirty. 2. Included bevel too narrow or root opening too close. 3. Contaminated electrode. 4. Long arc.	1. Clean base metal with stainless steel brush or chemicals. 2. Increase root opening or extend electrode to be closer to work. 3. Break off tungsten and reshape for polarity being used. 4. Move torch closer to work.
Porosity in Weld	1. Base metal dirty. 2. Trapped gases—travel speed too fast or wrong gas flow.	1. Clean base metal with stainless steel brush or chemicals. 2. Correct travel speed and increase shielding gas.
Tungsten Contamination on Work	1. Touching work with electrode. 2. Tungsten melting and mixing with base metal. 3. Shattering of tungsten from high heat input.	1. Use high frequency on Continuous for AC or Start Only for DC polarity. 2. Use less current or increase the diameter of the tungsten. 3. Make sure electrode ends are properly prepared or increase the diameter of the electrode to carry the high currents.

JOB 41: LAP WELD, HORIZONTAL POSITION, ALUMINUM

Equipment and Materials:

Standard TIG welding equipment
Personal hand tools
Protective clothing and gear

4 pieces, 1/8″ x 3″ x 6″ aluminum sheet
1/8″ 4043 filler wire
Argon shielding gas

PROCEDURE	KEY POINTS

1. Turn on the machine, the gas system and the water-cooling system. (If welder is so equipped.)

2. Set the machine for TIG aluminum welding.

    2. Try out the machine for horizontal welds on scrap.

3. Position the base metal and tack weld both samples, figure 41-2.

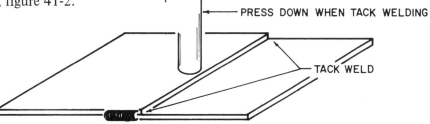

Fig. 41-2 Proper position of the tack welds for a lap weld

4. Wire brush each joint before welding.

5. Position one sample in the horizontal position and weld one side of the joint using the forehand motion and filler rod, figure 41-3.

    5. Cool samples in water as needed.

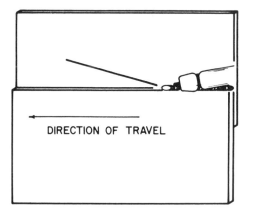

Fig. 41-3 Running a lap weld in the horizontal position

6. Turn the sample over and reposition for horizontal weld. Weld the joint, using forehand motion, without using filler material.

    6. Compare the joint to the one in which you used filler.

PROCEDURE	KEY POINTS
7. Position the other sample and repeat the procedure, using filler rod on both laps.	7. Repeat all procedures if more practice is needed.
8. Shut down all systems.	
9. Submit both samples to the instructor for comments and instructions.	

REVIEW QUESTIONS

Answer the following questions briefly.

1. Give the standard diameters of filler wire used for TIG welding aluminum.

2. List the four features needed from the filler wire to ensure good welds.

3. Why should the filler wire be of the same composition as the base metal being welded?

4. What possible corrections may be made if the electrode is used up too rapidly?

5. If the electrode is continuously oxidizing after every use, what can be done to eliminate the problem?

6. What may happen to the arc if the base metal is not properly cleaned?

7. What defect will be the result of trapping gases in the weld?

8. What condition may shatter the electrode when welding?

Unit 42 Four Layer Pad Buildup, Vertical Position, Aluminum

OBJECTIVES

After completing this unit the student will be able to:

- identify stringer beads and weave beads and describe their application to vertical position welding.

- discuss and apply correct torch and filler material angles for a vertical weld.

- satisfactorily TIG weld a four layer pad buildup in the vertical position using standard TIG welding procedures.

TIG Welding Aluminum in the Vertical Position

Fast and correct welds can be made in the vertical position when using the TIG process. Vertical welds are made by using certain bead motions and applying set angles to the TIG torch. Both the correct motions and the proper torch angle must be observed for quality welds.

Bead motion for vertical welds are of two groups, figure 42-1.

- The stringer bead motion.

- The weave bead motion.

The arc swing for the stringer bead is an upright "U" or an upside down "U." These motions are used with the vertical-up welding, from bottom to top, using a forehand torch angle. The bead motions used for weave beads are the same as for stringer beads, with wider torch movements, to increase the width of the bead. The width of weave beads should not exceed 6 times the diameter of the tungsten electrode being used.

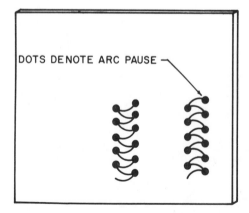

Fig. 42-1 Proper procedure for creating a vertical-up stringer bead

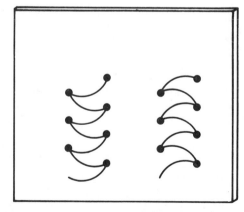

Fig. 42-2 Proper procedure for creating a vertical-up weave bead

The TIG torch should be positioned at 70 degrees from the bottom edge of the workpiece, with no side angle. The filler wire should be held at 20 degrees from the top edge of the workpiece and fed into the weld zone as needed, figure 42-2.

The torch body should be about parallel with the plate and should point upwards at 20 degrees. This position is used so that the arc force will hold the molten puddle up against the pull of gravity. The speed of movement for vertical welds must be a smooth, steady rate of travel.

JOB 42: FOUR LAYER PAD BUILDUP, VERTICAL POSITION, ALUMINUM

Equipment and Materials:

Standard TIG welding equipment
Personal hand tools
Protective clothing and gear
Stainless steel wire brush
1 piece, 1/8" x 5" x 5" aluminum sheet
1/8" 4043 aluminum filler wire
Argon shielding gas

PROCEDURE	KEY POINTS
1. Turn on the machine, the gas system and the cooling system (if applicable).	
2. Adjust the machine for vertical welding, using scrap aluminum.	
3. Place the plate vertical to the table. Begin at the bottom center of the plate and weld a bead all the way up the plate to the top, figure 42-3.	3. Wire brush the plate before welding.

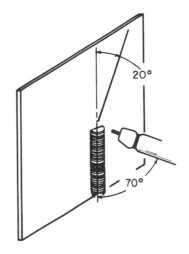

Fig. 42-3 Proper torch angles for a vertical-up weld

PROCEDURE	KEY POINTS
4. Follow the bead sequence shown, brushing off each pass, and fill the plate surface, figure 42-4.	4. Inspect each bead and make any adjustments necessary. 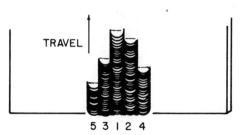 **Fig. 42-4 Proper bead sequence for a vertical pad buildup**
5. Turn the plate 1/4 turn and build another layer across the first.	5. The torch must slant upwards at 20 degrees and the filler rod is fed into the top of the puddle at the same angle from the top edge of the plate.
6. Cool the work as necessary. Rotate the plate until four layers have been completed.	6. After completion, shut down all systems.
7. If the tungsten electrode becomes contaminated, remove it and reshape it as needed.	7. Tungsten should have a rounded end for AC welding of aluminum.
8. Submit the pad to the instructor for comments and instructions.	

REVIEW QUESTIONS

Answer the following questions briefly.

1. Two methods of welding are used—forehand and backhand. Which is recommended for vertical aluminum TIG welding?

2. Vertical welds can be made vertical-up and vertical-down. Which is recommended for vertical aluminum TIG welding?

3. What are the two common bead motions used for TIG welding aluminum?

4. What is the maximum width of a weave bead recommended for TIG welding?

5. How much side angle should be used on the TIG torch in vertical welding?

Unit 43 Lap Weld, Vertical Position, Aluminum

OBJECTIVES

After completing this unit the student will be able to:

- describe the factors which affect the construction of a vertical lap weld.
- understand and explain the care of the TIG torch.
- name the parts of the TIG torch and explain the function of each part.
- successfully complete a lap weld, vertical position, using standard TIG welding procedures.

Lap Weld

The vertical position lap weld requires much practice before it can be successfully welded. The metal is placed in position and firmly tack welded. Pressure must be applied to hold the plates together while tacking, since any gap left between the two plates will cause welding problems. The torch must be directed slightly upwards and pointed into the corner junction of the plates. Gas flow will have to be increased slightly so that proper gas coverage is maintained over the molten puddle. A slight motion of the tungsten will help wash the melted metal from the upper plate down into the puddle and enough welding rod must be added to keep a full, rounded bead. A shallow bead will allow breakage of the weld if a strain is applied to the bottom sheet of the lap. Wherever possible, the lap should be welded on both sides of the joint.

Bead motions for this type of vertical weld are of two groups:

- The stringer bead motion, figure 43-1.
- The weave bead motion, figure 43-2.

It should be noted that the arc swing represents either an upright "U" or an upside down "U" configuration of the bead movement.

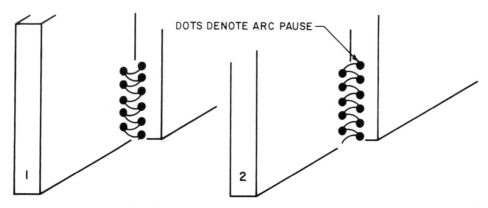

DOTS DENOTE ARC PAUSE

Fig. 43-1 Forehand, uphand stringer motions

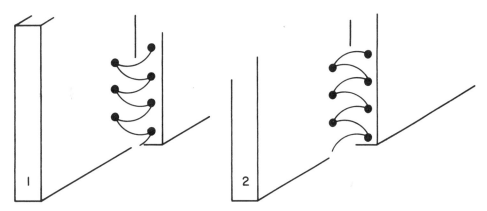

Fig. 43-2 Forehand, uphand weave motions

The two diagrammed bead motions are for uphand vertical welding, using a forehand travel procedure.

The torch should be pointed upwards at approximately 5 degrees, figure 43-3. The positioning of the torch will help the operator control the hot, molten puddle over the pull of gravity by using the arc force for pushing the puddle upwards. The angle of the filler rod in relationship to the torch and workpiece is shown in figure 43-4.

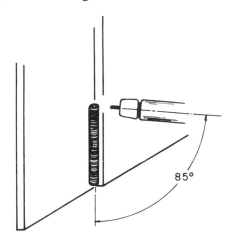

Fig. 43-3 The torch should be pointed up at a 5° angle.

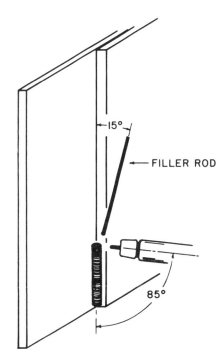

Fig. 43-4 The filler rod should be held at a 15° angle from the plates.

When welding in the vertical position, slight variations of arc pause, bead motion, torch angle, and filler material angle may be made to suit the individual operator.

Care of the TIG Torch

The special equipment used for TIG welding is expensive and requires proper care and usage.

Water-Cooled Torch. Correct handling of TIG welding equipment prolongs the life of the nozzle, collet, collet body, electrode, power cable, and hoses. Water-cooled torches are used where high production or high amperage are held through the welding operations. The water-cooled torch is made so that water is circulated through the torch body and the power cable to cool them, figure 43-5. The power cable is contained inside a hose (generally made of polyethylene) and the water inside flows around the cable, removing the heat as it returns to a water-supply tank. If a water-cooled torch is used where there is a lack of water, or no water at all, the power cable will heat and melt the power cable hose. The manufacturers' charts will show the number of gallons per minute which should circulate through the torch.

WATER COOLED TORCH

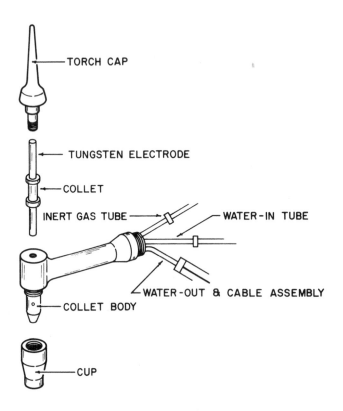

Fig. 43-5 Exploded view of water-cooled TIG torch

To prevent damage to the hose and power cable, a safety device known as a fuse assembly can be installed on the inlet side of the power cable. This installation contains a fuse link which is also cooled by the water flow. If there is no cooling water, or not enough, the fuse will melt preventing any damage. The fuse link can be easily replaced.

Air-Cooled Torch. Air-cooled torches are popular where low amperage operations can be used, figure 43-6. They require no cooling other than the surrounding atmosphere. The power cable for air-cooled torches is much heavier and carries the added heat; there is no fuse assembly. The power cable is usually built inside the inert-gas hose, because the gas shield will carry some of the heat to the weld zone and help to cool the torch. If welding heat over the rated capacity of the torch is used, hoses will be damaged and shielding gas can be lost.

Care During Welding Operations. The torch nozzle (cup) must not be pushed into the weld puddle. It is easily charred or broken. The proper distance between the nozzle and workpiece must be maintained and the nozzle must be kept free from contamination. Dirt on, or in, the nozzle interferes with the gas flow. When the ceramic nozzle or the tungsten electrode touches the workpiece, contamination of the weld may result. This contamination must not be cleaned from the nozzle by tapping the torch on any solid object. The torch nozzle, electrode, collet body, and collet are easily broken.

AIR COOLED TORCH

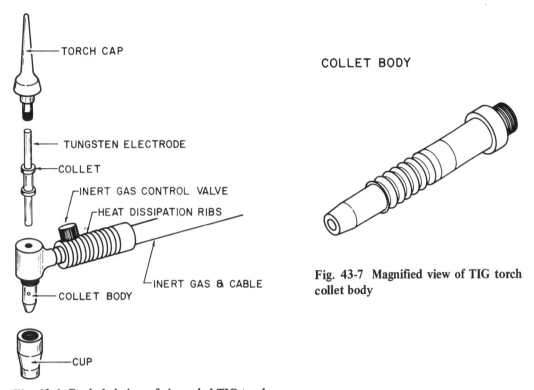

COLLET BODY

Fig. 43-7 Magnified view of TIG torch collet body

Fig. 43-6 Exploded view of air-cooled TIG torch

Disassembling the Torch. The torch is disassembled by unfastening the nozzle from the torch body. The collet body is threaded on the outside in two places, figure 43-7. These threads are easily damaged and must be handled with caution. The torch body has internal threads which the collet body screws into. The threads are fine and easily cross-threaded.

Maintenance Responsibility. It is the welder's responsibility to maintain the torch in good condition so that welds produced are of first quality. The nozzle and electrode must be examined frequently. Contamination must be removed or welds will not be satisfactory. The tungsten should be broken off and resharpened when it has been contaminated.

JOB 43: LAP WELD, VERTICAL POSITION, ALUMINUM

Equipment and Materials:

Standard TIG welding equipment
Personal hand tools
Protective clothing and gear
Stainless steel wire brush

4 pieces, 1/8″ x 3″ x 6″ aluminum sheet
1/8″ 4043 filler rod
Argon shielding gas

PROCEDURE	KEY POINTS
1. Turn on the machine, the gas system, and water system (if machine is so equipped).	
2. Set the machine for TIG welding of aluminum.	2. Use scrap metal and set the machine.
3. Check the tungsten electrode and put a round shape on the end for aluminum welding.	3. Tungsten electrodes are rounded on the end by striking the arc, on DC+ polarity, for a short time. Be careful not to contaminate the end of the electrode.
4. Position the base metal and tack weld both samples, figure 43-8.	

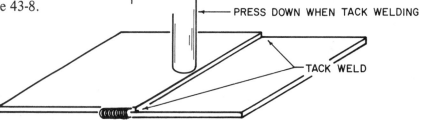

Fig. 43-8 Proper position of plates and tack welds for lap weld

5. Thoroughly wire brush each joint just before welding begins.	
6. Place one sample in the vertical position and weld one side, using the forehand motion, with filler rod.	6. The vertical lap joint requires a full, single pass fillet weld.
7. Weld the other side of the joint, using the forehand motion, without the use of filler rod.	7. Compare the filler wire joint and the fusion melt joint for appearance and strength.
8. Position the other sample and weld both sides of the joint, forehand, with filler rod.	
9. Shut down all systems.	
10. Submit the samples to the instructor for comments and instructions.	

REVIEW QUESTIONS

Answer the following questions briefly.

1. Name the parts of a standard water-cooled TIG torch.

2. Name the parts of a standard air-cooled TIG torch.

3. What is the purpose of the water flow in the water-cooled torch, during welding operations?

4. What is the function of a fuse assembly?

5. Why is the power cable heavier on the air-cooled torch than on the water-cooled torch?

6. Which TIG torch is used for high amperage, long term welding?

Unit 44 T Weld, Vertical Position, Aluminum

OBJECTIVES

After completing this unit the student will be able to:

- explain the technique of welding a T weld, including the machine adjustments which must be made.
- name and explain the makeup of the different types of welding joints used for TIG welding.
- list the factors which determine the use of different types of joints.
- successfully TIG weld a T joint in the vertical position using standard TIG welding procedures.

T Weld

The vertical position T weld (fillet) requires much practice before it can be completed successfully. The base material is placed in position and tack welded. The torch must be directed slightly upward and pointed into the corner of the two plates. Gas flow will have to be increased slightly so that a good gas shield is over the weld zone. A slight motion of the torch will result in equal legs on the fillet weld. The addition of filler rod must be sufficient to keep a full, round bead. A shallow bead will allow breakage of the weld if a strain is applied to the upright member of the T weld. Whenever increased strain is to be applied to the weld, it should be welded on both sides of the joint.

Types of Joints

The same types of joints used for other welding processes are used with TIG welding. The most common joints are:

- Lap Joint
- Corner Joint
- Butt Joint
- T Joint

Most welds made with the TIG process will consist of one, or a combination of, these joints. The proper joint design selection depends on the following factors:

- Service requirements
- Cost of joint preparation
- Cost of weld completion
- Type of base material
- Size of the parts to be welded
- Shape of the parts to be welded
- Appearance of the complete assembly

The filler materials, or TIG welding rods, do not have to be used if enough strength can be obtained without it. Many joints can be made strong enough by simply melting the base metal together, without the addition of a welding rod.

Cleaning

Proper cleaning of the base materials is a must, if quality welds are to be made. On small parts, hand cleaning with a stainless steel wire brush, steel wool, or chemical solvents is generally sufficient. For large parts, degreasing or solvent cleaning may be cheaper.

Precautions must be taken when chemical solvents are used for cleaning. The fumes from some solvents break down from the heat of the electric arc and form toxic gases. Good ventilating systems are needed to remove all the fumes and vapors from the work area.

JOB 44: T WELD, VERTICAL POSITION, ALUMINUM

Equipment and Materials:

Standard TIG welding equipment
Personal hand tools
Protective clothing and gear
Stainless steel wire brush
2 pieces, 1/8″ x 4″ x 6″ aluminum sheet
2 pieces, 1/8″ x 3″ x 6″ aluminum sheet
1/8″ 4043 filler rod
Argon shielding gas

PROCEDURE	KEY POINTS
1. Turn on the machine, the gas system and the cooling system (if so equipped).	
2. Set the machine for ACHF aluminum welding.	2. If necessary, prepare the tungsten electrode for aluminum welding.
3. Position the 3″ plates in the centers of the 4″ plates and tack weld both ends of each joint, figure 44-1.	

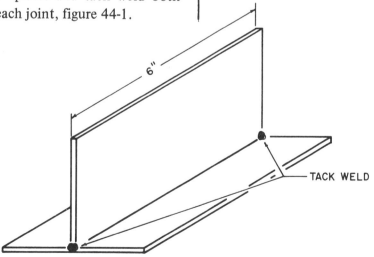

Fig. 44-1 Proper position of the plates and tack weld for a T joint

PROCEDURE	KEY POINTS

4. Place one sample in the vertical position and weld one side with a stringer bead motion. Use filler rod, figure 44-2.

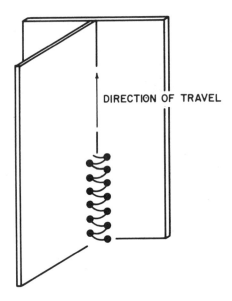

DIRECTION OF TRAVEL

Fig. 44-2 Proper bead motion for a vertical-up T weld

5. Rotate the joint and repeat the welding procedure, except run the bead down from top to bottom, figure 44-3.

5. Compare the vertical-up joint and the vertical-down weld.

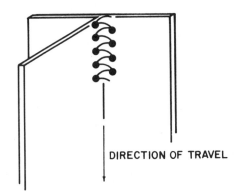

DIRECTION OF TRAVEL

Fig. 44-3 Proper bead motion for a vertical-down T weld

6. Position the other sample and weld both sides up the joint, using a forehand motion.

6. When the second side is welded, amperage may need to be increased to take care of the thicker metal.

7. Shut down all systems.

8. Submit both samples to the instructor for comments and instructions.

REVIEW QUESTIONS

Answer the following questions briefly.

1. Is it necessary to tack weld both ends of a T joint before welding it?

2. Where should the torch be pointed when welding the T joint?

3. Will it be necessary to increase the gas flow when welding in the vertical T joint?

4. What may happen if the bead is too shallow on a T joint fillet weld?

5. What types of joints are most often used in TIG welding?

6. Is it always necessary to use filler rod when TIG welding these joints?

SECTION 9: REVIEW

OBJECTIVE:

- Produce an aluminum T weld of proper penetration and quality in the vertical position using standard TIG welding procedures. Subject such a sample to a destructive test.

Equipment and Materials:

Standard TIG welding equipment
Personal hand tools
Protective clothing and gear
Stainless steel wire brush
1 piece, 1/8" x 5" x 6" aluminum sheet
1 piece, 1/8" x 2 1/2" x 6" aluminum sheet
3/32" 4043 filler rod
Argon shielding gas

PROCEDURE

1. Turn on the equipment and adjust for ACHF aluminum welding.

2. Position and tack weld the T joint as shown in figure R9-1.

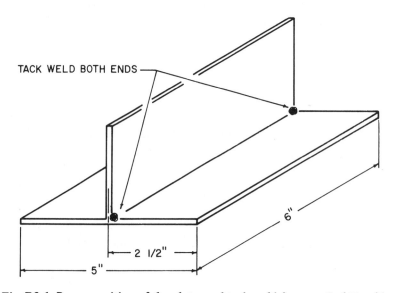

TACK WELD BOTH ENDS

6"

2 1/2"

5"

Fig. R9-1 Proper position of the plates and tack weld for a vertical T weld

3. Place the sample in the vertical position.

4. Complete one side of the joint using the bead sequence shown in figure R9-2, and forehand, vertical-up welding technique.

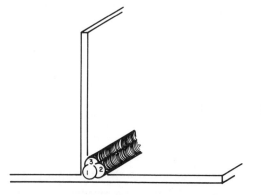

Fig. R9-2 Proper bead sequence for a vertical T weld

5. Perform the destructive test as shown in figure R9-3.

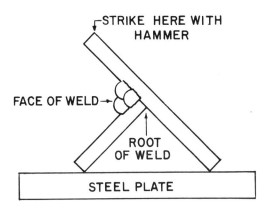

Fig. R9-3 Destructive test of a fillet weld

6. Submit the sample to the instructor for comments and further instructions.

SECTION 10:
Flat Position Welds of Stainless Steel, TIG Process

Unit 45 Four Layer Stringer Bead Pad Buildup, Flat Position, Stainless Steel

OBJECTIVES

After completing this unit the student will be able to:

- discuss the applications of stainless steel.

- understand the differences between welding stainless steel and aluminum with the TIG process.

- discuss the welding requirements for stainless steel.

- successfully complete the TIG welding of a four layer stringer bead pad buildup using standard TIG welding procedures.

Stainless Steel

Stainless steel is one of the most popular metal alloys and is used for many applications in the welding field. It is desirable for welded construction where strength and resistance to high temperatures, pressure, and corrosion are necessary. X-ray, quality welds needed for large, heavy wall piping systems used for nuclear power plants and other critical services are easily made with the TIG welding process. Stainless steel can be formed at the mill in all standard forms and shapes.

Stainless steel is the term used in reference to chromium-alloyed and nickel-alloyed steels. Some of these alloys are magnetic, others are not. There are a large number of these alloys and each type offers different properties where corrosion resistance and strength are required. A check of the manufacturers' charts is recommended when the welder is in doubt as to the makeup of any alloy.

Chromium-Nickel

The chromium-nickel steels are the ones most often used for welding. They are non-magnetic due to the large amount of nickel used in them. They are relatively easy to weld. Generally, the welding heat does not affect their strength or ductility. The filler rod used for welding should be closely matched to the composition of the base material. Manufacturers' charts are available at many welding supply houses. It is very important that the filler rod matches the base metal, and that the weld deposit and base metal have the properties required to successfully perform the job.

Heat Spread

The heat of welding will not spread out through the base metal of stainless steel as fast as it does on carbon steel. The spread of heat is about 50 percent less than on carbon steels, while the heat expansion rate is 50 percent greater. These two things tend to create warpage and shrinkage in thin gauge materials. Because stainless steel distorts to a greater degree than other materials during welding, tacking, holding jigs, and clamping fixtures are important.

Recommended Polarity

Stainless steel can be successfully welded on two polarities—direct current straight polarity (DC–) or alternating current with superimposed high frequency (ACHF). Direct current straight polarity allows the operator to produce welds of deeper penetration at high rates of speed. When welding operations are on this polarity the superimposed, high-frequency switch is set on the "start only" position. This allows the arc to be started without touching the tungsten electrode to the workpiece, eliminating contamination to the electrode.

Aluminum Welding and Stainless Steel Welding

The only major difference in welding these two alloys is that the welding of aluminum requires more heat input and a faster speed of travel. For stainless steel welding, travel speed is slower and lower amperage settings are selected on the machine.

TIG WELDING–STAINLESS STEEL							
Stock Thickness (inch)	Type of Joint	Amperes, DC Current–Straight Polarity			Electrode Diameter (inch)	Argon Flow 20 psi		Filler Rod Diameter (inch)
		Flat	Horizontal & Vertical	Overhead		lpm	cfh	
1/16	Butt	80-100	70-90	70-90	1/16	5	11	1/16
	Lap	100-120	80-100	80-100	1/16	5	11	1/16
	Corner	80-100	70-90	70-90	1/16	5	11	1/16
	Fillet	90-110	80-100	80-100	1/16	5	11	1/16
3/32	Butt	100-120	90-110	90-110	1/16	5	11	1/16
	Lap	110-130	100-120	100-120	1/16	5	11	1/16
	Corner	100-120	90-110	90-110	1/16	5	11	1/16
	Fillet	110-130	100-120	100-120	1/16	5	11	1/16
1/8	Butt	120-140	110-130	105-125	1/16	5	11	3/32
	Lap	130-150	120-140	120-140	1/16	5	11	3/32
	Corner	120-140	110-130	115-135	1/16	5	11	3/32
	Fillet	130-150	115-135	120-140	1/16	5	11	3/32
3/16	Butt	200-250	150-200	150-200	3/32	6	13	1/8
	Lap	225-275	175-225	175-225	3/32	6	13	1/8
	Corner	200-250	150-200	150-200	3/32	6	13	1/8
	Fillet	225-275	175-225	175-225	3/32	6	13	1/8
1/4	Butt	275-350	200-250	200-250	1/8	6	13	3/16
	Lap	300-375	225-275	225-275	1/8	6	13	3/16
	Corner	275-350	200-250	200-250	1/8	6	13	3/16
	Fillet	300-375	225-275	225-275	1/8	6	13	3/16

psi – pounds per square inch
lpm – liters per minute
cfh – cubic feet per hour

(Linde Co.)

Fig. 45-1 Recommendations for stainless steel TIG welding

Stainless Steel Recommendations

The preceding chart lists the recommendations for the various thicknesses of material, types of joints, and welding positions, figure 45-1, page 243. Included are the correct amperage, electrode diameter, gas flow, and filler rod material for each possibility.

JOB 45: FOUR LAYER STRINGER BEAD PAD BUILDUP, FLAT POSITION, STAINLESS STEEL

Equipment and Materials:

 Standard TIG welding equipment
 Personal hand tools
 Protective clothing and gear
 Stainless steel wire brush
 1 piece, 1/16″ x 5″ x 5″ stainless steel sheet
 1/16″ 308 filler wire
 Argon shielding gas

PROCEDURE	KEY POINTS
1. Remove the tungsten from the torch and prepare the end for DC– polarity welding, figure 45-2. Replace the electrode.	1. Install a #4 nozzle. 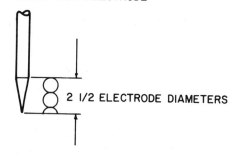 **Fig. 45-2 Electrode properly shaped for DC-polarity**
2. Shift the polarity switch to DC–.	
3. Turn on all systems.	
4. Set the inert-gas flow at 10 CFH.	4. Increase the flow if necessary.
5. Reduce the amperage setting for stainless steel welding between 80 to 100 amperes.	5. Adjust on scrap piece.
6. Shift the superimposed high-frequency switch to "start only."	
7. Position the plate for flat welding.	7. Set the metal up off the table, so that there is a space below it.
8. Wire brush the plate.	8. Use a clean stainless steel brush.

PROCEDURE	KEY POINTS
9. Beginning at the right side, run a bead across the center of the plate, using the forehand and stringer bead motions.	9. Brush each completed bead.
10. Follow the bead sequence shown and run beads until the plate is full, figure 45-3.	10. Cool the sample to remove heat build-up, as required.

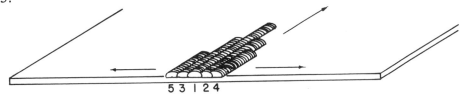

Fig. 45-3 Proper bead sequence for pad buildup

11. Show the completed layer to the welding instructor for comments.

12. Brush the beads.

13. Turn the sample 1/4 turn and repeat the procedures.

14. Continue to rotate the sample until four layers have been completed.

15. Submit the sample to the instructor for comments and instructions.

REVIEW QUESTIONS

Answer the following questions briefly.

1. What are the four reasons listed for the use of stainless steel in the welding field?

2. To what two steels does the term stainless steel refer?

3. Are chromium-nickel steels magnetic?

4. Does heat spread faster or slower in stainless steel than in carbon steel?

5. Where is the high-frequency switch set for DC- welding?

6. Should the end of the tungsten electrode be blunt or sharp for DC- welding?

Unit 46 Lap Weld, Flat Position, Stainless Steel

OBJECTIVES

After completing this unit the student will be able to:

- explain the striking of the arc and the addition of filler rod to the stainless steel lap weld.

- describe the necessary heat settings and arc length used for the lap weld.

- adjust the electrode extension for the weld.

- successfully TIG weld a lap joint in the flat position, on stainless steel, using standard TIG welding procedures.

Striking the Arc and Adding Filler Material to the Puddle

Four steps are required to run a bead when TIG welding:

(1) Bring the TIG electrode down close to the base metal, but not touching it. The superimposed high frequency will jump the air gap between the end of the tungsten electrode and the workpiece, establishing an arc. The inert-gas nozzle can be rested on the workpiece with little chance of touching the electrode to the work, figure 46-1. After the arc is started the nozzle must be lifted off the base metal. Move the TIG torch in a small circular motion to heat the puddle.

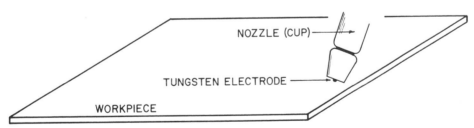

NOZZLE (CUP)

TUNGSTEN ELECTRODE

WORKPIECE

Fig. 46-1 Proper torch position for striking the arc

(2) Move the torch to the back edge of the puddle.

(3) Add filler rod to the leading edge of the puddle. Take care in the selection of filler rod size. If the filler rod is too large, it will subtract from the heat of the puddle, making the weld more difficult to perform.

(4) The filler rod is moved in and out of the front edge of the puddle as required for the correct amount of deposit. The torch and filler materials are continuously moved along the joint in this manner until the weld is completed. At the end of the weld, back up the torch and fill the crater to prevent crater cracks.

Practice will enable the welder to make all the necessary motions in a rhythmic pattern so the completed weld will have the appearance of a well-shingled roof, or fish scales.

Lap Joint

After establishing the arc, the weld puddle is formed so that the top edge of the over-lapping plate and the flat surface of the bottom plate flow together. Since the top edge of the overlapping plate will become molten before the flat surface, the angle and motion of the torch becomes important for puddle control. The top edge will also have a tendency to burn back or undercut. These two weld defects can be controlled by dipping the filler rod in and out of the puddle next to the top edge of the upper plate as it tries to melt away. Enough filler must be added to the puddle to fill the joint properly, figure 46-2. Finish the end of the weld by backing up the torch on the weld to fill the crater. The angle of the torch and filler material for lap welds is shown in figure 46-3.

Amperage and Arc Length

TIG welding amperage is set by the welder on the front panel of the machine, by shifting the current setting. Stainless steel will use less amperage than other materials of the same thickness because of its ability to retain the welding heat for a longer time. When the amperage is set correctly, the weld deposit can be run across the plate in a steady forward motion.

As a general rule, the arc length is held at one diameter of the electrode being used. This rule can be successfully used when the end of the electrode is rounded as it is for aluminum welding. When welding with direct current polarity, using a sharpened electrode, the arc length may be less than the electrode diameter. Each welder develops a personal style of welding which helps in making good welds.

Electrode Extension

Extensions of the tungsten will vary with each particular job. The stick-out distance may be from flush to the nozzle to possible 1/2" away from the nozzle. The longer the

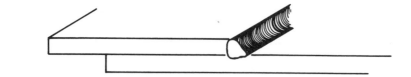

Fig. 46-2 A single-pass fillet bead of the proper thickness

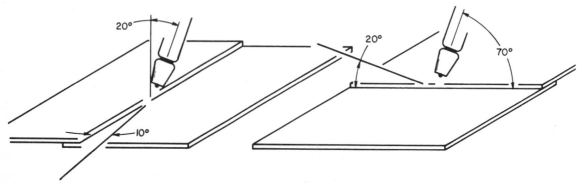

Fig. 46-3 Proper torch angles for a flat position lap weld

extension, the more likely it will be that the electrode will come in contact with the work-piece or filler material. These accidental contacts will consume the tungsten because contamination has to be removed and the end of the electrode has to be prepared. Start with the extension of one electrode diameter and make additional adjustments as required.

JOB 46: LAP WELD, FLAT POSITION, STAINLESS STEEL

Equipment and Materials:

> Standard TIG welding equipment
> Personal hand tools
> Protective clothing and gear
> Stainless steel wire brush
> 4 pieces, 1/16" x 2 1/2" x 6" stainless steel sheet
> 1/16" 308 filler rod
> Argon shielding gas

PROCEDURE	KEY POINTS
1. Turn on all systems and adjust the machine for stainless steel welding.	1. Visually check all systems for proper operations.
2. Position the stainless steel plates and tack weld, figure 46-4.	2. Brush each joint before welding begins.

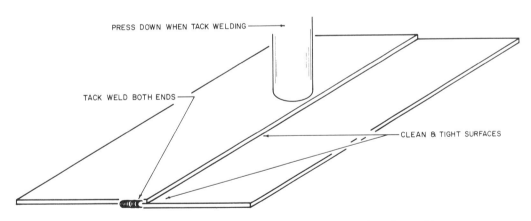

Fig. 46-4 Proper position of plates and tack welds for a lap weld

PROCEDURE	KEY POINTS
3. Position one tacked sample up off the table top and weld one side of the joint, using filler rod.	3. The lap weld requires a full, single pass fillet weld.
4. Inspect the weld and adjust the machine as needed.	
5. Turn the sample over and weld the other side without filler rod.	

PROCEDURE	KEY POINTS
6. Position the other sample and weld both sides of the joint, using a fore-hand motion with filler rod.	
7. Shut down all systems.	
8. Submit the samples to the instructor for comments and instructions.	

REVIEW QUESTIONS

Answer the following questions briefly.

1. List the four steps required to run a bead with TIG on stainless steel.

2. Which plate of a lap weld becomes molten first when heat is applied?

3. If the electrode is sharpened, how long should the arc length be?

4. Why is it more likely that the tungsten will become contaminated when the stick-out is set long?

Unit 47 T Weld, Flat Position, Stainless Steel

OBJECTIVES

After completing this unit the student will be able to:

- discuss the similarities between the lap joint and the T joint when TIG welding.
- discuss the use of backup bars and the effects of outside atmosphere on the stainless steel weld.
- successfully complete a T weld on stainless steel, in the flat position, using standard TIG welding procedures.

T Welds and Lap Joints

A T weld is a fillet weld made where two pieces of material meet at a 90 degree angle. A similar condition is found in the lap joint, where an edge (the upright plate) and a flat surface (the bottom plate) are to be welded. The edge of the upright member becomes molten before the flat surface and the angle and motion of the torch will play an important part in puddle control. At the upper edge of the weld puddle, either on the T weld or the lap joint, a tendency for undercutting exists. This condition can be corrected by adding the filler material next to the top edge of the weld zone, using a dipping in and out motion. Enough filler rod must be added to the puddle to build the bead at least three times the width of the electrode diameter, figure 47-1. Finish the end of a weld by backing up on the weld to fill the crater, and the fill will eliminate crater cracks.

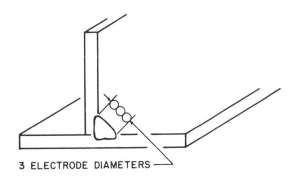

3 ELECTRODE DIAMETERS

Fig. 47-1 The bead width for a fillet weld should be three electrode diameters.

The angle of the torch and filler material for T joints directs more heat onto the flat surface, figure 47-2, page 252.

The electrode should be extended farther beyond the nozzle than in other types of welds so as to establish a shorter arc length. This is necessary for good fillet welds in the T joints.

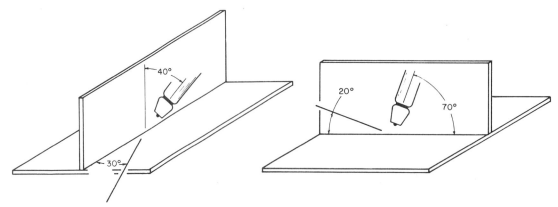

Fig. 47-2 Proper torch angles for a T weld

Joint Backup

The joint should be "backed up" on many TIG welding operations. Backing up the underside of the joint will protect the bottom side base metal from atmospheric contamination. The function of any shielding gas is to push the atmosphere away from the weld puddle. This permits welding to take place in a controlled environment. In shielded metal-arc welding, the coating placed on the stick electrode is melted by the heat of the arc and in doing so, releases a protective atmosphere.

In TIG welding, the inert-gas shield released through the torch nozzle has the same effect on the top side of the weld puddle. The bottom side, however, is unprotected without some backing. In many TIG welding operations, the joint is backed up by placing a backup bar under the meeting of the two pieces in the joint, or by allowing inert-gas to flow under the weld to protect it from contamination.

Atmosphere Contaminants

Those gases from the outside atmosphere (the air we breathe) which cause the most problems for a welder are oxygen, nitrogen, and hydrogen.

Oxygen. Oxygen combines quickly with other elements in the hot weld puddle to form oxides and gases. Alloying elements such as manganese and silicon can be used to counteract such reactions in carbon steels. When TIG welding, however, the only protection available to the molten puddle is the correct use of inert-gas shielding. With the right rate of flow of inert gas, the proper torch motion, and the correct speed of movement, the oxygen is kept out of the weld zone.

When incorrect inert-gas shields are used in TIG welding, free oxygen can combine with the carbon in stainless steel to form carbon monoxide. If carbon monoxide becomes trapped in the weld metal as it cools, it will collect in pockets which cause porosity in the weld.

Nitrogen. Nitrogen is the biggest problem encountered in the welding of steel materials because by volume there is so much of it in the surrounding air. Nitrogen forms nitrites (nitrogen salts) during the welding operations, causing hardness, a decrease of ductility, and lowered impact strength. These conditions often lead to material cracking in, and next to, the weld deposit. In addition, large amounts of nitrogen will cause porosity in the weld, These weld defects are easily eliminated by the correct shielding procedures.

Hydrogen. The presence of hydrogen during welding produces an erratic arc causing the arc to waver. This in turn affects the soundness of the weld deposit. As the weld puddle becomes solid, most of the hydrogen is pushed out of the liquid puddle but it may become trapped in pockets and cause stresses. These stresses may lead to small cracks in the weld which may later become larger. Trapped hydrogen may also cause under-bead cracking. Proper use of the shielding gas will correct these weld defects.

JOB 47: T WELD, FLAT POSITION, STAINLESS STEEL

Equipment and Materials:

Standard TIG welding equipment
Personal hand tools
Protective clothing and gear
Stainless steel wire brush
2 pieces, 1/16″ x 5″ x 6″ stainless steel sheet
2 pieces, 1/16″ x 2 1/2″ x 6″ stainless steel sheet
1/16″ 308 filler rod
Argon shielding gas

PROCEDURE	KEY POINTS
1. Turn on all systems.	
2. Set the machine for stainless steel welding. The high frequency switch is to be set at "start only."	2. Adjust the welding of the machine on stainless steel of the same thickness as that used for the job.
3. Position the 2 1/2-inch metal in the centers of the other pieces, 90 degrees to the base sheets for a T weld, and tack both ends of each joint.	3. Tack weld these joints with the same procedures used in other jobs.
4. Place one sample in the flat position (up off the welding bench) and weld one side with a stringer bead and forehand motion, using filler rod.	4. Prevent excessive penetration with correct amperage and speed of movement of the torch.
5. Cool the sample.	
6. Inspect the back of the joint for honeycombing.	6. Honeycombing will be present if the weld has burned completely through the joint. It is caused by contamination of the metal by the atmosphere.

PROCEDURE	KEY POINTS
7. Lower the heat setting if honeycombing is present.	7. An advantage of the TIG process is the ability to weld thin sections with complete control of penetration quality.
8. Position the other sample and weld it, using filler rod and the forehand motion.	
9. Shut down all systems.	
10. Submit both samples to the instructor for comments and instructions.	

REVIEW QUESTIONS

Answer the following questions briefly.

1. How is the lap weld similar to a T joint in puddle control?

2. What is joint "backup"?

3. Can inert gas be used as a joint backup?

4. What three gases from the atmosphere cause the most welding difficulties?

5. What is formed when oxygen combines with melting metal?

6. What is formed when free oxygen combines with the carbon in stainless steel?

7. What three conditions may occur as a result of the formation of nitrites?

8. What happens to the arc when hydrogen is present during welding?

9. What specific type of cracking may occur when hydrogen becomes trapped in the hot metal of the weld?

SECTION 10: REVIEW A

OBJECTIVE:

- Produce a stainless steel butt weld of proper penetration and quality in the flat position using standard TIG welding procedures. Subject such a sample to a guided-bend test.

Equipment and Materials:

Standard TIG welding equipment
Personal hand tools
Protective clothing and gear
Stainless steel wire brush
Carbon-arc cutting equipment or hacksaw
2 pieces, 1/8" x 4" x 6" stainless steel sheet
1/8" 308 filler rod
Argon shielding gas

PROCEDURE

1. Turn on the equipment and adjust for stainless steel welding.

2. Position and tack weld the joint as shown in figure R10-1.

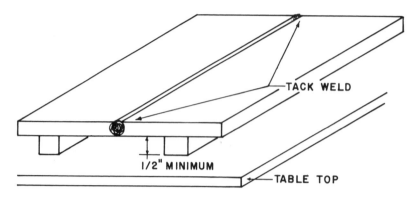

Fig. R10-1 Proper position of the work and the tack welds

3. Weld the sample using a forehand technique and a stringer bead motion, figure R10-2.

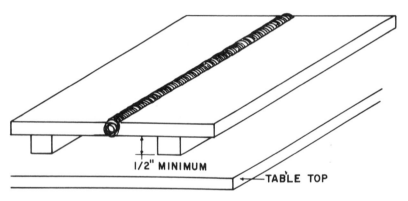

Fig. R10-2 A stringer bead is run down the joint.

4. Lay out and cut the sample into sections as shown in figure R10-3.

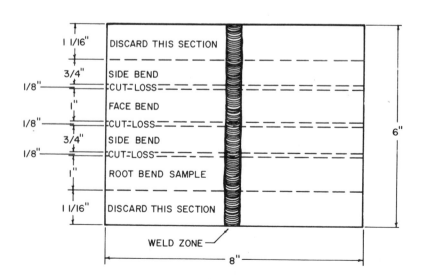

Fig. R10-3 Proper layout of sample for specimens to be used in a guided-bend test

5. Perform a guided-bend test on the face and root of the various strips.

6. Examine the various strips. There should be no separation of base metal from the weld and no porosity.

7. If the welded samples show defects, repeat steps 1-6.

8. When the samples are satisfactory, submit them to the instructor for comments and further instructions.

SECTION 10: REVIEW B

OBJECTIVE:

- Produce a stainless steel inside corner weld of proper penetration and quality in the flat position using standard TIG welding procedures. Subject such a sample to a destructive test.

Equipment and Materials:

Standard TIG welding equipment
Personal hand tools
Protective clothing and gear
Stainless steel wire brush

2 pieces, 1/16″ x 2 1/2″ x 6″ stainless steel sheet
1/16″ 308 filler rod
Argon shielding gas

PROCEDURE

1. Turn on the equipment and adjust for stainless steel welding.

2. Set the high frequency switch for "start only."

3. Position the plates for an inside corner weld and tack weld the ends and the center of the joint, figure R10-4.

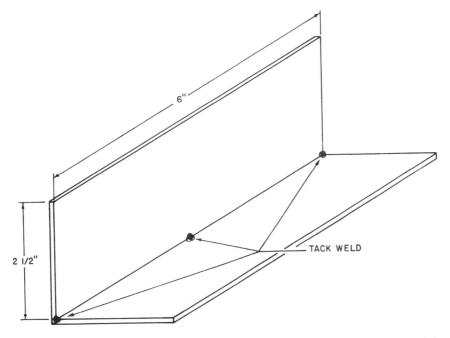

Fig. R10-4 Proper position of the plates and the tack welds for an inside corner joint

4. Place the sample in the flat position, up off the table, and weld the inside corner joint. Use filler rod and the T-joint torch motions.

5. Perform a destructive test as shown in figure R10-5.

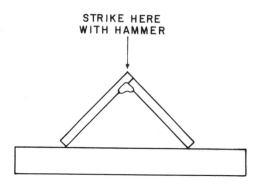

Fig. R10-5 Performing a destructive test on an inside corner weld

6. Submit the sample to the instructor for comments and further instructions.

SECTION 11:
Horizontal and Vertical Welds of Stainless Steel, TIG Process

Unit 48 Four Layer Pad Buildup, Horizontal Position, Stainless Steel

OBJECTIVES

After completing this unit the student will be able to:

- discuss the selection of filler wires for stainless steel welding.

- understand the importance of proper welding speeds and heat inputs for stainless steel welding.

- inspect and identify welding faults and make the corrections necessary to produce quality welds.

- successfully complete a four layer pad buildup in the flat position on stainless steel, using standard TIG welding procedures.

Filler Rod

Corrosion-resistant steel (stainless steel) must be welded with the filler rod which is suited to the base metal. The rod diameter of the filler material should be the smallest possible of a manufacturer's recommended size. Therefore a lower welding amperage can be used to make the weld deposit. If the rod diameter is too large, it will soak up a good deal of the welding heat which will make welding more difficult.

Heat Inputs

Stainless steel will distort much more than either carbon steel or aluminum and proper clamping of the base metal is required for a great many weldments. Travel speeds are much slower than those recommended for the other materials. The recommended welding procedures for thin sections calls for low heat inputs. With low amperage, it will take longer to make a good weld. More heat and faster travel speeds will make a difficult job harder to do and are not recommended except for the experienced welder. With high amperage and travel speeds, the operator must proceed with perfect rhythm. Even the smallest hesitation will cause the weld puddle to overheat and burn through in the form of excessive penetration. If such burn-through occurs, the weld will suck the atmosphere into the molten puddle, making a brittle, unusable weld. Only if backing is used on the weld can this situation be prevented.

Stainless Steel Weld Color

The color of a finished stainless steel weld will indicate whether too much heat has been used during welding. The student will have to learn to "read" the weld puddle and understand what it tells about penetration and weld bead contour.

A dark, purple color or a dark purple-blue bead with very few ripples or ripples that stretch into a long V indicate too much heat.

A good stainless steel TIG weld made with the correct heat is light red or a mixture of light reds and light purples.

Welding Inspection and Correction of Defects

VISUAL WELD INSPECTION

Defect	Reason	
Poor penetration; Lack of fusion	Poor joint preparation; welding amperage too low	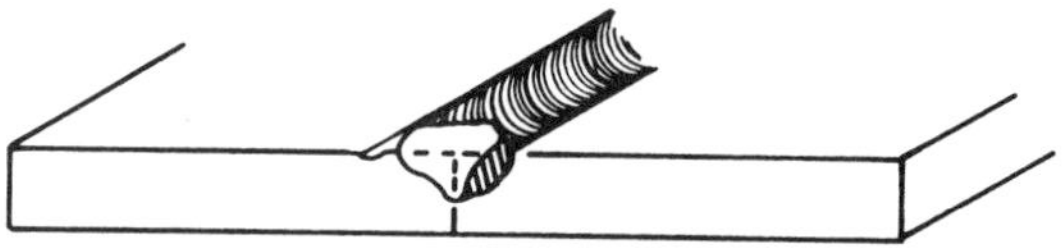
Bead too wide; Too much buildup Too much penetration	Welding speed too slow	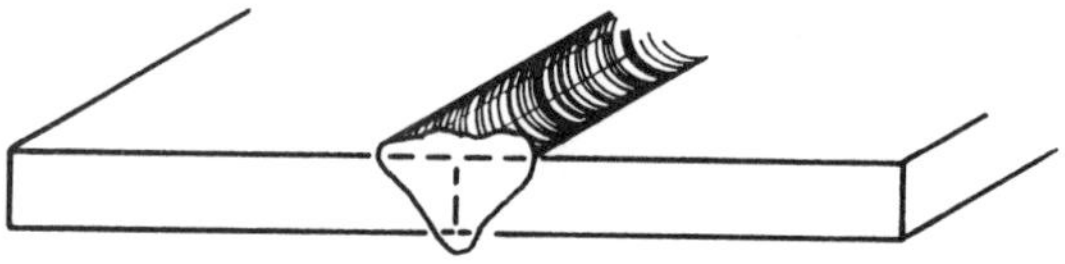
Undercut: Penetration not even; Not enough weld deposit	Amperage too high; Filler rod not added from right place	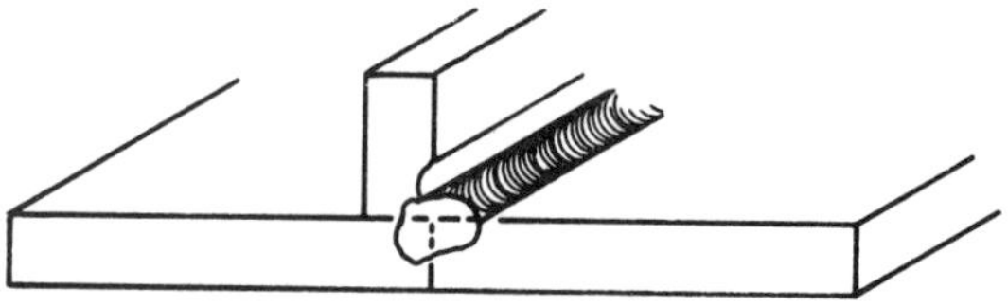
Bead too high; Penetration shallow	Low welding amperage	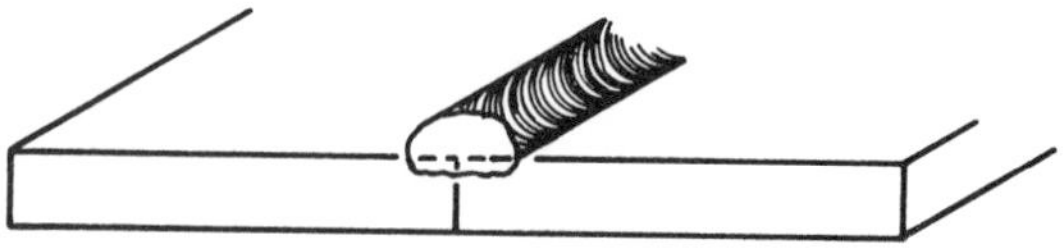
Undercut at edge; Too much penetration	Too much welding amperage	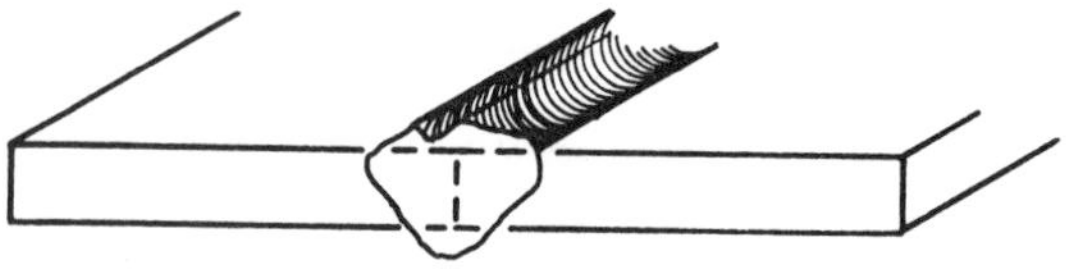
Bead too small; No penetration	Travel too fast	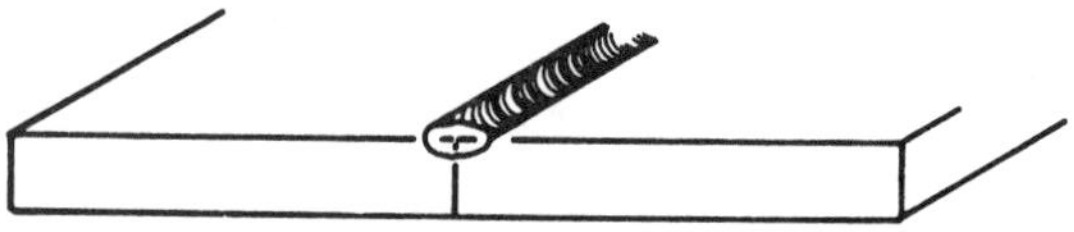

Fig. 48-1

JOB 48: FOUR LAYER PAD BUILDUP, HORIZONTAL POSITION, STAINLESS STEEL

Equipment and Materials:

Standard TIG welding equipment
Personal hand tools
Protective clothing and gear
Stainless steel wire brush

1 piece, 1/8" x 5" x 5" stainless steel
3/32" 308 filler rod
Argon shielding gas

PROCEDURE	KEY POINTS
1. Turn on all systems.	1. Visually check all systems for proper operation.
2. Set the machine for stainless steel welding.	
3. Adjust the machine by welding on stainless steel scrap.	3. Use only the forehand and stringer bead techniques for horizontal welds.
4. Place the plate vertical to the table. Begin at the bottom of the plate and weld a bead across, using a forehand motion, figure 48-2.	4. The TIG torch should slant upward in the direction of travel about 5 degrees. The torch angle and rod angle should be adjusted according to the skill of the operator.

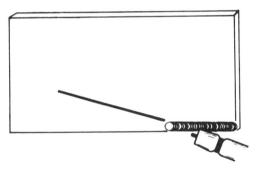

Fig. 48-2 Running a horizontal bead across the bottom of the plate

5. Check the beads as they are completed and make necessary corrections.	5. Wire brush each bead.
6. Follow the bead sequence shown in other horizontal welding jobs.	
7. After the first layer is completed, rotate the plate 1/4 turn so that the next layer of beads crosses the first.	
8. Keep turning the plate as each layer is completed until four layers have been welded on the plate.	8. Horizontal welds will have a tendency to undercut and sag. Proper torch and rod manipulation must be maintained.

REVIEW QUESTIONS

Answer the following questions briefly.

1. When TIG welding with less heat, should the welding wire be smaller or larger?

2. How does stainless steel react differently than carbon steel or aluminum when heat is applied?

3. What happens to a stainless steel weld when burn-through occurs and there is no backing for the weld?

4. What colors should appear on a good stainless steel TIG weld?

Unit 49 Lap Weld, Horizontal Position, Stainless Steel

OBJECTIVES

After completing this unit the student will be able to:

- describe the lap weld and the amount of penetration required for successful welding of the fillet weld.

- understand the care and storage requirements for stainless steel filler wires.

- successfully complete a lap weld in the horizontal position on stainless steel using standard TIG welding procedures.

Lap Welds on Stainless Steel Sheet

A lap weld requires a fillet weld to give strength to the joint. Lapping one piece of metal over another gives an edge which is 90 degrees with the other plate. Fillet welds are one of the major weld designs because the plates forming the joint need little preweld preparation. Most fillet welds require a closer arc than other types of joints and penetration on fillets must be maximum whether the joint is welded from one side or both sides.

Care and Storage of Stainless Steel Filler Rod

Sound quality welds can only be made from filler materials that have had proper care during use and storage. If the filler metal has been handled improperly, the weld deposit may be poor. Effective welds on stainless steel have to be made with filler rod designed for TIG welding. The wire must have the correct diameter and be matched in alloy and composition to the base metal. In general, the diameter of the filler wire should about equal the thickness of the metal to be welded.

Filler wire demands good care. It should be covered and stored in a dry place and the temperature should be kept constant. Filler rod should not be left out when it will not be used for a long time. Such rod should be taken from welding stations and stored in the original containers and the carton should be resealed. If rod ovens are available, the filler rod should be stored in them to prevent moisture from entering the boxes.

Identification of the base metal is necessary to select the right filler. Check all sources of information to make sure the base metal alloy is known.

The following factors must be considered in selecting filler rod for each welding job:

- Base metal strength

- Base metal composition

- Position of welding

- Amperage to be used

- Joint fitup and design
- Thickness of base metal
- Shape of the base metal
- Conditions of service for the completed weldment
- Production efficiency
- The work conditions for the job

The following chart, figure 49-1, lists the proper filler wire for various stainless steel alloys.

FILLER MATERIAL	FILLER COMPOSITION					ALLOYS WHICH CAN BE WELDED
	Carbon	Manganese	Silicon	Chromium	Nickel	
308L*	.025	1.8	.40	20.6	9.7	304, 308, 321, 347
308 Hi Sil**	.025	1.8	.85	20.6	9.7	301, 304
309L	.025	1.8	.40	24.0	13.5	309, and non-heat-treatable straight chromium
310	.12	1.8	.45	26.0	21.0	310, 304-clad and overlay
316L	.025	1.80	.35	19.5	13.0	316L
317L	.025	1.8	.40	19.0	12.5	317

* "L" signifies low carbon content
** "Hi Sil" signifies higher than normal content of silicon

Fig. 49-1 Recommended filler materials for various alloys of stainless steel

JOB 49: LAP WELD, HORIZONTAL POSITION, STAINLESS STEEL

Equipment and Materials:

Standard TIG welding equipment
Personal hand tools
Protective clothing and gear
Stainless steel wire brush

4 pieces, 1/8" x 3" x 6" stainless steel sheet
1/8" 308 filler rod
Argon shielding gas

PROCEDURE	KEY POINTS
1. Turn on all systems.	
2. Set the machine for TIG stainless steel welding.	2. Adjust the machine on scrap metal for horizontal welds.
3. Position the metal and tack weld both samples, figure 49-2.	

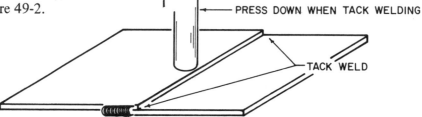

PRESS DOWN WHEN TACK WELDING

TACK WELD

Fig. 49-2 Proper position of the plates and the tack welds for a lap joint

PROCEDURE	KEY POINTS
4. Position one sample in the horizontal plane and weld one side of one joint, using the forehand motion and filler rod, figure 49-3.	4. Wire brush each joint before welding.

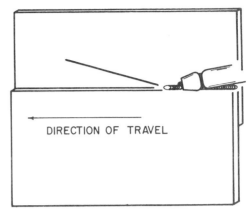

DIRECTION OF TRAVEL

Fig. 49-3 Proper position of the filler rod and the torch for a horizontal lap weld

PROCEDURE	KEY POINTS
5. Turn the sample over and weld the joint, using forehand motion without filler rod.	5. Adjustments of amperage may be necessary when welding without rod.
6. Position the other sample and repeat all procedures except that filler rod must be used for both welds.	6. Compare all the welds.
7. Shut down all systems.	
8. Submit the samples to the instructor for comments and instructions.	

REVIEW QUESTIONS

Answer the following questions briefly.

1. Name one primary reason why fillet welds are often used in industry.

2. Do fillet welds require closer arcs than other types of joints?

3. To make good welds with the TIG process, the wire must have the correct _____ and it must be matched in _____ and _____ to the base metal.

4. Where should filler wire be stored when not in use?

Unit 50 Four Layer Stringer Bead Pad Buildup, Vertical Position, Stainless Steel

OBJECTIVES

After completing this unit the student will be able to:

- discuss the recognized welding techniques used in the TIG welding industry.
- know and practice good workmanship standards as listed in the text.
- successfully weld a four layer pad buildup in the vertical position, using standard TIG welding procedures.

Stainless Steel Buildup, Vertical Position

This type of weld is very similar to other welds made in the vertical position. Pad buildup is normally done with a stringer bead motion and an uphand technique.

The torch should be pointed up at a 20 degree angle to help hold the puddle up against the force of gravity.

Recommended TIG Welding Techniques

The object of all welding processes is to complete the welding without creating internal stresses in the metal surrounding the weld. In order to gain this goal, the amperage (welding heat) should be set at the lowest practical level for each welding application. The right amperage must be set in the correct place for the right length of time.

When welding is done on thin metal or stainless steel, or when the heat is directed in a small area, the tungsten electrode should be sharpened. Sharpening ensures that the arc will occur and jump off the electrode at the same point each time the arc is struck. The arcing will also stay in the same spot during the arcing time.

A short arc should be set and held. The short arc will hold the heat in the area directly around the weld zone. The arc must be directed to the exact place where it will melt the base metal for the job. Using too much heat will apply it to larger areas and increase stress and distortion.

The arc must be directed on the joint to be welded until a molten puddle shows. After the puddle appears, welding speed of travel should move at a speed which allows both members of the joint to melt under the arc.

When the puddle is slow in forming, a slight circular motion sometimes speeds the melting. Good gas coverage is of extreme importance and the travel speed should not be so slow that it remelts already fused metal.

When filler rod is used, the rod diameter should be about equal to the thickness of the base metal.

The rod is dipped in and out of the weld zone to make a uniform deposit of weld. Correct motions and angles are very important.

The length of distance from the end of the electrode or the nozzle is related to the shape of the weldment and the type of joint. The longer the extension of the electrode, the less effective the gas shield and the length of the electrode may make it easier to become contaminated by contact with the base metal or weld puddle. When the electrode is extended far from the nozzle, the gas shield must be increased to cover the larger area.

Workmanship Standards

To produce clean, high-quality welds, the following rules should be put into effect:

- Clean the surface of all materials to be welded.
- Carefully prepare all bevels, lands, and root openings.
- Preheat and interpass heats should be controlled.
- Filler material must match the base material and should be of the correct diameter for the job.
- Welding procedures and bead sequences must be followed.
- Defective tack welds must be removed and replaced before final welding.
- Defective welds must be removed and replaced before final welding.
- Surface porosity must be removed and rewelded.
- Weld-end craters must be filled.
- All beads must be brushed.
- All undercutting must be controlled.

JOB 50: FOUR LAYER STRINGER BEAD PAD BUILDUP, VERTICAL POSITION, STAINLESS STEEL

Equipment and Materials:

Standard TIG welding equipment
Personal hand tools
Protective clothing and gear
Stainless steel wire brush
1 piece, 1/8" x 5" x 5" stainless steel sheet
3/32" 308 filler wire
Argon shielding gas

PROCEDURE	KEY POINTS
1. Turn on all the systems.	
2. Adjust the machine for vertical welds.	2. Always use scrap stainless steel and set the welding machine.
3. Place the plate vertical to the table. Begin at the bottom and weld a bead all the way up to the top, figure 50-1.	3. Wire brush the plate before welding.

PROCEDURE

KEY POINTS

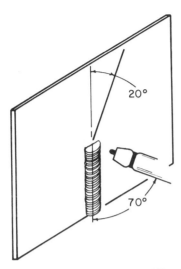

Fig. 50-1 Proper torch angle and filler rod position for a vertical-up weld

4. Complete the first layer using an alternating bead sequence and overlap each bead by 1/3.

5. Turn the plate 1/4 turn and build another layer across the first, figure 50-2.

4. Brush and inspect each bead. Make machine adjustments as needed.

5. The torch must slant upwards at 20 degrees and the filler rod is fed into the top of the puddle at the same angle from the top.

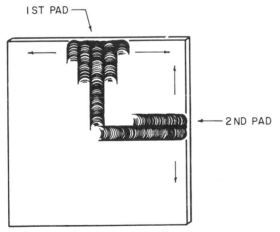

Fig. 50-2 Rotate the plate 90° after each layer until four layers are completed.

6. Rotate the plate until four layers have been deposited.

7. Submit the pad to the instructor for comments and instructions.

6. If the electrode becomes dirty, remove it and reshape it.

7. After completing the exercise, shut down all systems.

REVIEW QUESTIONS

Answer the following questions briefly.

1. What is the stated object of all welding processes?

2. How should the electrode be prepared for welding thin stainless steel?

3. What type of arc is recommended for use on stainless steel sheet?

4. What amount of amperage may increase distortion of stainless steel sheets?

5. What size diameter filler rod should be used in comparison with the thickness of the sheet to be welded?

6. How much shielding gas should be used when the tungsten electrode is extended far from the nozzle?

Unit 51 Lap Weld, Vertical Position, Stainless Steel

OBJECTIVES

After completing this unit the student will be able to:

- discuss the composition and applications of striped electrodes.
- explain the basic reasons for poor arc connection in TIG welding and ways to eliminate the condition.
- successfully make a lap weld, vertical position, on stainless steel, using standard TIG welding procedures.

Striped Tungsten

A recently developed tungsten electrode, called a striped electrode, has been introduced into the TIG field. The striped tungsten differs from other electrodes by having a solid strip of 2 percent thorium placed in a wedge the full length of the tungsten, figure 51-1.

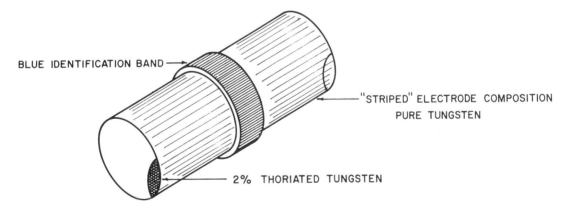

BLUE IDENTIFICATION BAND

"STRIPED" ELECTRODE COMPOSITION
PURE TUNGSTEN

2% THORIATED TUNGSTEN

Fig. 51-1 TIG electrode composition

The combination of thorium and tungsten as separate elements creates different arcing characteristics.

The striped electrode is in great demand because of certain advantages over regular tungsten electrodes:

- Higher currents can be used with smaller diameter electrodes.
- Arc stability improves appearance of the bead.
- Good heat resistance
- Lower electrical resistance
- Good shaping characteristics
- Easy arc starting

- Effective on all base materials

- Better performance than pure tungsten

- Longer electrode life

- Less contamination to the weld puddle

The strip of thorium equals about 1/4 of the cross section of the electrode. It cannot be seen as a definite division. It combines the shaping and stability of pure tungsten, with the arc starting and amperage capability of thoriated tungsten.

The striped electrode works well on AC polarity on aluminum and magnesium. It is also good for some DC applications. The striped electrode is not recommended for welding where the electrode must be ground since this may remove the thorium and leave pure tungsten at the point.

The striped electrode is color coded with a blue band around the center of the electrode.

TIG Arc Characteristics

Starting the arc is somewhat more difficult with the TIG process than with some other operations. A poor arc occurring while TIG welding may be the result of one or more of the following:

- Wrong diameter nozzle

- Dirty tungsten electrode

- Split electrode

- Burned electrode

- Wrong setting on the high frequency

- Wrong inert-gas flow rate

- Dirty base material

- Wrong amperage

- Defective equipment

- Poor ground on base material

Each of these conditions should be checked and corrections made. Three methods are used to establish an arc for TIG welding:

1. High frequency, continuous

2. High frequency, start only

3. Touch starting

When welding with AC polarity, and the superimposed high frequency, the tungsten does not have to touch the work to strike the arc. The high-frequency current makes a path for the amperage to follow by jumping the air gap between the tungsten and the base metal.

For most welding jobs, the arc length is about one tungsten diameter. When using argon as a shielding gas, the length of the arc may vary.

The high frequency, "start only," can be used with either polarity. On DC settings, if the machine is equipped with a high-frequency unit, it may be used to advantage. On the "start only," the high frequency automatically turns off after the arc is started.

The touch starting is used with DC without the high frequency to start the arc. The electrode has to touch the workpiece in order to arc. When this method is used, the electrode should be raised approximately 1/8″ above the weld zone to keep from contaminating the electrode. The arc can be struck on a separate sheet of copper, aluminum, or carbon steel and then carried over to the base metal to be welded.

JOB 51: LAP WELD, VERTICAL POSITION, STAINLESS STEEL

Equipment and Materials:

Standard TIG welding equipment
Personal hand tools
Protective clothing and gear
Stainless steel wire brush
4 pieces, 1/16″ x 2 1/2″ x 6″ stainless steel sheet
1/16″ 308 filler rod
Argon shielding gas

PROCEDURE	KEY POINTS
1. Turn on all systems.	
2. Set the machine for stainless steel welding.	2. Use scrap metal for setting the machine.
3. Position the stainless steel sheets and tack weld, figure 51-2.	3. Brush each joint before welding.

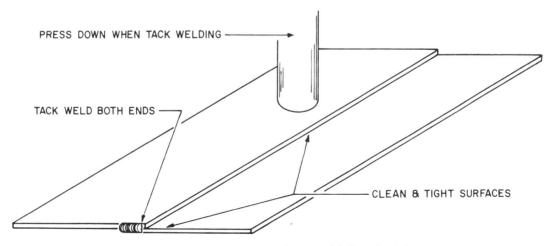

PRESS DOWN WHEN TACK WELDING

TACK WELD BOTH ENDS

CLEAN & TIGHT SURFACES

Fig. 51-2 Proper position of tack welds for a lap joint

PROCEDURE	KEY POINTS
4. Position one tacked sample in the vertical plane and weld the joint with the forehand motion, using filler rod, figure 51-3.	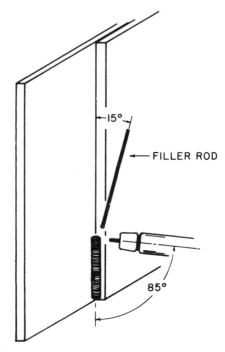

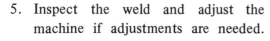

Fig. 51-3 Proper position of the torch and rod for a vertical-up weld

5. Inspect the weld and adjust the machine if adjustments are needed.	
6. Rotate the sample and weld the other side, without filler rod.	
7. Position the other sample and weld both sides using a forehand motion, with a filler rod.	7. Compare the joint welded without rod against the other.
8. Shut down all systems.	
9. Submit both samples to the instructor for comments and instructions.	

REVIEW QUESTIONS

Answer the following questions briefly.

1. How is the striped tungsten different from regular tungstens?

2. What type of currents can be used with smaller diameter striped tungstens?

3. How does the striped tungsten electrode life compare to ordinary electrodes?

4. Can the strip of thorium be seen in the striped tungsten electrode?

5. Name the three methods of starting an arc in TIG welding.

6. Does defective equipment cause a poor arc?

7. Can the wrong diameter nozzle cause poor arc starting?

8. Does a split electrode have an effect on the arc?

Unit 52 T Weld, Vertical Position, Stainless Steel

OBJECTIVES

After completing this unit the student will be able to:

- describe the proper position of the torch and filler rod for a vertical T weld.
- successfully TIG weld a T weld on stainless steel in the vertical position using standard TIG welding procedures.

Stainless Steel T Weld, Vertical Position

The vertical T weld is a difficult weld to master. The torch must be directed slightly upward and pointed into the corner of the two plates. The gas flow must be increased to ensure an ample gas shield over the weld zone.

The addition of filler rod must be sufficient enough to keep a full, rounded bead. A shallow bead will not produce adequate penetration and allows the weld to break if any force is applied to the upright member of the T weld.

A vertical-up motion combined with a stringer bead technique is the proper method of travel, figure 52-1.

JOB 52: T WELD, VERTICAL POSITION, STAINLESS STEEL

Equipment and Materials:

Standard TIG welding equipment
Personal hand tools
Protective clothing and gear
Stainless steel wire brush
2 pieces, 1/16" x 4" x 6" stainless steel sheet
2 pieces, 1/16" x 3" x 6" stainless steel sheet
1/16" 308 filler rod
Argon shielding gas

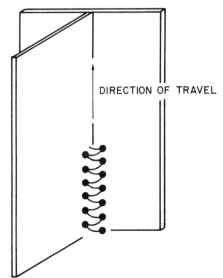

DIRECTION OF TRAVEL

Fig. 52-1 **Proper direction of travel and bead motion for a vertical-up weld**

PROCEDURE	KEY POINTS
1. Turn on all systems.	1. Visually check all systems.
2. Set the machine for stainless steel welding.	2. Adjust the welding machine by making sample welds on scrap metal.

PROCEDURE	KEY POINTS
3. Extend the tungsten electrode if necessary.	
4. Position the 3-inch sheets in the center of the 4-inch sheets and tack weld both ends of the joint, figure 52-2.	4. Tack weld the joints with the same procedures used in other vertical T welds.

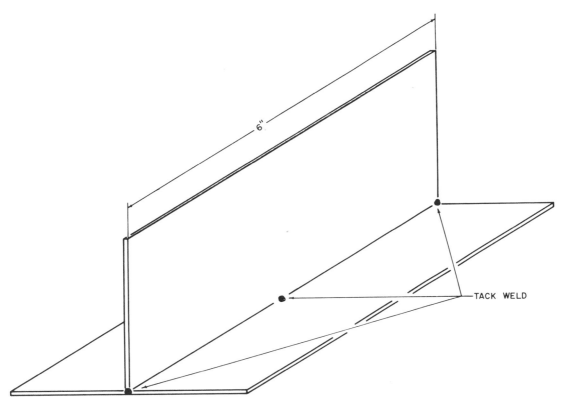

Fig. 52-2 **Proper position of the plates and tack welds for a T weld**

5. Place one sample in the vertical position and weld one side with a stringer bead motion, using filler rod.

6. Cool the sample.

7. Inspect the back of the joint for weld defects.

8. Readjust the amperage if honeycombing appears on the back of the joint.

9. Position the other sample and weld it using filler rod.

10. Cool the sample.

PROCEDURE	KEY POINTS
11. Shut down all systems.	
12. Submit the samples to the instructor for comments and instructions.	

REVIEW QUESTIONS

Answer the following questions briefly.

1. What motion is recommended for the vertical T weld?

2. What filler rod is recommended for the vertical T joint in Job 52?

SECTION 11: REVIEW

OBJECTIVE:

- Produce a stainless steel outside corner weld of proper penetration and quality in the vertical position using standard TIG welding procedures. Subject such a sample to a destructive test.

Equipment and Materials:

Standard TIG welding equipment
Personal hand tools
Protective clothing and gear
Stainless steel wire brush
2 pieces, 1/16" x 2 1/2" x 6"
    stainless steel sheet
Argon shielding gas

PROCEDURE

1. Turn on the equipment and adjust for stainless steel welding.

2. Tack weld the corner joint as shown, figure R11-1.

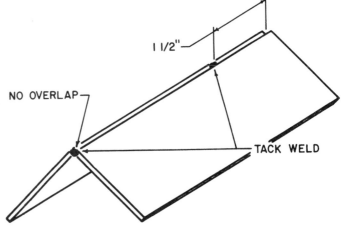

Fig. R11-1 **Proper position of the plates and tack welds for an outside corner weld**

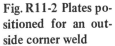

Fig. R11-2 Plates positioned for an outside corner weld

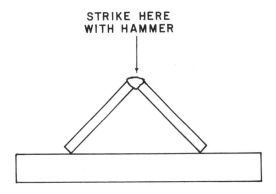

Fig. R11-3   Performing a destructive test on an outside corner weld

3. Position the plates in the vertical position, figure R11-2.

4. Weld the sample using a forehand technique and a stringer bead motion.

5. Perform the destructive test as shown in figure R11-3.

6. Submit the sample to the instructor for comments and instructions.

Glossary

AC (Alternating Current): Electricity which reverses its direction of electron flow.

Alloy: A mixture of two or more metals.

Anode: Positive pole — part of an electrical circuit towards which the electrons are flowing.

Arc: The flow of electricity through an air gap. The arc flowing through the air produces high temperatures.

Backing: Material placed on the back side of a weld to keep the penetration even.

Base Metal: The metal to be welded.

Bead: The weld; used to describe the neat ripples formed by semiliquid metal.

Bevel: An edge which has been ground at an angle.

Butt Joint: The joint made when two pieces are in the same plane, with the edge of one piece touching the other.

Cathode: Negative pole.

Crater: A depression in the weld, usually at the end.

Crown: The surface of the finished bead.

DC (Direct Current): Electricity which flows in only one direction.

Direct (Straight) Polarity: Electrical current flowing from the electrode to the base metal. Electrode negative and base metal positive.

Electrode: The material which carries the electricity up to the point where the arc is formed. (Arc, rod, tungsten)

Filler Rod: Metal wire which is melted into the puddle of the weld. (Welding rod)

Fillet Weld: A weld in the corner of the angle formed by two or more pieces of metal.

Flash: The impact of the electric arc against the human eye.

Flat Position Weld: A weld made horizontally on the horizontal surface of the parent metal.

Fusion: Mixing of melted metal.

Gas Metal-Arc Welding: A weld made using a wire feed of continuous metal and a shielding gas. (MIG welding) (Metallic Inert Gas Welding)

Gas Tungsten-Arc Welding: A weld made using a tungsten (non-consumable) electrode and a shielding gas. (TIG welding) (Tungsten Inert-Gas Welding)

Helmet: A protective hood which fits over the head. In welding it contains a dark lens through which the light of the arc may be watched.

Horizontal Position: A weld made on a horizontal seam, generally with the surface of the metal in a vertical position.

Inert-Gas Arc Welding: Surrounding the arc with a gas which has no reaction with the metal and keeps the atmosphere away from the molten puddle.

Infrared Rays: Dangerous rays produced by the light of arc welding — injurious to the eyes and skin.

Lap Joint: A joint where the pieces to be welded overlap each other.

MIG: Gas metallic arc welding. (Metallic inert gas)

Pass: One progression across the metal; one bead of weld.

Penetration: The depth of fusion into the metal being welded.

Porosity: Gas pockets or voids in metal which has been welded.

Puddle: The molten part of the weld, where the arc is supplied.

Reverse Polarity: Causing the electrodes to flow from the base metal to the electrode; electrode positive.

Sequence: The order in which the beads (passes) are welded on the joint.

Short Arc: Gas-metal arc welding (MIG) which uses low voltage; the arc is interrupted as the metal bridges the gap of the arc.

Spray Arc: Gas-metal arc welding (MIG) which uses an arc voltage high enough to transfer the electrode metal across the arc in small globules.

Straight Polarity: Causing the electrodes to flow from the electrode to the base metal; electrode negative.

Tack, Tack Weld: A small weld used to hold any assembly together before it is fully welded.

T Joint: A joint made by putting one piece of metal against another at a 90 degree angle.

TIG: Gas tungsten arc welding. (Tungsten inert gas)

Tungsten: TIG welding electrode.

Ultraviolet Rays: Harmful energy waves given off by the arc dangerous to the eyes and skin.

Vertical Position: A joint where the welding is done on a vertical seam on a vertical surface.

NITROGEN, ARGON, HELIUM
AND CARBON DIOXIDE SAFETY
PRECAUTIONS

Nitrogen, argon, helium, and carbon dioxide are inert, colorless, odorless, and tasteless gases. Nitrogen makes up about 78 per cent of the atmosphere, argon about 1 per cent.

WARNING:

NITROGEN, ARGON, HELIUM, AND CARBON DIOXIDE CAN CAUSE ASPHYXIATION AND DEATH IN CONFINED, POORLY VENTILATED AREAS.

NITROGEN, ARGON AND HELIUM AS LIQUIDS OR COLD GASES AND CARBON DIOXIDE AS COLD GAS CAN CAUSE SEVERE FROSTBITE TO THE EYES OR SKIN. DO NOT TOUCH FROSTED PIPES OR VALVES. IF ACCIDENTAL EXPOSURE TO THESE GASES OCCURS, CONSULT A PHYSICIAN AT ONCE. IF A PHYSICIAN IS NOT READILY AVAILABLE, WARM THE AREAS AFFECTED BY FROSTBITE WITH WATER THAT IS NEAR BODY TEMPERATURE.

USE A PRESSURE-REDUCING REGULATOR WHEN WITHDRAWING GASEOUS NITROGEN, ARGON, HELIUM, OR CARBON DIOXIDE FROM A CYLINDER OR OTHER HIGH-PRESSURE SOURCE.

KEEP EQUIPMENT AREA WELL VENTILATED.

Nitrogen, argon, and helium are non-toxic, but they can cause asphyxiation in a confined area that does not have adequate ventilation. Any atmosphere which does not contain enough oxygen for breathing (at least 18 per cent) can cause dizziness, unconsciousness, or even death. When there is doubt about the adequacy of ventilation, use an oxygen analyzer with a 0 to 25% scale to check for oxygen.

Carbon dioxide affects the important acid-base balance in the body. Carbon dioxide is formed in normal functioning within the body, but the body can tolerate increased amounts of carbon dioxide only in limited concentration. This is recognized in OSHA standards where a Threshold Limit Value of 5,000 parts per million by volume (0.5 per cent concentration) has been adopted. For safety, concentrations above this level should not be permitted; increased concentrations can cause bodily harm or death. Additionally, carbon dioxide can cause asphyxiation by displacing oxygen resulting in dizziness, unconsciousness or death.

Nitrogen, argon, helium and carbon dioxide cannot be detected by the human senses and will be inhaled like air. If adequate ventilation is not provided, these gases may displace normal air without warning that a life-depriving atmosphere is developing. Store containers outdoors or in other well ventilated areas. Never enter any tank, pit or other confined area where these gases may be present until purged with air and tested for a breathable atmosphere, using a gas analyzer.

NITROGEN, ARGON, HELIUM, AND CARBON DIOXIDE CAN BE EXTREMELY COLD.

(About 100 to 450 degrees F. below zero).
COVER EYES AND SKIN.

Accidental contact of liquid or cold gas with the eyes or skin may cause severe frostbite. Handle liquid so that it will not splash or spill. Protect your eyes with safety goggles or face shield, and cover the skin to prevent contact with the liquid or cold gas, or with cold pipes and equipment. Protective gloves that can be quickly and easily removed and long sleeves are recommended for arm protection. Wear cuffless trousers outside boots or over high-top shoes to shed spilled liquid.

KEEP AIR AND OTHER GASES AWAY FROM LIQUID HELIUM.

The low temperature of liquid helium (452 deg. F. below zero) can solidify any gas. Such solidified gases can plug pressure-relief passages and foul relief valves. Plugged passages are hazardous because of the need to relieve excess pressure produced as heat leaks into the continually evaporating liquid. Always store and handle liquid helium under positive pressure and in closed systems to prevent infiltration and solidification of air or other gases.

KEEP EXTERIOR SURFACES OF LIQUID HELIUM EQUIPMENT CLEAN TO PREVENT COMBUSTION.

Air will condense on exposed helium liquid or cold-gas surfaces, such as vaporizers and piping. Nitrogen, having a lower boiling point than oxygen, will evaporate first, leaving an oxygen-enriched condensation on the surface. To prevent the possible ignition of grease, oil, or other combustible materials on such surfaces, all areas of possible air condensation should be kept free of these materials.

NEVER USE CONTAINERS, EQUIPMENT, OR REPLACEMENT PARTS OTHER THAN THOSE SPECIFICALLY DESIGNATED FOR USE IN NITROGEN, ARGON, HELIUM OR CARBON DIOXIDE SERVICE.

For further safety information, refer to Linde Form 9888, "Precautions and Safe Practices – Liquefied Atmospheric Gases," and Form 11-785, "Safe and Efficient Handling of Liquid Helium."

IF NECESSARY TO DISPOSE OF WASTE GAS OR LIQUID, EXERCISE CAUTION.

Gaseous nitrogen, argon, helium, or carbon dioxide should be released only in an open outdoor area. Liquid nitrogen, argon or helium should be dumped into an outdoor pit filled with clean, grease-free and oil-free gravel, where it will evaporate rapidly and safely.

HYDROGEN SAFETY PRECAUTIONS

Hydrogen is a colorless, odorless, and tasteless gas. It is the lightest of all elements.

WARNING:

HYDROGEN IS A FLAMMABLE GAS. A MIXTURE OF HYDROGEN WITH OXYGEN OR AIR IN A CONFINED AREA WILL EXPLODE IF IGNITED BY A SPARK, FLAME OR OTHER SOURCE OF IGNITION. HYDROGEN FLAME IS VIRTUALLY INVISIBLE.

HYDROGEN AS A LIQUID OR COLD GAS MAY CAUSE SEVERE FROSTBITE TO THE EYES OR SKIN. DO NOT TOUCH FROSTED PIPES OR VALVES. IF ACCIDENTAL EXPOSURE TO LIQUID HYDROGEN OCCURS, CONSULT A PHYSICIAN AT ONCE. IF A PHYSICIAN IS NOT READILY AVAILABLE, WARM THE AREAS AFFECTED BY FROSTBITE WITH WATER THAT IS NEAR BODY TEMPERATURE.

USE A PRESSURE-REDUCING REGULATOR WHEN WITHDRAWING GASEOUS HYDROGEN FROM A CYLINDER OR OTHER HIGH-PRESSURE SOURCE. TAKE EVERY PRECAUTION AGAINST HYDROGEN LEAKS. ESCAPING HYDROGEN CANNOT BE DETECTED BY SIGHT, SMELL OR TASTE. BECAUSE OF ITS LIGHTNESS, IT HAS A TENDENCY TO ACCUMULATE IN THE UPPER PORTIONS OF CONFINED AREAS.

KEEP HYDROGEN AWAY FROM SOURCES OF IGNITION, AND DO NOT PERMIT ANY ACCUMULATION OF GAS.

Concentrations of hydrogen between 4 per cent and 75 per cent by volume in air are relatively easy to ignite by a low-energy spark and may cause an explosion. Smoking, open flames, unapproved electrical equipment, and other ignition sources must not be permitted in hydrogen areas. Hydrogen, when ignited, will burn with an almost invisible flame. Store containers outdoors or in other well-ventilated areas and away from heat sources, such as furnaces, ovens, hot-metal ladles, and radiators and away from flammable materials, such as gasoline, kerosene, oil and combustible solids.

KEEP AIR AND OTHER GASES AWAY FROM LIQUEFIED HYDROGEN.

The low temperature of liquid hydrogen can solidify any gas except helium. Such solidified gases can plug pressure-relief passages and foul relief valves. Plugged passages are hazardous because of the need to relieve excess pressure produced as heat leaks into the continually evaporating liquid. Air must be kept out of contact with liquid hydrogen to prevent accumulation of potentially explosive concentrations. Therefore, always store and handle liquid hydrogen under positive pressure and in closed systems to prevent infiltration and solidification of air or other gases.

KEEP EXTERIOR SURFACES OF LIQUID HYDROGEN EQUIPMENT CLEAN TO PREVENT COMBUSTION.

Air will condense on exposed hydrogen liquid or cold-gas surfaces, such as vaporizers and piping. Nitrogen, having a lower boiling point than oxygen, will evaporate first, leaving an oxygen-enriched condensation on the surface. To prevent the possible ignition of grease, oil, or other combustible materials on such surfaces, all areas of possible air condensation should be kept free of these materials.

LIQUID HYDROGEN IS EXTREMELY COLD.

(about 420 deg. F. below zero).

Accidental contact of liquid or cold gas with the eyes or skin may cause severe frostbite. Handle liquid hydrogen so that it will not splash or spill. Protect your eyes with safety goggles or face shield, and cover the skin to prevent contact with the liquid or cold gas, or with cold pipes and equipment. Clean, protective gloves that can be quickly and easily removed, and long sleeves are recommended for arm protection. Cuffless trousers should be worn outside boots or over high-top shoes to shed spilled liquid.

KEEP EQUIPMENT AREA WELL VENTILATED.

Although hydrogen is non-toxic, it can cause asphyxiation in a confined area that does not have adequate ventilation. Any atmosphere which does not contain enough oxygen for breathing (at least 18 per cent) can cause dizziness, unconsciousness, or even death. Hydrogen gas cannot be detected by human senses and will be inhaled like air. If adequate ventilation is not provided, it may displace normal air without warning. Store containers outdoors, or in other well ventilated areas. Never enter any tank, pit or other confined area where hydrogen may be present until purged first with nitrogen, then with air, and checked.

NEVER USE CONTAINERS, EQUIPMENT, OR REPLACEMENT PARTS OTHER THAN THOSE SPECIFICALLY DESIGNATED FOR USE IN HYDROGEN SERVICE.

OBSERVE ALL APPLICABLE SAFETY CODES WHEN INSTALLING HYDROGEN EQUIPMENT.

Before marking any installation become thoroughly familiar with NFPA (National Fire Protection Association) Standards No. 50-A, "Standard for Gaseous Hydrogen Systems at Consumer Sites," and 50-B, "Standard for Liquefied Hydrogen Systems at Consumer Sites," and with all local safety codes. For further safety information, refer to Linde Form 9914, "Precautions and Safe Practices – Liquid Hydrogen."

IF NECESSARY TO DISPOSE OF WASTE GAS OR LIQUID, EXERCISE CAUTION.

Liquid and gaseous hydrogen must be disposed of outdoors in isolated, well-ventilated areas away from personnel, combustible materials, and ignition sources. Liquid hydrogen for disposal should be completely vaporized and the vapor vented in a safe manner. Keep in mind that a flammable mixture will exist for some distance downwind of the disposal area. An aluminum pan, 3-feet square by 2-in. deep makes a suitable flash evaporator for disposal of moderately small quantities of liquid hydrogen.

WELDING PROCEDURES DESIGN AND SPECIAL PROCESSES

by Frank R. Schell
   Bill Matlock

Preface

The design of weldments and the communication of this design to others is as important to the welding industry as the actual welding processes. Metal is the most used of any broad group of materials in existence and welding is a practical way to join metals. However, the changes that take place in metal during the heating and cooling caused by welding must be understood if weldments are to give satisfactory service. The designers of welded objects must communicate their carefully engineered designs to the welders who will fabricate these objects. Therefore, it is important for those who design, weld, or use fabricated metal products to understand basic metallurgy, welding design, and symbols used on welding drawings.

Section one of this book discusses materials for welding, metallurgy, welding design, and communication of these designs in a concise manner. Only information which is of particular importance to the welder has been included. At the end of each of the four units is a series of review questions which provides the student with an opportunity for self-evaluation.

The second section of WELDING PROCEDURES: DESIGN, AND SPECIAL PROCESSES covers welding processes which, although they are not available in all school situations, are of particular importance to the welding industry. The carbon-arc process is actually a variation of electric-arc welding process discussed in book two. As in shielded metal-arc welding, the carbon-arc process relies on heat generated by an electric current as it jumps the gap from an electrode to the base metal. The air carbon-arc process is one of growing importance to the welding industry and no welder should enter the field without being familiar with this process.

Section two provides the student welder with an opportunity to gain some experience in air carbon-arc cutting and to become familiar with other special welding processes. There are jobs at the end of each unit in this section which provide hands-on application opportunities for experiencing the content discussed within the unit.

Contents

SECTION 1:
Welding Design, Symbols, and Metallurgy

Unit 1 Steel Shapes and Ordering Information

OBJECTIVES

Upon completion of this unit, the student will be able to:

- identify the standard steel shapes.
- demonstrate the correct procedures for ordering steel.
- list the proper symbols of various structural steel shapes used in the welding trade.

Angle Iron

One of the most commonly welded structural shapes is angle iron, figure 1-1. Four measurements are used to order angle iron:

- the width of one leg.
- the width of the other leg.
- the thickness of the legs.
- the length of the piece.

The angle iron shown in figure 1-1 is ordered in this way:

1 piece (or pc.) 2″ x 2″ x 1/4″ by 2″ angle

The symbol used in the trade to represent angle iron resembles the structural shape. The symbol is a capital letter L.

Angle-iron leg sizes are always measured from the outside corner to the outside of the leg, figure 1-1.

Angle iron can be ordered with a variety of leg sizes. The legs need not have the same measurement. Angle iron may be ordered with a 3-inch leg on one side and a 2-inch leg on the other, or with any number of combinations.

The complete ordering formula is: W x W x T x L
Example: L 2″ x 2″ x 1/4″ x 20′

Channel Iron

Another structural shape used in the welding field is channel iron, figure 1-2, page 2.

Two different methods may be used to order channel iron. The first method is expressed as: D x F x T x L, where

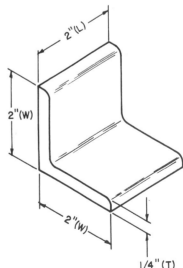

Fig. 1-1 Angle iron

1

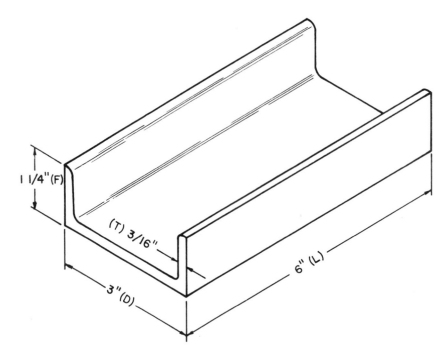

Fig. 1-2 Channel iron

- D equals the depth of the channel in inches;
- F equals the width of the flange in inches;
- T equals the thickness of the web in inches; and
- L equals the length in inches or feet.

This formula is complex and ordinarily not required in the field. The most commonly used ordering formula is:

- D (depth) x Pounds per Foot.

Generally, channel iron can be bought by ordering it by size and pounds per foot, since standard channel iron will weigh a certain number of pounds per foot. This weight governs the size of the flange and the thickness of the web.

The depth of the channel iron in figure 1-2 is 3 inches and each foot weighs 8 pounds. The symbol for channel iron is C. Therefore, the complete ordering formula is D x Pounds per Foot x Length.

Example: 1 piece (or pc.) C 6" x 8# x 18'

American Standard Beam

A third structural shape is the American standard beam, figure 1-3. The name is used to specify this particular shape. An American standard beam is ordered with the formula D x L where:

- D equals the depth in inches;
- L equals the length in feet and inches.

American standard beams may also be specified in pounds per foot when a certain strength factor is needed. In this case, the formula would read:

- D (depth) x L (length) x Pounds per Foot.

2

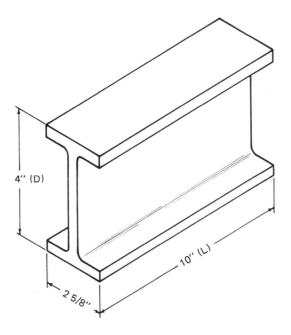

Fig. 1-3 American standard beam

The symbol for American standard beams is S.

    Example: 1 piece (or pc.) S 4″ x 12′ x 7.7 pounds per foot

Wide-Flange Beam

    Another type of beam is the wide-flange beam. The wide-flange beam has parallel flanges, figure 1-4.

    Wide-flange beams are ordered with the formula: D x F x L. They can also be ordered with the specified pounds per foot. Either of the following methods may be used:

- D x F x L; or

- D x Pounds per Foot x Length.

    The symbol for the wide-flange beams is WF.

    The depth of the wide flange in figure 1-4 is 6 inches, the width of the flange is 4 inches, and the length is 10 inches. The weight per foot is 16 pounds.

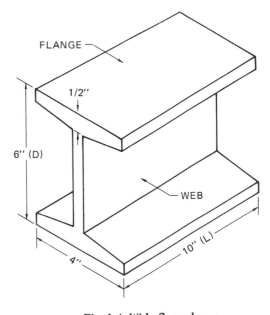

Fig. 1-4 Wide-flange beam

The formula is:

WF 6" x 4" x 10" or WF 6" x 16# x 10"

Strips, Bands, and Sheet Metal

Steel rolled into sheets 1/8 inch thick or less is known as *sheet metal.* When the thickness is over 1/8 inch, it is not called sheet metal, but is designated as *plates* or *flats.*

Most sheet metal is designated in gauges, rather than in fractional measurements. The most common designation used for sheet metal is the U. S. Standard Sheet Metal Gauge, figure 1-5.

When sheet metal is ordered, the formula used is T (thickness) x W (width) x L (length), figure 1-6.

	UNITED STATES STANDARD GAUGE FOR SHEET STEEL				
Number of Gauge	Approximate Thickness, in Fractions of an Inch	Approximate Thickness, in Decimal Parts of an Inch	Approximate Thickness, in Millimeters	Weight per Square Foot, in Pounds	Weight per Square Meter in Kilograms
11	1/8	.125	3.175	5.0	24.41
12	7/64	.109	2.778	4.375	21.36
13	3/32	.094	2.381	3.75	18.31
14	5/64	.078	1.984	3.125	15.26
16	1/16	.062	1.587	2.5	12.21
18	1/20	.05	1.270	2.0	9.76
20	3/80	.0375	0.9525	1.5	7.32
22	1/32	.03125	0.7937	1.25	6.10
24	1/40	.025	0.635	1.00	4.88
26	3/160	.01875	0.476	0.75	3.66
28	1/64	.0156	0.396	0.625	3.05
30	1/80	.0125	0.3175	0.5	2.44

Fig. 1-5

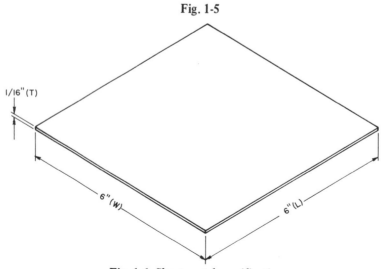

Fig. 1-6 Sheet metal specifications

The complete ordering formula for the piece in figure 1-6 is:

1 piece (or pc.) 16 gauge x 6" x 6"

Flats and Plates

Flats are furnished in different widths and generally in either 18- or 20- foot lengths. Some companies designate the material as a flat until it exceeds 6 inches in width, and others furnish it in 8-inch or 10-inch widths. Sheet metal that is wider than these widths and over 1/8 inch in thickness, is generally called *plate*. The structural steel symbol for a flat bar is FB. The structural steel symbol for a plate is PL.

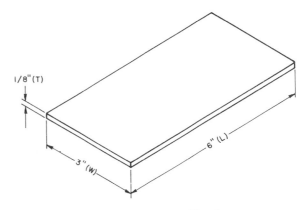

Fig. 1-7 Flat bar specifications

The formula for ordering flats is the same as for sheet metal:

T x W x L.

The ordering formula for the steel bar illustrated in figure 1-7 reads:

1 piece (or pc.) FB 3/16" x 3" x 6"

The complete ordering formula for the piece illustrated in figure 1-8 is:

1 piece (or pc.) PL 1/4" x 12" x 18".

For strips, bands, sheets, flats, and plates the order is always made with the symbol first, the thickness second, the width third, and the length fourth.

Square, Round, and Hexagon Bars and Pipe and Tubing

The other most important structural shapes are square, round, and hexagon bars and pipe and tubing.

Most kinds of steel are either hot rolled or cold rolled. *Hot-rolled steel* is made from white-hot steel, rolled through various sized rollers until the desired thickness and width are

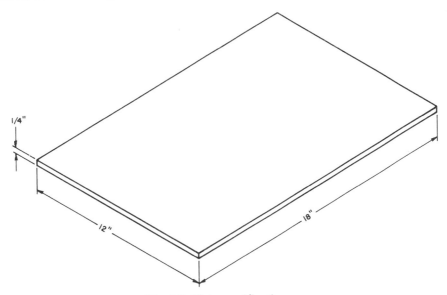

Fig. 1-8 Plate specifications

5

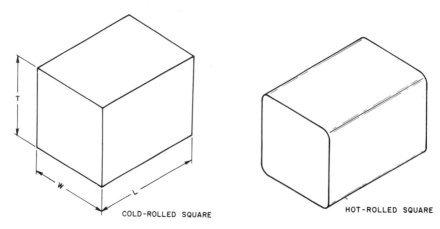

Fig. 1-9 Square bar stock

achieved. *Cold-rolled steel* is rolled from cold steel, allowing the size to be kept within close specifications. In all cases, cold-rolled steel is much closer to exact size than hot-rolled steel, which may vary a small amount from size specifications.

Cold-rolled square bars have sharp, square corners, while hot-rolled square bars have rounded corners, figure 1-9.

Round bar, ordered in cold-rolled steel is exact in size and diameter. Ordered in hot-rolled steel, it is generally a small amount oversized but cheaper in price. Cold-rolled steel has a shiny, nickel-like appearance; hot-rolled steel is dark and has some mill scale on it.

The structural symbol for square bars is □ or ⌼ .

Square bars are ordered in this manner:

T (thickness) x W (width) x L (length).

Round bars are ordered in this manner:

D (diameter) x L (length). The symbol is ○ or ⌀ and the ordering measurement is followed by the designation cold rolled (CR) or hot rolled (HR), figure 1-10.

Hexagon (six-sided) bars are ordered by measuring the distance across the flats, figure 1-11.

The ordering formula reads:

F (distance across the flats) x L (length). The hexagon shape must be specified in the order.

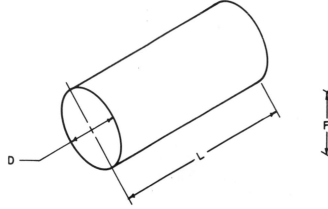

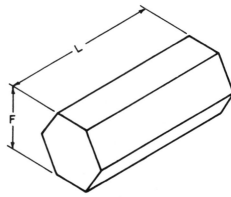

Fig. 1-10 Round bar stock   **Fig. 1-11 Hexagon bar stock**

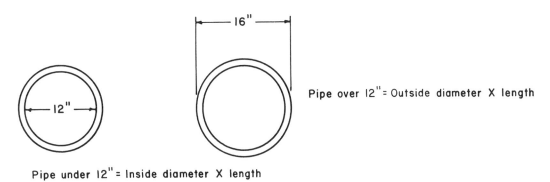

Pipe over 12" = Outside diameter X length

Pipe under 12" = Inside diameter X length

Fig. 1-12

Pipe and tubing are standard structural shapes used in welding.

Pipe is ordered by the *inside diameter* (ID) times the length, unless over 12 inches in diameter. If over 12 inches in diameter, it is ordered by the *outside diameter* (OD). Therefore, if a piece of 2-inch pipe is requred, it has a 2-inch inside measurement, but will be larger than 2 inches on the outside because of the wall thickness, figure 1-12.

Figure 1-12 refers to pipe having a standard wall thickness. Pipe may also be ordered by schedules. *Schedules* are established so that heavy-duty wall thicknesses may be ordered.

The formula for ordering pipe:

ID x L unless over 12 inches, when it becomes OD x L (L being the given length).

The formula for ordering tubing is:

OD (outside diameter) x L (length)

Figure 1-13 shows pipe sizes and outside diameters. Subtracting the pipe size from the outside diameter indicates the thickness of the two walls. If the result of the subtraction is divided by 2, the wall thickness will be known.

Example: Pipe size = 1"; Outside Diameter (OD) = 1.315"
1.315" – 1" = .315 (wall thickness, combined)
.315 ÷ 2 = .1575" or 5/32" wall thickness of standard 1" pipe.

If pipe other than standard wall thickness is desired, it may be ordered by referring to various schedules. Figure 1-14, page 8, illustrates schedules of pipe. Schedules 30 and 40 are considered to be standard pipe.

Nominal Pipe Size	Outside Diameter of Pipe
1/8 ID	0.405
1/4 ID	0.540
3/8 ID	0.675
1/2 ID	0.840
3/4 ID	1.050
1 ID	1.315
1 1/4 ID	1.660
1 1/2 ID	1.900
2 ID	2.375
2 1/2 ID	2.875
3 ID	3.500
3 1/2 ID	4.000
4 ID	4.500
5 ID	5.563
6 ID	6.625
8 ID	8.625
10 ID	10.750
12 ID	12.750
14 OD	14.000
16 OD	16.000
18 OD	18.000
20 OD	20.000
24 OD	24.000

Fig. 1-13

Nominal Pipe Size	Nominal Wall Thickness									
	Sched. 10	Sched. 20	Sched. 30	Sched. 40	Sched. 60	Sched. 80	Sched. 100	Sched. 120	Sched. 140	Sched. 160
1/8 ID	. . . .	. . . .	. . . .	0.068	. . . .	0.095	. . . .	. . . .	. . . .	. . . .
1/4 ID	. . . .	. . . .	. . . .	0.088	. . . .	0.119	. . . .	. . . .	. . . .	. . . .
3/8 ID	. . . .	. . . .	. . . .	0.091	. . . .	0.126	. . . .	. . . .	. . . .	. . . .
1/2 ID	. . . .	. . . .	. . . .	0.109	. . . .	0.147	. . . .	. . . .	. . . .	0.188
3/4 ID	. . . .	. . . .	. . . .	0.113	. . . .	0.154	. . . .	. . . .	. . . .	0.219
1 ID	. . . .	. . . .	. . . .	0.133	. . . .	0.179	. . . .	. . . .	. . . .	0.250
1 1/4 ID	. . . .	. . . .	. . . .	0.140	. . . .	0.191	. . . .	. . . .	. . . .	0.250
1 1/2 ID	. . . .	. . . .	. . . .	0.145	. . . .	0.200	. . . .	. . . .	. . . .	0.281
2 ID	. . . .	. . . .	. . . .	0.154	. . . .	0.218	. . . .	. . . .	. . . .	0.344
2 1/2 ID	. . . .	. . . .	. . . .	0.203	. . . .	0.276	. . . .	. . . .	. . . .	0.375
3 ID	. . . .	. . . .	. . . .	0.216	. . . .	0.300	. . . .	. . . .	. . . .	0.438
3 1/2 ID	. . . .	. . . .	. . . .	0.226	. . . .	0.318	. . . .	. . . .	. . . .	. . . .
4 ID	. . . .	. . . .	. . . .	0.237	. . . .	0.337	. . . .	0.438	. . . .	0.531
5 ID	. . . .	. . . .	. . . .	0.258	. . . .	0.375	. . . .	0.500	. . . .	0.625
6 ID	. . . .	. . . .	. . . .	0.280	. . . .	0.432	. . . .	0.562	. . . .	0.719
8 ID	. . . .	0.250	0.277	0.322	0.406	0.500	0.594	0.719	0.812	0.906
10 ID	. . . .	0.250	0.307	0.365	0.500	0.594	0.719	0.844	1.000	1.125
12 ID	. . . .	0.250	0.330	0.406	0.562	0.688	0.844	1.000	1.125	1.312
14 OD	0.250	0.312	0.375	0.438	0.594	0.750	0.938	1.094	1.250	1.406
16 OD	0.250	0.312	0.375	0.500	0.656	0.844	1.031	1.219	1.438	1.594
18 OD	0.250	0.312	0.438	0.562	0.750	0.938	1.156	1.375	1.562	1.781
20 OD	0.250	0.375	0.500	0.594	0.812	1.031	1.281	1.500	1.750	1.969
24 OD	0.250	0.375	0.562	0.688	0.969	1.219	1.531	1.812	2.062	2.344

Fig. 1-14

REVIEW QUESTIONS

Answer the following questions briefly.

1. What is the structural symbol for angle iron?

2. How is angle iron ordered?

3. What is the structural symbol for channel iron?

4. How is channel iron ordered?

5. What is the American standard beam structural symbol?

6. How is an American standard beam ordered?

7. What is the wide flange beam structural symbol?

8. How is wide-flange beam ordered?

9. What is the structural symbol for plate?

10. How is plate ordered?

11. What is the structural symbol for flat?

12. How is flat steel ordered?

13. How is square stock ordered?

14. How is hexagon bar ordered?

15. How is round stock ordered?

16. What is the difference between cold-rolled steel and hot-rolled steel?

17. How is pipe ordered? Are there any exceptions?

18. How is tubing ordered?

19. What does the schedule of pipe indicate?

20. When measuring angle-iron legs, where are the measurements taken?

Unit 2 Welding Joint Design

OBJECTIVES

Upon completion of this unit, the student will be able to:

- explain the application of static and dynamic loads.
- define the five basic joint designs.
- describe the variations of application of the five basic designs.

Selection of the correct joint design is very important. The considerations in joint design include safety, welding codes, and cost.

Joint designs and welding procedures are the responsibility of an engineering department. However, this does not mean that the welder should not be concerned with these selections. Regardless of who is involved in the selection, the major function of both operations is to achieve the maximum strength at the least expense.

Not only must the product have sufficient strength to perform well under the load conditions expected, but it must be pleasing in appearance. A welder who has a thorough understanding of the basic welding joints will be in a position to give practical assistance to an engineering department.

Load Performance

Correct selection of joint design is necessary if welded members are to perform within the load service and safety requirements. One of the first considerations for joint design must be its ability to withstand the loads applied. These can come in the form of static loads or dynamic loads.

Static Loads

A *static load* is a steady load which does not increase or decrease under service conditions. An example of a static load is an upright supporting steel beam in a steel building. In this case, once the beam is loaded, the small increases or decreases of the load will not be considered as factors contributing to failure.

Dynamic Loads

A *dynamic load* is a continually changing load which increases or decreases under certain conditions. An example of a dynamic load is an upright, supporting steel beam in a highway bridge. In this case, the beam is supporting increases and decreases of loads as traffic increases or decreases. The effects of these load conditions must be considered as a contributing factor to failure under such service conditions.

10

Types of Joints

There are five basic joint designs. Some variations of the basic designs will be described in terms of their use, preparation, advantages, and disadvantages. The basic joints are:

- butt joint
- lap joint
- T joint
- corner joint
- edge joint

Square Butt Joints

The design of a welding joint is considered on the basis of the preparation cost, the position of the weld, the type of service load to be applied, and the cost of completion of the welded joint. The square butt weld is the most economical joint design because of the lack of plate preparation, but it must be confined to products made of light materials.

Square butt joints can be of two designs: the square closed butt joint or the open butt joint. These joints are fairly strong in a static load condition but they are not recommended when tension developed by bending is concentrated on the root of the weld.

Square Closed Butt Joint

On metal 1/8 inch thick, or thinner, complete penetration is possible by welding from only one side. On metal over 1/8 inch thick and up to 3/16 inch thick, complete penetration is possible by welding from both sides. This joint is satisfactory for all normal service load conditions. Preparation of the square closed butt joint is simple and inexpensive since the joint requires only that the edges of the plates be clean and match up in the design of the joint, figure 2-1.

Square Open Butt Joint

On metal over 1/8 inch thick and up to 1/4 inch thick, complete penetration is possible by welding from both sides. This joint is satisfactory for all normal service load conditions. However, if complete (100 percent) penetration is not achieved, the open square butt joint is no stronger than the closed square butt joint. The plate preparation is the same as for the closed square butt joint. At times the joint welding is more time consuming because the root opening must be set the correct distance along the complete length of the joint according to instructions given by the drawing, figure 2-2, page 12.

Grooved Butt Joint

Whenever steel is beveled, the costs of the completed joint increase. This is because plate preparation increases the costs of labor, equipment, operational time, and filler material.

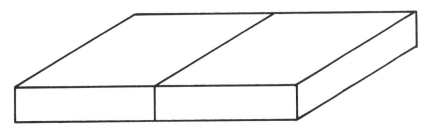

Fig. 2-1 Square closed butt joint

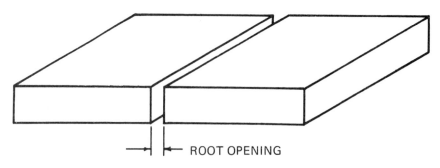

Fig. 2-2 Square open butt joint

expenditures. When heavy plates are to be welded, they must be beveled to insure complete penetration. Without proper plate preparation, complete penetration is impossible and inadequate penetration will result in a weak joint.

Single-V Butt Joint

The single-V butt joints can be of the closed or open design. They are stronger than square butt joints and are used where greater stresses exist. On metal 1/4 inch to 5/8 inch in thickness, 100 percent penetration can be achieved. Plate preparation costs will increase over the square butt joint and greater amounts of filler material will be used to fill the beveled joint. The degree of plate bevels can vary from 15 degrees to 45 degrees on plates prepared for the single-V joint. If welding can be completed from one side only, full penetration to the root of the weld must be obtained. These joints are stronger than square butt joints in a static load condition, but they are subject to failure if severe bending tensions are developed on the root of the weld. Single-V joints that are welded from both sides will provide full strength and meet the requirements for code welding, figure 2-3.

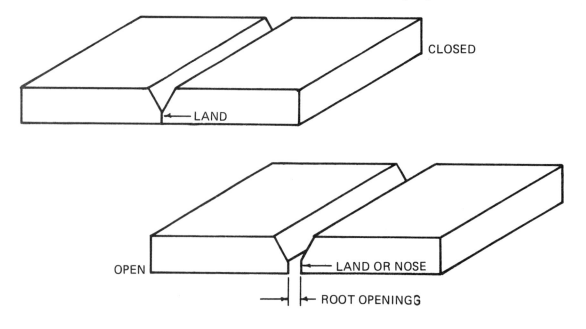

Fig. 2-3 Single-V butt joints

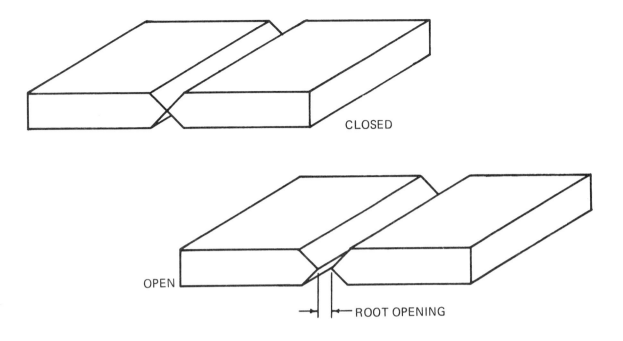

CLOSED

OPEN

ROOT OPENING

Fig. 2-4 Double-V butt joints

Double-V Butt Joint

The double-V butt joint can also be of the closed or open design. It is stronger than the single-V butt joint and is used where greater load stresses exist. On metal over 5/8 inch in thickness and up to 1 1/2 inches thick, 100 percent penetration can be achieved. Plate preparation costs will increase over the single-V butt joint. The degree of plate bevels can vary from 30 degrees to 60 degrees of included angle. When welding the double-V butt joint, full penetration to the root of the weld must be obtained. These joints are stronger than the single-V butt joints and are best for all load conditions. These joints can be used only where welding can be completed from both sides of the joint. To keep joint distortion and warpage to a minimum, the welding beads must be alternated from one side to the other, following prescribed bead sequences. Double-V joints that have been welded correctly will provide full strength and meet the requirements for code welding, figure 2-4.

Single-Bevel and Double-Bevel Butt Joints

The single-bevel and double-bevel butt joints are recommended for the open root design. They are used when load stresses are greater than can be met by the square butt joints, but less than the value required by the V butt joints. The cost of preparing and completing the joint will be less than for the V butt joints because only one plate edge is beveled. The amount of bevel will vary from 15 degrees to 45 degrees. This joint is difficult to fuse and penetration into the square side of the matching plate of the joint may be hard to achieve, figure 2-5, page 14.

Single-U Butt Joints

The single-U butt joint can be of the closed or open design. It is strong and readily meets all usual load stresses. On metal from 1/2 inch to 3/4 inch in thickness, 100 percent

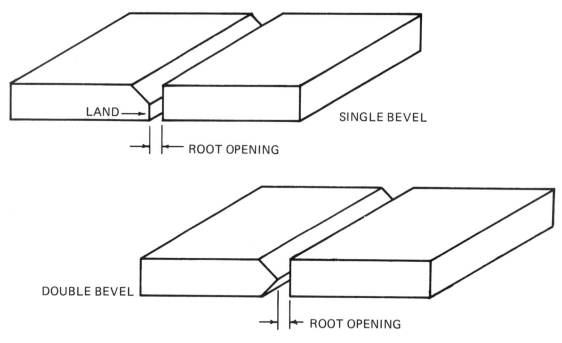

Fig. 2-5 Bevel butt joints

penetration can be achieved. The joints are used where high quality is required. Plate preparation costs will increase over the bevel and V joints, but there will be a decrease in filler material. The radius of the plate grooves will vary, depending on the position of the plates when welded. A general radius on plates 1/2 inch to 1 inch in thickness will be 1/4 inch at the bottom of the groove. Complete penetration is required for the single-U groove design to give satisfactory service under load conditions. It is easier for the welder to achieve complete penetration of single-U joints when welding can be completed from both sides of the joint, figure 2-6.

Double-U Butt Joints

The double-U butt joint can be of the closed or open design. It is of the maximum strength classification and readily meets all load stresses. On metal 3/4 inch thick and thicker, 100 percent penetration can be achieved. The joint is used where high quality is required and where welding can be completed from both sides of the joint. Plate preparation costs will increase over the single-U joints, but less filler metal is required, figure 2-7.

A general radius on plates 3/4 inch thick and thicker will be 1/4 inch at the bottom of the groove. Variations of included degree angles of both the single- and double-U groove design will be noted for the horizontal position of welding. Complete penetration is required for the double-U groove design to give satisfactory service under load conditions. Welding from both sides of the double-U joint will give the operator better control over stresses and distortion created from welding heat input. The choice between the double-U groove design should be made on the basis of completed welding costs and load conditions.

Single-J Butt Joint

Single-J butt joints are used on work similar to that requiring single-U joints. However, they are used when load conditions are not as great. The single-J butt joint can be of closed

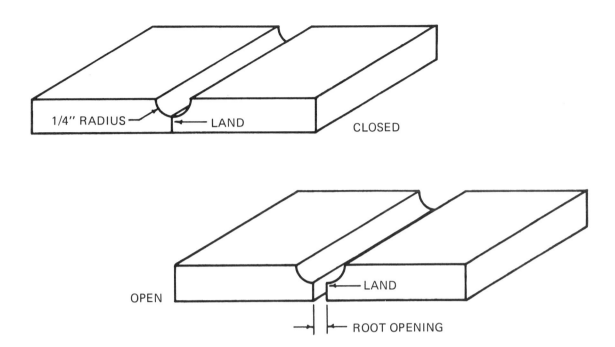

Fig. 2-6 Single-U butt joints

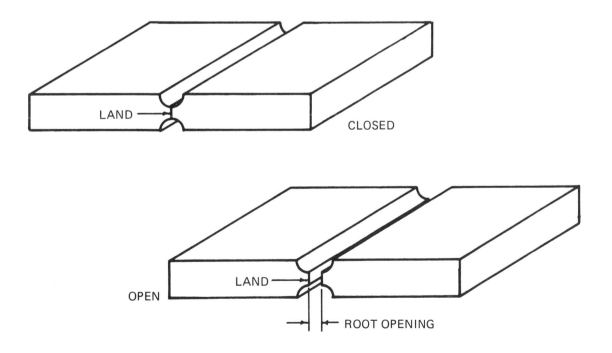

Fig. 2-7 Double-U butt joints

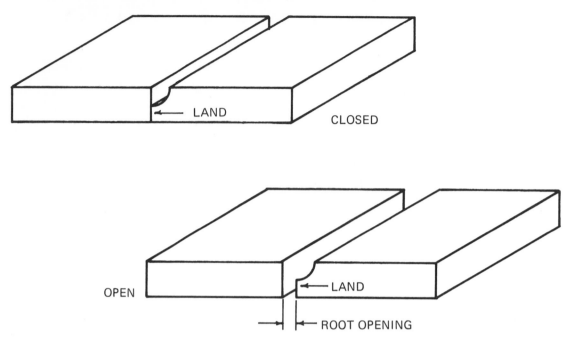

Fig. 2-8 Single-J butt joints

or open design. On metal 1/2 inch to 1 1/2 inches thick 100 percent penetration can be achieved. The cost of plate preparation is less than for double groove designs because only one plate must be grooved. The cost of filler material is also less since only one groove needs to be filled. A general radius on plates 1/2 inch to 1 1/2 inches thick will be 1/2 inch at the bottom of the groove. Complete penetration is required for the single-J groove design to give satisfactory service under load stresses. These joints are subject to failure if severe tensions are developed on the root area of the weld zone. It may be difficult for the welder to achieve complete fusion in the joint because the matching plate has a straight edge. To ensure adequate penetration, the joint should be welded from both sides, figure 2-8.

Double-J Butt Joint

Double-J butt joints are used on work similar to that requiring double-U joints, but where load stresses are not as demanding. The double-J butt joint can be of the closed or open design, but filler material cost will decrease because of a decrease in cross-sectional area. Both sides of the joint must be exposed for welding. A general radius on plates 3/4 inch to 1 1/2 inches thick will be 1/2 inch at the bottom of the groove. Complete fusion and penetration are required for this design to give strength under load conditions. Care must be taken to be certain that proper fusion occurs in the straight edge of the matching plate, figure 2-9.

Lap Joint

The lap joint is one of the most commonly used joints in the welding field. It is simple, inexpensive, and easily welded, since no beveling or matching is needed for joint preparation. The only preparations needed for a successful weld are clean base materials

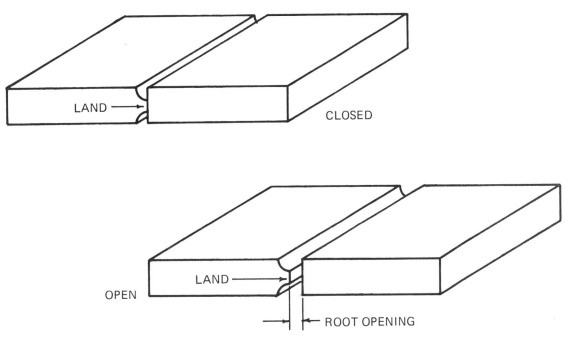

Fig. 2-9 Double-J butt joints

that are evenly aligned. The joint consists of lapping one plate over the other and welding. The amount of plate overlap will depend on the thickness of the material and the strength needed from the joint under load stresses. Generally, thicker plates require more overlap than do thinner sections. The increased overlap is used for added support of the finished joint. When the weld is subject to heavy stresses, the lap joint should be welded on both sides. In the single- or double-lap joint, penetration to the root of the joint is required. The double-lap joint has greater strength than the single lap but it requires twice as much welding as would the simpler and stronger butt weld, figure 2-10.

Plug and Slot Joints

Plug and slot joints resemble lap joints in that one plate must be placed directly over the other. In this case, one of the plates has a hole or slot cut in it and placed overlapping the other plate. Figure 2-11, page 18, shows both a plug and slot joint.

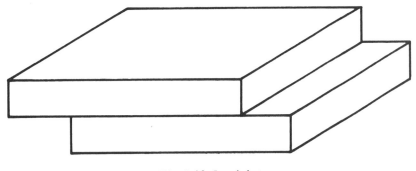

Fig. 2-10 Lap joint

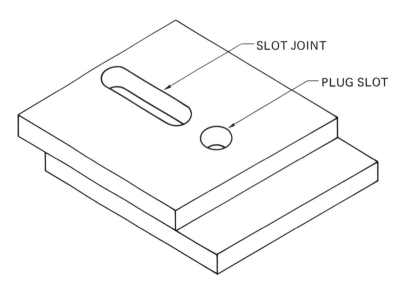

Fig. 2-11 Plug and slot joints

Square T Joint

The square T joint can be used on plate thicknesses up to 1/2 inch. It is simple and inexpensive since no beveling or machining is needed for joint preparation. The only preparation needed is clean base material evenly positioned. The amount of weld deposit will depend on the thickness of the material and the strength needed. Generally, the thicker plates require welding from both sides of the joint, especially when heavy load stresses will be applied, figure 2-12.

The single-fillet T joint will not withstand bending tensions on the root of the weld and should be used with caution. In the single or double square joint, penetration to the root of the joint is required. The double square T joint has greater strength than the single square

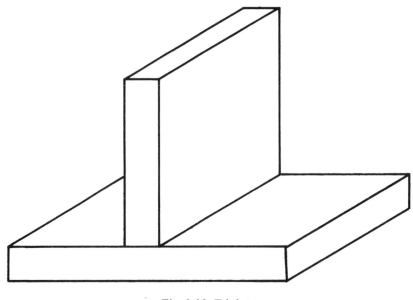

Fig. 2-12 T joint

T joint but requires twice as much welding. T-joint designs are ones that welders tend to overweld. Overwelding increases completed costs along with an increase in warpage and distortion stresses.

Single-Bevel T Joint

Because of the increased strength of the single-bevel T joint, it is able to withstand heavier load stresses than the square T joint. The single-bevel T joint can be used on plate thicknesses from 3/8 inch to 5/8 inch. The cost of preparation will be more than for the square T joint because the upright member must be beveled and the fit-up time is greater. The joint consists of positioning one plate on the other and welding. The amount of weld deposit will depend on the thickness of the plate and the degree of bevel preparation. The amount of filler material needed is less than that used on the square T joint. When heavy load stresses are applied the thicker plate should be welded from both sides. In the single-bevel T joint, penetration to the root of the joint is required. If welding can be completed from both sides of the joint, the resistance to load stresses will be greatly increased. Tacking or jigging necessary for holding the plates in correct position for welding should be completed before welding operations begin, figure 2-13.

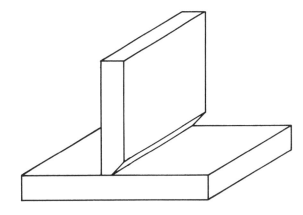

Fig. 2-13 Single-bevel T joint

Double-Bevel T Joint

Because of the increased strength of the double-bevel T joint, it is able to withstand heavier load stresses than the single-bevel T joint. The double-bevel T joint can be used on plate thicknesses from 1/2 inch to 1 inch. The costs of preparation will be more than for the single bevel because the upright member must be beveled on both sides. The fit-up time will be about the same as for the single-bevel joint. The amount of weld deposit will depend on the thickness of the plate and the degree of bevel preparation, figure 2-14.

This joint must be welded from both sides, with adequate root fusion and penetration. The double-bevel T joint will perform well under heavy stress loads when the welder makes certain that proper fusion has occurred in both the vertical and flat plates which make up the joint.

Single-J T Joint

The single-J T joint can be used for heavier load stresses than the square T joint, but not for loads that the double-bevel T joint can support. The single-J T joint can be used on plate thicknesses of 1 inch or more. The costs of preparation

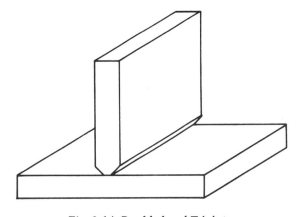

Fig. 2-14 Double-bevel T joint

will increase over the bevel T joint because the upright member must be machined with special tools designed to make the J groove. The fit-up time will be about the same as for single- or double-bevel T joints. Because of the edge preparation of the J groove, filler material used will decrease from that used on the bevel T joint. If welding can be completed from only one side, care must be taken to insure complete fusion and penetration in the root zone area. If welding from both sides is possible, the strength of the joint will increase greatly when a backup bead is placed on the side opposite the J groove, figure 2-15.

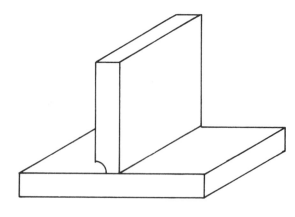

Fig. 2-15 Single-J T joint

Double-J T Joint

The double-J T joint can be used for the most severe load stresses encountered in the T joint design. The double J can be used on plate thicknesses of 1 1/4 inch or thicker. The costs of preparation will be more than for the single J because the upright member must be machined on both sides with special tools designed to prepare the J groove. Often, the J grooves will be of different sizes on opposite sides of the joint. The fit-up time remains about the same as for the single bevel or single J joint. These joints are used where welding can be accomplished from both sides. The utmost care must be exercised so that complete fusion and penetration is gained into both the upright and flat members of this joint, figure 2-16.

Flush Corner Joints

On metal 1/8 inch thick, or thinner, complete fusion and penetration are possible on square corner joints. Metal over 1/8 inch thick and up to 3/16 inch thick may be completely fused by welding from both sides. Often, it is impossible to weld from both sides, therefore, this joint is somewhat restricted to thinner gauge materials. Joint preparations are simple and inexpensive since it requires only that the edges be clean and that they match up on the joint design. Care must be used in welding the square corner joint in order to make complete fusion and penetration into both members of the design. This joint should be used cautiously where bending tensions will develop on the root side of the joint. The square corner is easy to weld, but subject

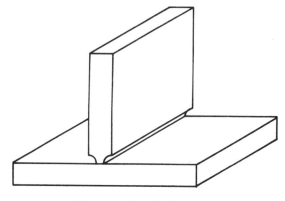

Fig. 2-16 Double-J T joint

to warping and distortion, because of heat input into lightweight materials, figure 2-17.

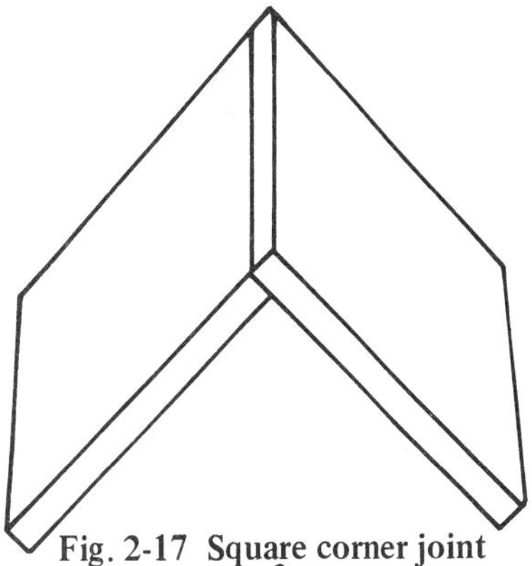

Half-Open Corner Joints

The half-open corner joint may be used on metal from 3/16 inch to 3/8 inch thick. Penetration and fusion are easily achieved and joint preparation is simple and inexpensive. The edges of the metal should be clean and positioned to match the design specified. Clamping or holding devices can be used to position the joint members. The design of the joint forms a groove and permits the welder to achieve complete fusion and penetration to the root zone of the weld. This joint will support more than the flush corner joint but is not recommended where severe bending tensions develop at the root of the joint, figure 2-18.

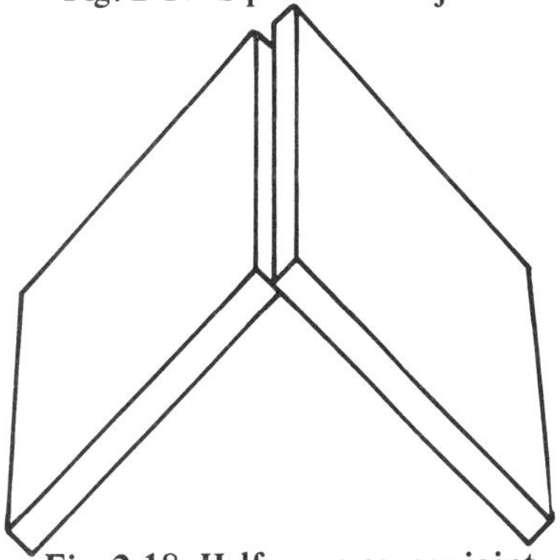

Full-Open Corner Joint

The full-open corner joint may be used on metal of 3/8 inch to 1 inch thick. Penetration and fusion are easily achieved and joint preparations are simple and inexpensive, since only cleaning and aligning are necessary. Holding clamps or jigs are necessary because of the heavier metal used. If welding is completed from only one side, root fusion and penetration must be made into both members of the joint. When welding can be done from both sides, the joint will support severe loading stresses. Costs of filler material are greater than for flush and half-open corner joints because of the larger cross section of the weld zone, figure 2-19.

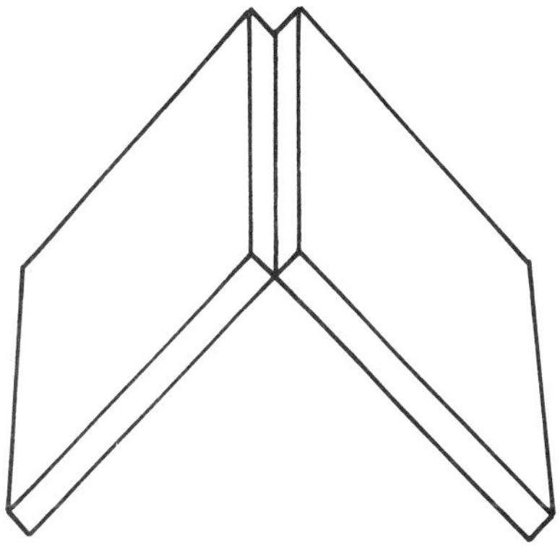

Fig. 2-17 Square corner joint

Fig. 2-18 Half-open corner joint

Fig. 2-19 Full-open corner joint

Edge Joint

The edge joint is suitable for metal thicknesses of 1/8 inch or less. Complete fusion and penetration through the joint are impossible due to the design. This joint preparation is somewhat costly and time consuming, since the edges must be clean and many times the members must be bent to meet the design. Filler material cost is less because less filler is needed on flat surfaces which are not grooved. The edge joint will support only a minimum

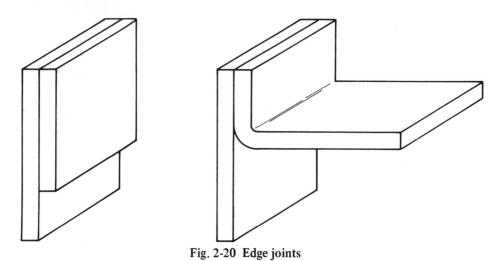

Fig. 2-20 Edge joints

of load stress and should not be used where heavy tension develops on the root side of the joint. This joint, figure 2-20, is subject to warpage and distortion.

Summary

The information included in this unit indicates the importance of correct joint designs. In general, the strength of the work will depend on the type of joint design and the ease with which the welder can complete the product. Proper joint designs depend on such factors as material thickness, position of welding, ease of welding, fusion and penetration needed, and load stresses to be applied. Correct joint design and welding procedures will keep warpage and distortion to a minimum, reduce cracking defects from weld zone shrinkage, and make it easier for the welder to produce quality welds with good appearance at the lowest possible cost.

REVIEW QUESTIONS

Answer the following questions briefly

1. Define static load.

2. Define a dynamic load.

3. What are the five basic joint designs?

4. What is the difference between an open butt joint and a closed butt joint?

5. Why is it necessary to bevel heavier metal before it is welded?

6. What is a bevel?

7. Make simple sketches of the following joints:
   - A. square open butt
   - B. single-bevel butt
   - C. double-bevel butt
   - D. lap joint
   - E. flush corner joint

Unit 3
Welding Symbols

OBJECTIVES

Upon completion of this unit, the student will be able to:

- recognize the symbols used to identify the various types of welding joints.
- explain and properly use all of the various welding symbols.

Basic Welding Symbol

All personnel involved with the field of welding must be able to read and interpret various welding symbols. These symbols are used on drawings and shop sketches to specify the information needed to complete a particular weld.

The basic welding symbol is made up of a reference line, an arrow, and in some cases a tail, figure 3-1. Other information is added to this basic symbol in the form of weld symbols, dimensions, abbreviations, notes, and other supplementary symbols.

In certain cases, when notes, specifications, or other reference information must be added to the symbol, a tail is used. This information is then placed within the tail.

Most of the information is placed on the reference line of the welding symbol. Where a symbol is placed on the reference line it has great significance. The bottom side of the reference line is known as the *arrow side,* the top is known as the *other side,* figure 3-2.

This can best be illustrated by the following examples. Figure 3-3 shows the weld symbol for a square butt weld placed on the arrow side of the reference line and its resultant weld. Figure 3-4, page 24, shows the result of the same symbol being placed on the other side of the reference line. If the weld symbol were placed on both sides of the reference line, a resulting weld as shown in figure 3-5, page 24, would be constructed. If the reference line is drawn to be vertical, the right side of the line is considered the arrow side while the left side is the other side.

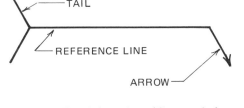

Fig. 3-1 Basic welding symbol

Fig. 3-2 The bottom of the reference line is known as the arrow side while the top is known as the other side.

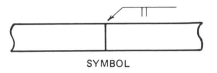

SYMBOL

SIGNIFICANCE

Fig. 3-3 A square butt weld on the arrow side of the joint

Fig. 3-4 A square butt weld on the other side of the joint

Fig. 3-5 A square butt weld on each side of the joint

Some welds require only one plate to be prepared by beveling or gouging. In this case, the arrow breaks in the direction of the plate to be prepared, figure 3-6.

In addition to a particular weld symbol, a large amount of other information can be displayed as part of a complete welding symbol. A complete welding symbol and all of its possible elements is shown in figure 3-7. The welding symbol and all of its elements are designated by the American Welding Society.

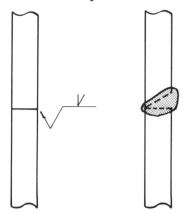

Fig. 3-6 The arrow breaks in the direction of the plate to be prepared.

Weld Symbols

Weld symbols are added to the basic welding symbol to represent the various types of joint designs which may be required. The weld symbols are shown in figure 3-8. Figure 3-9, page 26, shows the basic welding symbols and their location significance.

Dimensions

Dimensions are added to the welding symbol to indicate such things as size, depth, length, and center-to-center spacing of welds. Figure 3-10, page 27, illustrates the use of a dimension to specify the size of a fillet weld with equal legs. If a fillet weld has unequal legs, the symbol is labeled as shown in figure 3-11, page 27.

The effective throat of a weld is specified as shown in figure 3-12, page 27. The *effective throat* is measured from the surface of the weldment to the bottom of the weld.

Figure 3-13, page 27, shows the procedure used to specify the length of a weld. Length is often used along with a dimension indicating the center-to-center spacing for intermittent

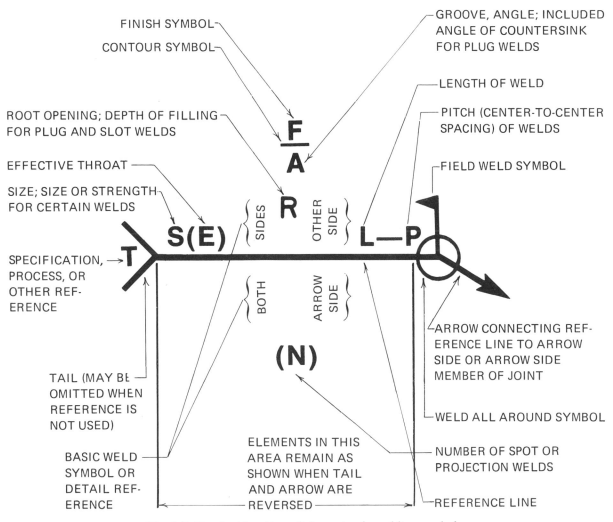

FINISH SYMBOL

CONTOUR SYMBOL

GROOVE, ANGLE; INCLUDED ANGLE OF COUNTERSINK FOR PLUG WELDS

ROOT OPENING; DEPTH OF FILLING FOR PLUG AND SLOT WELDS

LENGTH OF WELD

PITCH (CENTER-TO-CENTER SPACING) OF WELDS

EFFECTIVE THROAT

FIELD WELD SYMBOL

SIZE; SIZE OR STRENGTH FOR CERTAIN WELDS

$\dfrac{F}{A}$

R

S(E) SIDES OTHER SIDE L—P

SPECIFICATION, PROCESS, OR OTHER REFERENCE

T

BOTH ARROW SIDE

ARROW CONNECTING REFERENCE LINE TO ARROW SIDE OR ARROW SIDE MEMBER OF JOINT

(N)

TAIL (MAY BE OMITTED WHEN REFERENCE IS NOT USED)

WELD ALL AROUND SYMBOL

BASIC WELD SYMBOL OR DETAIL REFERENCE

ELEMENTS IN THIS AREA REMAIN AS SHOWN WHEN TAIL AND ARROW ARE REVERSED

NUMBER OF SPOT OR PROJECTION WELDS

REFERENCE LINE

Fig. 3-7 Standard location of elements of a welding symbol

Groove						
Square	V	Bevel	U	J	Flare-V	Flare-bevel
‖	V	V	Y	Ρ	Y	⌐

Fillet	Plug or slot	Seam	Back or Backing	Surfacing	Flange	
					Edge	Corner
◺	▭	⊕	◡	◡◡	⌡⌊	⌊⌊

Fig. 3-8 Basic weld symbols

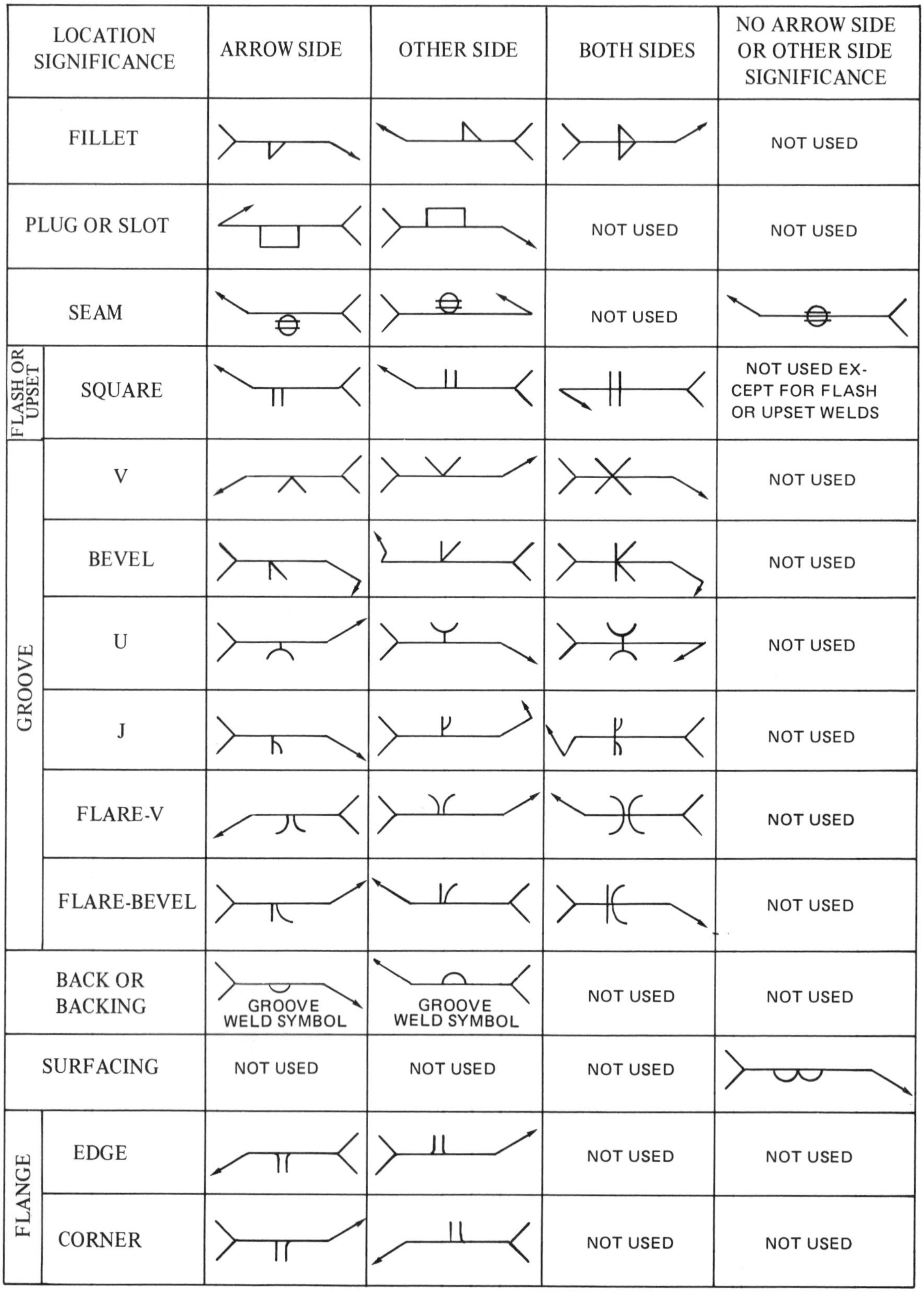

LOCATION SIGNIFICANCE		ARROW SIDE	OTHER SIDE	BOTH SIDES	NO ARROW SIDE OR OTHER SIDE SIGNIFICANCE
FILLET					NOT USED
PLUG OR SLOT				NOT USED	NOT USED
SEAM				NOT USED	
FLASH OR UPSET	SQUARE				NOT USED EX-CEPT FOR FLASH OR UPSET WELDS
GROOVE	V				NOT USED
	BEVEL				NOT USED
	U				NOT USED
	J				NOT USED
	FLARE-V				NOT USED
	FLARE-BEVEL				NOT USED
	BACK OR BACKING	GROOVE WELD SYMBOL	GROOVE WELD SYMBOL	NOT USED	NOT USED
	SURFACING	NOT USED	NOT USED	NOT USED	
FLANGE	EDGE			NOT USED	NOT USED
	CORNER			NOT USED	NOT USED

Fig. 3-9 Basic welding symbols and their location significance

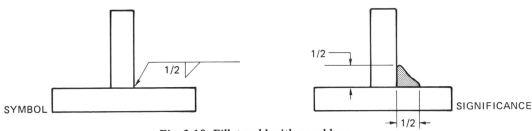

Fig. 3-10 Fillet weld with equal legs

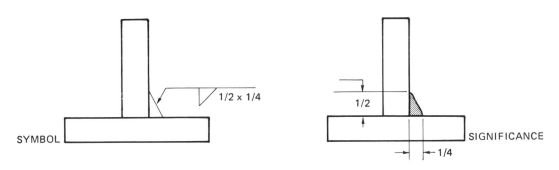

Fig. 3-11 Fillet weld with unequal legs

Fig. 3-12 Specifying the effective throat of a square butt weld

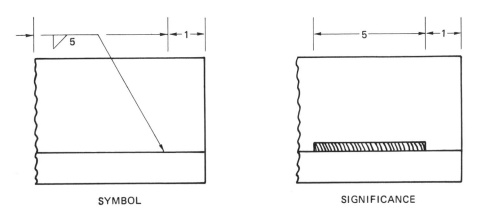

Fig. 3-13 Specifying the length of a fillet weld

welds, figure 3-14. When the weld symbols are directly opposite each other, as in figure 3-14, the welds are also formed directly opposite each other as in figure 3-15. The resultant weld is staggered as shown.

Another dimension which may be given on a welding symbol is the root opening. The application of this dimension is shown in figure 3-16.

The included angle of a joint with prepared edges may be specified as shown in figure 3-17.

Supplementary Symbols

Supplementary symbols are also added to the basic welding symbol as required. These symbols are shown in figure 3-18 with their respective interpretations.

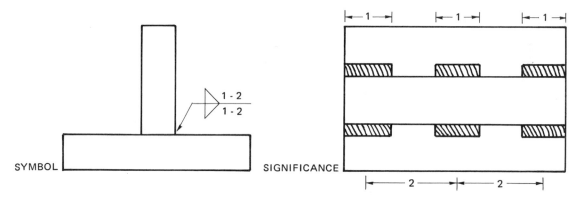

Fig. 3-14 Specifying the length and center-to-center spacing of fillet welds on both sides of the joint

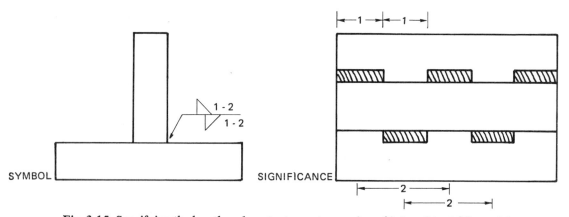

Fig. 3-15 Specifying the length and center-to-center spacing of intermittent fillet welds

Fig. 3-16 Specifying the root opening of a V weld

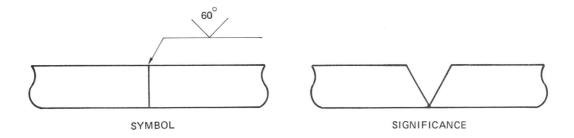

Fig. 3-17 Specifying the included angle of a V joint

Weld all around	Field weld	Melt-thru	Contour		
			Flush	Convex	Concave
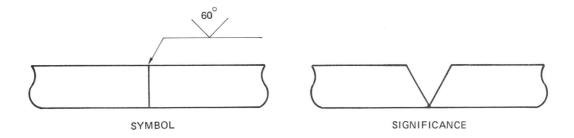					

Fig. 3-18 Supplementary symbols

Weld-All-Around Symbol. The weld-all-around symbol instructs the welder to completely encircle a particular joint. Figure 3-19 shows a weld-all-around symbol with a fillet weld symbol and the resultant weld.

Field Weld Symbol. Certain weldments must be constructed on the job due to their size or to meet special conditions. Welds to be made at the job location are indicated by the field weld symbol.

Melt-Thru Symbol. The melt-thru symbol indicates the need for a particular weld to have complete penetration. Figure 3-20 shows the melt-thru symbol used with a butt weld symbol and the resultant weld.

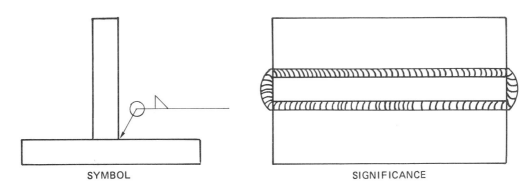

Fig. 3-19 Specifying a joint to be welded all around

Fig. 3-20 Specifying the melt-thru of a square butt weld

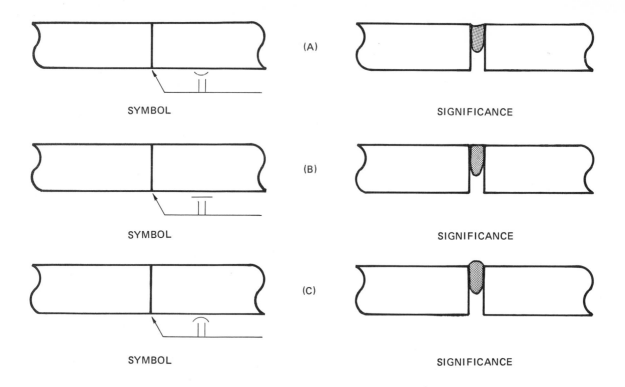

Fig. 3-21 Specifying the contour of a weld bead (A) Concave (B) Flush (C) Convex

Symbol	Interpretation
C	Chipping
G	Grinding
M	Machining
R	Rolling
H	Hammering

Fig. 3-22 The various finish symbols

Contour Symbols. Contour symbols indicate the resultant surface shape of a particular weld. Figure 3-21 shows each of the contour symbols used with a butt weld symbol and the resultant welds.

Finish Symbols

In certain cases, the surface finish of a weld bead may require that a specific method be used to shape the desired contour. The method used is specified by the use of a particular letter symbol. These letters and their designations are shown in figure 3-22.

Abbreviations

Many abbreviations are used as part of notes or specifications added to the basic welding symbol through the use of a tail. Many of the abbreviations used in the field of welding are shown in figure 3-23.

Letter Designation	Welding Process	Letter Designation	Welding Process
AAW	Air-Acetylene Welding	IRB	Infrared Brazing
AHW	Atomic Hydrogen Welding	IW	Induction Welding
BB	Block Brazing	LBW	Laser Beam Welding
BMAW	Bare Metal-Arc Welding	NTW	Nonpressure Thermit Welding
CAW	Carbon-Arc Welding	OAW	Oxyacetylene Welding
CW	Cold Welding	OHW	Oxyhydrogen Welding
DB	Dip Brazing	PAW	Plasma-Arc Welding
DFW	Diffusion Welding	PEW	Percussion Welding
DW	Die Welding	PGW	Pressure Gas Welding
EBW	Electron Beam Welding	PTW	Pressure Thermit Welding
EW	Electroslag Welding	RB	Resistance Brazing
EXW	Explosion Welding	RPW	Projection Welding
FB	Furnace Brazing	RSEW	Resistance-Seam Welding
FCAW	Flux Cored Arc Welding	RSW	Resistance-Spot Welding
FLB	Flow Brazing	RW	Roll Welding
FLOW	Flow Welding	SAW	Submerged Arc Welding
FOW	Forge Welding	SCAW	Shielded Carbon-Arc Welding
FRW	Friction Welding	SMAW	Shielded Metal-Arc Welding
FW	Flash Welding	SW	Stud Welding
GCAW	Gas Carbon-Arc Welding	TB	Torch Brazing
GMAW	Gas Metal-Arc Welding	TCAB	Twin-Carbon Arc Brazing
GSSW	Gas-Shielded Stud Welding	TCAW	Twin-Carbon Arc Welding
GTAW	Gas Tungsten-Arc Welding	TW	Thermit Welding
HW	Hammer Welding	USW	Ultrasonic Welding
IB	Induction Brazing	UW	Upset Welding

Fig. 3-23 Welding abbreviations

REVIEW QUESTIONS

1.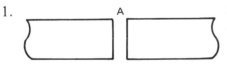

   a. What type of joint is pictured?

   b. What type of opening is signified by the letter (A)?

   c. Draw the welding symbol which would apply to this joint.

2.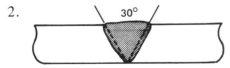

   a. What type of joint is pictured?

   b. What does the "30" signify?

   c. Draw the welding symbol which would apply to this joint.

3.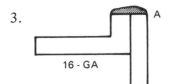

   a. What type of joint is pictured?

   b. Draw the welding symbol which would apply to the weld shown at (A).

4.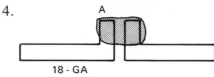

   a. What type of joint is pictured?

   b. Draw the welding symbol which would apply to the weld shown at Ⓐ.

5.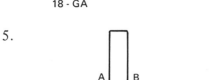

   a. What type of joint is pictured?

   b. Draw the welding symbol which would apply to welds Ⓐ and Ⓑ if Ⓐ is the arrow side of the joint.

6.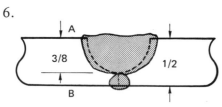

   a. What type of joint is pictured?

   b. Draw the welding symbol which would apply to welds Ⓐ and Ⓑ if Ⓐ is the arrow side of the joint.

   c. What is the significance of the "3/8" shown on the drawing?

7.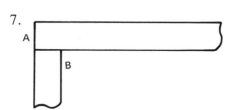

   a. What type of weld will apply to B?

   b. What type of weld will apply to A?

   c. Draw the welding symbol which would apply to Ⓐ and Ⓑ if Ⓑ is the arrow side of the joint.

8.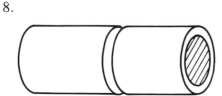

   a. What type of joint is pictured?

   b. Draw the welding symbol which would apply to the joint.

   c. Draw the welding symbol which would apply to this joint if a 1/8" root opening is needed to weld it.

9.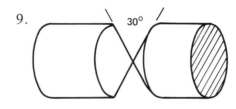

   a. What significance does the 30" have on this joint?

   b. What type of joint is pictured?

   c. Draw the weld symbol which would apply to this joint.

10.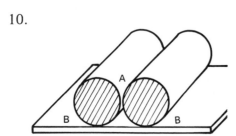

   a. What type of joint is pictured at Ⓐ?

   b. What type of joint is pictured at Ⓑ?

   c. Would you use more than one symbol to show the welding of this joint?

   d. Draw the necessary symbol(s) for these joints.

11.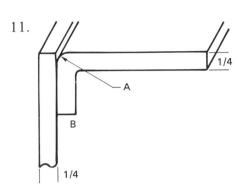

a. What type of joint is pictured?

b. Draw the welding symbol which would apply to this joint, if Ⓐ is the arrow side of the joint.

12.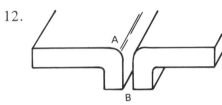

a. What type of joint is pictured?

b. Draw the welding symbol for this joint, if Ⓑ is the arrow side.

c. Draw the welding symbol for this joint if a 1/8" root opening is required at Ⓐ.

13. Draw each of these symbols as they would appear on the arrow side of a welding symbol. Identify each of the symbols you draw by its proper name.

(a) V

(b) Y

(c) ⊢/

(d) ◺

(e) ‖

(f) ⊩

(g) ⊔⊩

(h) ⊩⊢

(i) ⊤⊢

(j) V

(k) ◯

(l) ▶

Unit 4 Physical
and Mechanical Metallurgy

OBJECTIVES

Upon completion of this unit, the student will be able to:

- explain the application of metallurgy in the welding field.
- define the physical and mechanical properties of metal.
- recognize and correct weld faults resulting from metallurgical problems.
- explain welding electrode classifications.

Metallurgy

Metallurgy is the study of methods used for taking metal out of ore, refining it, and preparing it for use. The field of metallurgy also includes the study of metals with attention to their structure, composition, physical properties, and behavior under different use conditions.

Process Metallurgy

The heating and melting of metal when a weld is made involves process metallurgy. *Process metallurgy* is the study of a metal in a producing function. When iron ore is melted in the steel mills and made into different usable steel shapes, the process can be compared to the making of a weld. The pouring of molten metal into molds at the steel plant is similar to melting a welding rod and mixing it into the base metal being welded. Both of these processes are producing metal, figure 4-1.

Physical Metallurgy

As metal cools from a liquid state into a solid, or when a weldment is shaped or machined, it is undergoing a process of *physical metallurgy*. When weld metal cools, it becomes solid. The weldment may be used as welded, it may be machined, shaped, or flattened. To meet

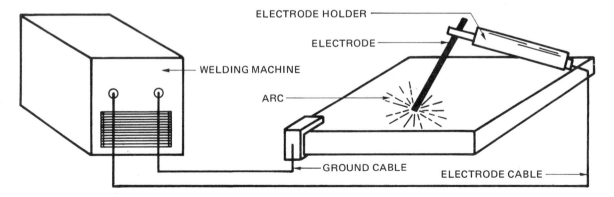

Fig. 4-1 Striking the arc creates heat for melting the parent metal. This is a form of process metallurgy.

the requirements for the welded part, the weld may be worked hot or cold. Physical metallurgy is the study of a metal working function, figure 4-2.

Mechanical Properties of Metal

The mechanical properties of metal are those qualities which help it resist a force which may be applied to it. Included in the mechanical properties affecting welding are tensile strength, compressive strength, ductility, hardness, fatigue strength, and others, figure 4-3.

All metals have mechanical properties, but they may vary considerably.

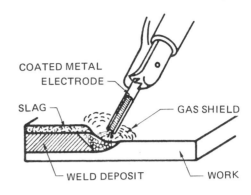

Fig. 4-2 When the liquid puddle of the weld cools, the metal progresses through a series of physical changes. The study of these changes is called physical metallurgy.

Some metals are designed to withstand a great deal of tensile pull, but are not capable of resisting a strong compressive force. Metals must be matched so they contain the mechanical properties which are needed for the particular job they are to do.

Physical Properties of Metal

The *physical* properties of metal are the characteristics which are not related to its ability to withstand a force applied to it. Included in the physical properties are corrosion resistance, electrical conductivity, thermal (heat) conductivity, heat expansion and contraction, and hardness, figure 4-4, page 36.

Parent Metal in Welding

The material to be welded is called *parent* or *base metal.* When the physical and mechanical properties of the parent metal are known, a welding joint can be designed to withstand different types of loads. The electrode to be used in welding the joint must be selected so that the weld is equal to, or exceeds, the strength of the parent metal.

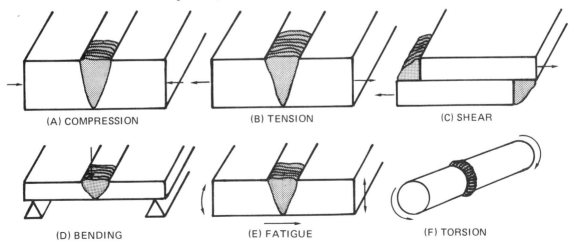

Fig. 4-3 One, or all, of the illustrated forces may be applied to a weld at the same time.

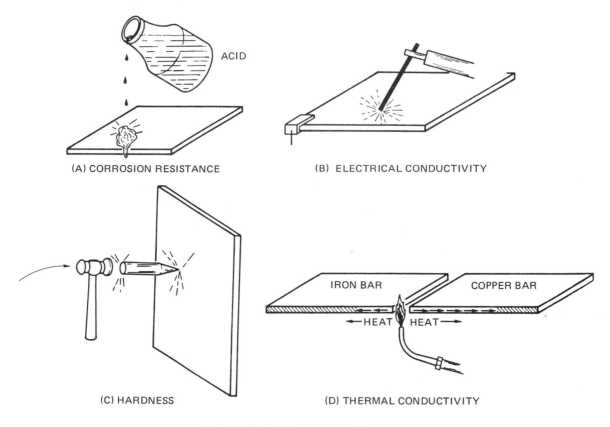

Fig. 4-4 Physical properties of metal

Load Failure

Any metal will fail if the load on it is great enough. As previously mentioned, metals are designed to resist different kinds of loads, such as repeated bends, pulls, pressures, twists, and impact. Any welded structure must be engineered to be strong enough to withstand the kind of load which might be applied after it is put into service, figure 4-5.

Tensile Strength

Tensile strength is measured by actually measuring the amount of force used to pull a piece of metal apart. The steel industry has developed standards for the tensile strength of different metals, figure 4-6.

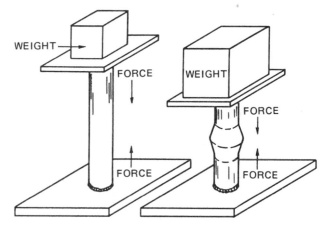

Fig. 4-5 When a structure is overloaded it is subject to load failure. Metal must be designed to withstand maximum loads which may be put on the product.

Yield Point and Elastic Limit of Metal

Low-carbon or medium-carbon steel can be pulled to a certain point where the load on the pull does not have to be increased to stretch the metal. This is called the *yield point*

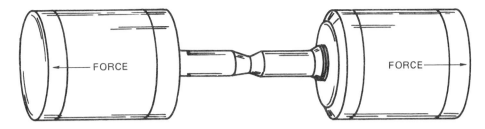

Fig. 4-6 Tensile strength is determined by measuring the amount of pull which is necessary to pull apart a piece of metal.

Fig. 4-7 A pulling force gradually applied to a piece of steel causes it to neck-in and stretch. Before it breaks, when the metal has stretched to where it does not return to the original size, it has passed the elastic limit.

of the metal. A piece of metal can be pulled and stretched and still return to the original length when the pull is removed. However, when it is pulled past the point where it returns to its original length, it has reached its *elastic limit.* The amount of pull used can be measured so that the steel can be designated for the kind of job it will be best for, figure 4-7.

Ductility of Metal

Some metals are designed so that they can be bent into various shapes without breaking. When this quality is built into the metal, it is said to be *ductile.* A good example of a ductile material is the metal used for automobile fenders. The metal can be formed into the shape needed without cracking or breaking and it retains the shape of the bend. Highly ductile material will stretch before breaking, figure 4-8.

Fig. 4-8 Ductile metal can be bent into various shapes without breaking and will stay in the bent shape.

Brittleness of Metal

When a metal, such as cast iron, cannot be bent without breaking, it is said to be *brittle.* Brittle metal has no elastic qualities. It will not return to its original shape when a load is removed, but will break at the point of overloading, figure 4-9, page 38.

Metal Fatigue

If a piece of metal is bent back and forth enough times it will fail. The break is due to *metal fatigue,* which means that the metal fails to return to its original shape. Other types

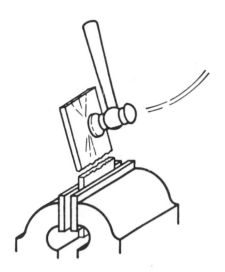

Fig. 4-9 Metal which breaks easily instead of bending is brittle.

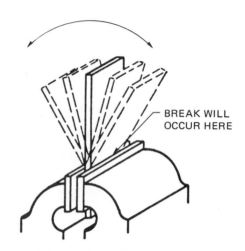

Fig. 4-10 When a piece of metal fails under repeated loads or bending, it is said to have metal fatigue.

of metal fatigue can result from applying and removing a load many times at the same spot. A grinding wheel used on a weld often leaves small scratches across the surface. This section of the metal tends to fail at this point before it fails on the unground surface of the parent metal, figure 4-10.

Effects of Temperature on Metal

Some metals will fail under a load if the temperature of the metal changes. Low temperatures tend to make metal more brittle unless it has been designed to withstand sudden loads or impacts at low temperatures. High temperatures generally make metal more ductile. This means that it can bend more readily without fracturing. Excessive heat will tend to deform the metal so it does not return to its normal size after a load has been applied to it, figure 4-11.

Thermal Conductivity

Some metals conduct heat faster than others, that is, the heat spreads through them faster. The ability of a metal to transfer heat is called *thermal conductivity*, figure 4-12. Generally, if a metal conducts heat rapidly, it is a tough metal.

Compressive Strength

Any metal which can resist crushing is said to have *compressive strength*. If a metal deforms easily under a load, its compressive strength is low.

Hardness in Metal

Hardness of metal is defined as a metal's ability to resist penetration. Some metals become harder with the

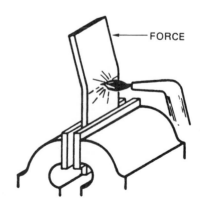

Fig. 4-11 High temperatures generally make metal bend easily when force is applied.

Metal	Thermal Conductivity Cal./cm/°C/Sec.
Copper	0.93
Aluminum	0.57
Tungsten	0.40
Magnesium	0.37
Nickel	0.22
Iron	0.18
Lead	0.08

Fig. 4-12 Some metals conduct heat faster than others. This table lists some common metals in order of their thermal conductivity.

Metal	Electrical Conductivity (silver = 100)
Silver	100
Copper	97.61
Aluminum	63.00
Magnesium	39.44
Steel	31.98
Zinc	29.57
Iron (wrought)	14.57
Tin	14.39
Nickel	12.89
Lead	8.42

Fig. 4-13 Electrical conductivity of metals

application of heat. The welder must know what effect the application of welding heat will have on the material being welded. This avoids shrinkage and distortion and the tendency of the metal to crack as it cools.

Electrical Resistance and Conductivity

Metal which resists the flow of electricity tends to heat slower than that which has lower resistance. Therefore, it requires more voltage and loses energy. The electrical conductivity of the material is related to the resistance. Metals which do not resist the flow of heat have high electrical conductivity, figure 4-13.

Melting Points

Melting points of metals vary. The melting point of a metal indicates its *fusibility* (ability to be melted and fused), figure 4-14.

Commonly Used Steels

When carbon is added to steel, the hardness of the material increases but the ductility decreases. In most cases, the higher the hardness, the lower the melting point.

Carbon steels are generally put into groups according to the amount of carbon they contain. The Society of Automotive Engineers has devised a steel numbering system which is used to identify steels and their content, figure 4-15, page 40.

As a rule, a steel with a carbon range between 0.05 percent and 0.30 percent is called *low-carbon steel*, or *mild steel.* Mild steel is most commonly used for welding and appears in most welded structures. *Medium-carbon steels* have a

Metal	Melting Temperatures °F	°C
Aluminum	1217	659
Bronze	1562–1832	850–1000
Copper	1981	1083
Iron	2786	1530
Lead	621	327
Mild Steel	2462–2786	1350–1530
Nickel	2646	1452
Silver	1761	960
Tin	450	232
Zinc	786	419

Fig. 4-14 Melting points of commonly used metals

Type of Steel	Numerals (and Digits)
Carbon Steels	1xxx
Plain Carbon	10xx
Free Cutting (Screw Stock)	11xx
Free Cutting, Manganese	X13xx
High-Manganese Steels	T13xx
Nickel Steels	2xxx
0.50% Nickel	20xx
1.50% Nickel	21xx
3.50% Nickel	23xx
5.00% Nickel	25xx
Nickel-Chromium Steels	3xxx
1.25% Nickel, 0.60% Chromium	31xx
1.75% Nickel, 1.00% Chromium	32xx
3.50% Nickel, 1.50% Chromium	33xx
3.00% Nickel, 0.80% Chromium	34xx
Corrosion and Heat Resisting Steels	30xxx
Molybdenum Steels	4xxx
Chromium	41xx
Chromium-Nickel	43xx
Nickel	46xx and 48xx
Chromium Steels	5xxx
Low-Chromium	51xx
Medium-Chromium	52xxx
Corrosion and Heat Resisting	51xxx
Chromium-Vanadium Steels	6xxx
Tungsten Steels	7xxx and 7xxxx
Silicon-Manganese Steels	9xxx

Fig. 4-15 SAE Steel Numbering System

carbon range from 0.30 percent to 0.45 percent. This type of steel is strong and hard and does not weld as easily as low carbon steel. *High-carbon steels* have a carbon range from 0.45 percent to 0.75 percent. Both the medium-carbon and high-carbon steels can be heat-treated and hardened readily. Special welding techniques and electrodes are needed to weld these steels. Alloy steels are those which have had special elements added to make them applicable to special needs, figure 4-16.

Cast Iron

Cast iron is an alloy of iron, carbon and silicon. The most common type, gray cast iron, is widely used for farm implement or machinery castings. Cast iron has high strength, but it is very brittle.

Cast Steel

Steel castings are made from molten steel poured into molds of the shape to be used. Steel castings are strong and tough and used for precision equipment.

Carbon Class	Carbon Range %	Typical Uses
Low	0.05–0.15	Chain, nails, pipe, rivets, screws, sheets for pressing and stamping, wire.
	0.15–0.30	Bars, plates, structural shapes.
Medium	0.30–0.45	Axles, connecting rods, shafting.
High	0.45–0.60	Crankshafts, scraper blades.
	0.60–0.75	Automobile springs, anvils, bandsaws, drop hammer dies.
Very High	0.75–0.90	Chisels, punches, sand tools.
	0.90–1.00	Knives, shear blades, springs.
	1.00–1.10	Milling cutters, dies, taps.
	1.10–1.20	Lathe tools, woodworking tools.
	1.20–1.30	Files, reamers.
	1.30–1.40	Dies for wire drawing.
	1.40–1.50	Metal cutting saws.

Fig. 4-16 Uses for steel by carbon content

Stainless Steel

Corrosion-resistant steels (stainless steels) are grouped into three sections. The chromium nickel steels contain 0.25 percent or less carbon, 16 percent to 25 percent chromium, and 6 percent to 22 percent nickel. The 18-8 stainless steel (18 percent chromium, 8 percent nickel) is the most commonly used in the welding field. Chromium-nickel steels do not increase in strength from heat treatment. However, cold-working these materials will increase the strength of the metal. Hardenable chromium steels contain 12 percent to 18 percent chromium and carbon from 0.15 percent to 1.20 percent. Chromium-nickel steels within these percentages will harden when heat treated.

Nonferrous Metals

Metals which do not contain iron are called *nonferrous metals.* These include copper, nickel, zinc, magnesium, lead, and aluminum.

Nonferrous metals require special welding operations. This is due to the rapidly developing oxidation which results when nonferrous metals are heated.

Electrode Selection Based on Joint Requirements

Many factors are involved in the selection of the proper welding electrode. Most important are the joint design, position, and material. The following chart should be used when electric arc welding of mild steel, figure 4-17, pages 42 and 43.

Metallurgy in Welding

Heat affects the parent metal during and after a welding operation. Generally, four things take place to make a weld.

Full strength welds on mild steel can usually be made with any of a variety of different electrodes. Selection of the best electrode for maximum welding efficiency should be based on joint requirements. Here is a method for considering joint requirements:

1. Classify the joint as freeze, fill, follow, or a combination of these.

2. Choose the electrode group — fast-freeze, fast-fill, fill-freeze (fast-follow), or Low Hydrogen (fill-freeze) — from the information below and the sketches.

FREEZE JOINTS

Joints welded vertically and overhead are freeze joints. The weld metal must freeze quickly to keep the molten metal from spilling out of the joint.

For 3/16″ to 5/8″ plate, use fast-freeze electrodes.

For 5/8″ and thicker plate, the fill-freeze low hydrogen electrodes are more economical because deposit rates are higher and they make welds with fewer large beads so overall cleaning time is reduced.

AWS Numbering System

a. The prefix "E" designates arc welding electrode.

b. The first two digits of 4 digit numbers and the first three digits of 5 digit numbers indicate minimum tensile strength:

E60xx . . . . . 60,000 psi Tensile Strength
E70xx . . . . . 70,000 psi Tensile Strength
E110xx . . . 110,000 psi Tensile Strength

c. The next-to-last digit indicates position:

Exx1x . . . . . All Positions
Exx2x . . . . . Flat position and horizontal fillets

d. The last two digits together indicate the type of coating and the current to be used (See "Electrode Groups" below).

e. The suffix (Example: EXXXX-A1) indicate the approximate alloy in the deposit.

-A1 . . . . 1/2% Mo
-B1 . . . . 1/2% Cr, 1/2% Mo
-B2 . . . . 1 1/4% Cr, 1/2% Mo
-B3 . . . . 2 1/4% Cr, 1% Mo
-C1 . . . . 2 1/2% Ni
-C2 . . . . 3 1/4% Ni
-C3 . . . . 1% Ni, .35% Mo, .15% Cr
-D1 & D2 . . . . 25-.45% Mo, 1.25-2.00% Mn
-G . . . . .50 min Ni, .30 min Cr, .20 min Mo, .10 min V
(Only one of the listed elements is required)
E11018-M 1.3-1.8 Mn, 1.25-2.50 Ni, .40 Cr., .25-.50 Mo, .05 max. V.

FILL JOINTS

Groove, flat and horizontal fillets, and lap welds in plate over 3/16″ thick are fill joints. They primarily require fast-fill electrodes with high deposit rates to fill the joint in the shortest time.

Fast-fill electrodes only weld level or slightly downhill (15° max.) joints. More steeply inclined fill joints are best welded with fill-freeze electrodes.

For the required tight fitup, plates are butted tight, a back-up strip is used, or a stringer bead is made with fill-freeze electrodes.

FOLLOW JOINTS (Sheet Steel)

Welding sheet metal under 3/16″ thick requires electrodes that weld at high travel speeds with minimum skips, misses, slag entrapment and undercut.

Fillets and laps in all positions are best welded with EXX12 or EXX13 fill-freeze electrodes because they have excellent fast-follow ability.

Other types of joints are best welded with fast-freeze electrodes because they have good follow-freeze ability.

Index to Electrode Groups

The Fast-freeze Group

Exx10 . . . Organic . . . . DC only
Exx11 . . . Organic . . . . AC or DC

The Fast-fill Group

Exx24 . . . Rutile, approx. 50% iron powder . . . . AC or DC
Exx27 . . . Mineral & approx. 50% iron powder . . . . AC or DC

The Fill-freeze ('Fast-Follow') Group

Exx14 . . . Rutile & approx. 30% iron powder . . . . DC or AC
Exx12 . . . Rutile (Fast-Follow) . . . . DC or AC
Exx13 . . . Rutile (Fast-Follow) . . . . AC or DC

The Low Hydrogen Group

with Fill-freeze characteristics

Exx18 . . . Low Hydrogen & approx. 30% iron powder . . . . . . DC or AC

with Fast-fill characteristics

Exx28 . . . Low Hydrogen & approx. 50% iron powder . . . . . . AC or DC

The Alloy Steel Group

Exxx18 . . Low Hydrogen & approx. 30% iron powder . . . . . . DC or AC

Fig. 4-17 Electrode selection based on joint requirements

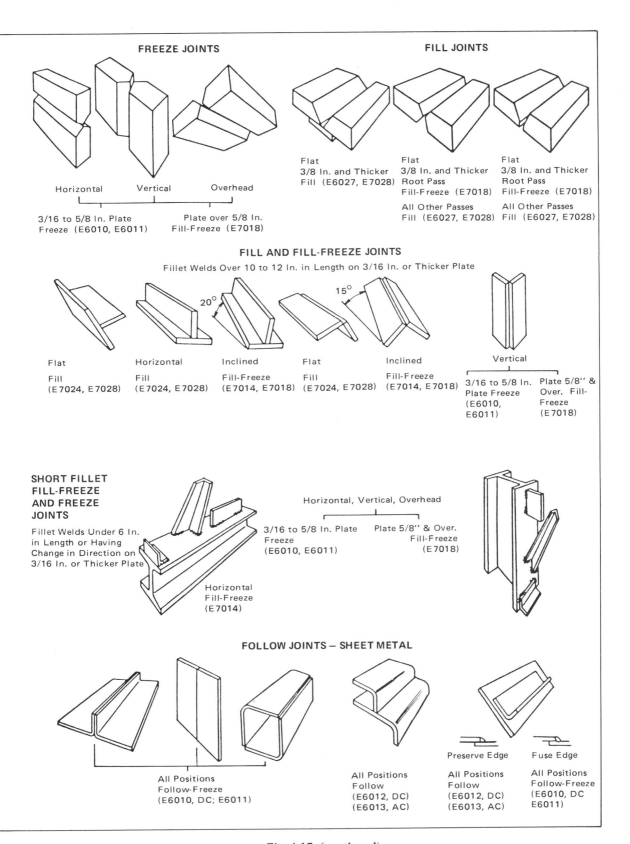

Fig. 4-17 (continued)

- The metal is heated;

- The heat is worked on the parent metal to deposit the weld;

- The weld and the parent metal cool;

- Under certain conditions, the metal and weld must be reheated to relieve the stresses which can be set up in the metal and the weld.

During welding, there are various stages of heat. At the point of the weld, the metal is molten. The part already welded is cooling, and the metal ahead of the weld is in a generally cool state, figure 4-18.

Grain structures in all of the areas shown will be different, and the differences in heat and cooling are called the *heat-affected zone.* Since the changes in grain structure have a direct bearing on the strength of the weld, the welder must know how heat has affected the operation.

Electrodes are coated with flux which melts over the molten puddle and covers the weld just made in order to protect the weld from outside atmosphere. The flux also forms a gaseous shield over the weld zone. Molten metal attracts oxygen and nitrogen and will become contaminated very rapidly if it is not protected. The addition of oxygen to molten metal weakens the weld and cuts down on the impact strength of the weld. When nitrogen is added to molten metal, the weld can become brittle, figure 4-19. The flux protects the weld from these contaminants.

Welding with the oxyacetylene torch protects the molten puddle from these impurities by covering the weld with the envelope of the flame. The envelope serves the same purpose as the flux coating on an electric arc welding electrode. Other forms of welding use an inert gas for this purpose.

Crater Cracks

When the crater of the weld at the end of the bead cools too rapidly and the bead cools too slowly, a condition known as *crater cracks* appears. Crater cracks occur because of hot

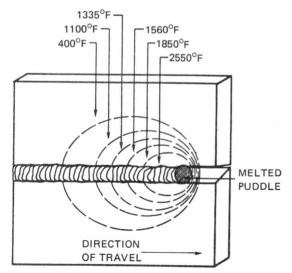

Fig. 4-18 Heat affected zones during welding

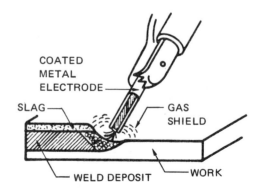

Fig. 4-19 The flux on coated arc welding rods melts to form slag on top of the molten weld and protects the metal from oxygen and nitrogen which are present in the atmosphere.

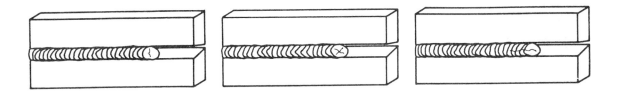

Fig. 4-20 Crater cracks at the end of the weld are caused by hot shrinkage.

shrinkage. The crater at the end of a weld must be filled when the weld is made since crater cracks can cause failure under stress, figure 4-20.

Expansion and Contraction of Metal

The addition of heat to metal makes it expand. Cooling has the opposite effect and contracts the metal. If a piece of metal is heated in one spot, for example, in the center, the surrounding cooler metal will force the molten metal to buckle, pushing it up into a bump, figure 4-21.

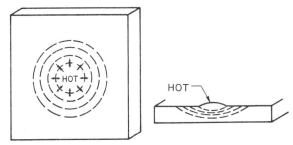

Fig. 4-21 Heat application on metal causes it to buckle because it expands and the colder metal surrounding the heated area does not expand.

The expansion and contraction of metal during welding must be controlled. This is done by using holding fixtures (jigs); by intentionally having the members of the weldment out of line so the heat can pull them into line; by not over-welding (adding too much heat to the joint); and various other methods.

Trouble Shooting Weld Problems Related to Metallurgy

Problem	Cause	Remedy
Warpage	Wrong welding sequence	Follow proper sequence.
	Shrinkage	Clamp or tack-weld parts; use faster welding procedures.
	Improper set-up of parts before welding	Align parts and clamp them in place.
	Poor control of welding heat	Use a skip motion to weld; preheat parts.
Cracking	Wrong welding procedures	Welding should allow open ends of joint to move. Preheat the joint.
	Base metal not weldable	Metal may contain too much sulphur or phosphorous—use low hydrogen electrodes.
	Welds too small	Use welds to equal the size of the metal in strength.
	Weld joint too rigid	Eliminate all rigid joints by redesigning.
Brittleness	Excessive welding heat	Use proper heat.
	Improper joint preparation	Check base metal analysis and weld using proper procedures for carbon content. May need preheating and postheating.
	Wrong electrodes	Select electrodes that will mix with parent metal.
Porosity	Poor base metal	Be sure base metal is correct.
	Wrong welding procedures	Use lower electrical currents; keep the puddle fluid and stay in the puddle long enough to let the gas escape. Use a weave motion.
	Poor electrode selection	Use electrodes recommended for material being welded.

Fig. 4-22

REVIEW QUESTIONS

Answer the following questions briefly.

1. What is metallurgy?

2. What is process metallurgy?

3. What is physical metallurgy?

4. Name five mechanical properties of metal.

5. Name five physical properties of metal.

6. What is parent metal?

7. How is tensile strength measured?

8. What is the yield point of metal?

9. Define ductility.

10. Define brittleness.

11. What is metal fatigue?

12. What effect does temperature change have on metal which is put under a load?

13. What is thermal conductivity?

14. What is compressive strength?

15. Define hardness.

16. If metal resists the flow of electricity through it, will it take more or less voltage to heat it?

17. What is fusibility?

18. Define low-carbon steel.

19. Define medium-carbon steel.

20. Define high-carbon steel.

21. What is cast iron?

22. What is the carbon content of chromium-nickel steels?

23. What are the chromium and nickel contents of the 18-8 stainless steels?

24. What is nonferrous metal?

25. What is the tensile strength of an E-6011 arc electrode in pounds?

26. In which of the following positions can an E-6011 arc electrode be used: horizontal position, vertical position, overhead position?

27. Explain the term heat affected zone.

28. What is a crater crack?

29. Give three reasons why a welded joint should not be over-welded.

30. Name three possible causes of porosity in a weld.

SECTION 2:
Special Arc Processes

Unit 5 Carbon-Arc Equipment

OBJECTIVES

Upon completion of this unit, the student will be able to:

- describe the equipment used for the process.
- gouge a plate.

The Carbon-Arc Cutting Process

The metal to be cut or gouged is melted with the electric arc and blown out of the cut with a jet of compressed air. The carbon-arc method does not oxidize the metal as the oxyacetylene torch does. It can be used to cut or gouge metals which do not oxidize easily as the action is primarily one of melting and blowing-away the melted metal.

Carbon-Arc Cutting Equipment

The equipment used for carbon-arc cutting consists of:

1. A special torch which resembles a regular arc welding electrode holder except that it has air attachments for supplying compressed air along the electrode, figure 5-1.

2. A standard air compressor, capable of delivering compressed air at over 100 psi. A regulator for the air delivery is not generally necessary because the torch operates between 60 and 100 pounds per square inch of air and the adjustment is not critical.

3. Air lines to supply air to the torch.

4. Carbon-arc electrodes which are a composition of carbon and graphite. A copper-cladding added to the electrode prolongs its life and will make the groove more uniform in size.

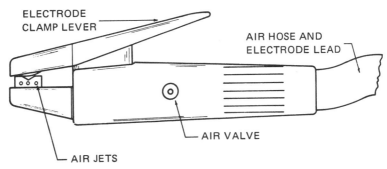

Fig. 5-1 Air carbon-arc cutting torch

5. A power source capable of delivering from 190 to 210 amperes for a sustained period of time. The power source should be able to deliver up to 600 amperes. The high current used in arc gouging will burn out the arc welding machine if the machine is not built for heavy-duty production.

Cutting Polarity

For general purposes, direct current reverse polarity is the most frequently used. However, carbon arc electrodes are available which act well employing alternating current. The alternating current electrode has some advantages not found in the DCRP electrode. The DCRP electrode produces a slag which does not blow off of cast iron readily, while the AC electrode, through its action of producing minute explosions, tends to blow off the cast iron slag more readily and make for faster cutting speed and cleaner grooves.

Protective Clothing and Arc Lens

Proper protective clothing is very necessary when the carbon-arc cutting process is used. The high temperatures achieved and the excessive blowing of the slag can be dangerous. Care must be taken that the area is free of inflammable materials. Since the high amperage creates a bright light, a lens a shade or two darker than the regular arc welding lens should be used if the cutting job is to be sustained for a long period of time.

JOB 5 APPLICATION OF AIR CARBON-ARC CUTTING EQUIPMENT

Equipment and Materials:

Arc welding power supply	Copper-coated carbon electrodes
Air carbon-arc cutting torch	1/4-inch to 1/2-inch mild steel plate
Compressed air power supply	

PROCEDURE	KEY POINTS
1. Attach an air carbon-arc torch to a welding electrode holder lead.	
2. Connect the air lines to the torch.	
3. Adjust the welder to 190 to 210 amperes.	3. Set the current selector switch to DCRP.
4. Install a carbon electrode in the torch.	4. There are two small air orifices on one side of the carbon arc electrode clamp. Make sure these orifices are on the underside of the electrode (the side closer to the plate being gouged).
5. Adjust the electrode so that 3 to 3 1/2 inches extend from the torch.	
6. Lower electrode holder to about 1/2 inch from the plate, drop the welding hood over the face, depress the air trigger, and contact the electrode to the plate.	6. Heavy protective clothing is necessary. The lens in the arc hood should be one to two shades heavier than for ordinary arc welding.

7. After the arc is established, progress across the plate with a slow, steady travel.

8. Keep the electrode angle at about 5 degrees up from the plate as shown in figure 5-2.

9. Practice gouging furrows in the metal until the gouging is satisfactory.

7. Keep the carbon electrode about 1/8 of an inch into the plate and a small furrow will be dug. The depth and width of the gouge is regulated by the speed of travel, the angle of the electrode, and the depth the electrode is pushed into the plate.

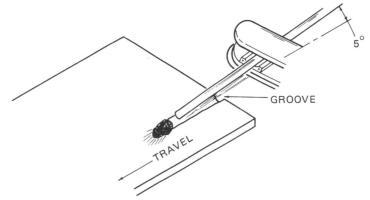

Fig. 5-2 **Gouging plate with an air carbon-arc torch**

REVIEW QUESTIONS

Answer the following questions briefly.

1. What is the purpose of the compressed air which is used in the carbon-arc gouging process?

2. How much air pressure is recommended for satisfactory gouging with the air carbon-arc process?

3. What is the purpose of the copper-cladding which is used on the carbon electrode?

4. What is the recommended polarity setting of the welding machine for carbon-arc gouging?

5. What length of stick-out is recommended for the carbon electrode when gouging steel plate?

6. What three things regulate the depth and width of the gouge made by the carbon-arc process?

Unit 6 Beveling Mild Steel Plate with the Air Carbon-Arc Torch

OBJECTIVES

Upon completion of this unit, the student will be able to:

- explain the procedures used for beveling steel plate with the air carbon-arc torch.
- bevel a plate.

Beveling

Beveling is the cutting of a slope on the edge of a piece of metal. Welding joints made from material which is over 3/16 inch thick require some method of assuring penetration through the joint. Beveling the edges of the metal and leaving a space between the pieces (root gap) allows the welder to completely penetrate the joint on the first welding pass.

Included Angle

The *included angle* of a welding joint is the sum of the degrees of bevel on the two plates of the joint. A 60-degree included angle indicates that each plate must be beveled 30 degrees.

Cross-sectional Area

The edge of a plate which has been square cut (90 degrees) will have less cross-sectional area than an edge which is cut on a bevel, since the bevel increases the width of the thickness. It will therefore require more heat to completely bring such an area up to the melting point.

JOB 6: BEVELING MILD STEEL PLATE WITH THE AIR CARBON-ARC TORCH

Equipment and Materials:

Standard air carbon-arc cutting equipment
3/16-inch, 1/4-inch, and 3/8-inch scrap mild steel plate

PROCEDURE	KEY POINTS
1. Connect torch and air lines, and adjust the welder to 190 to 210 amperes.	1. Install a carbon electrode in the torch.
2. Adjust the electrode to 3 to 3 1/2 inches stickout.	
3. After the arc is struck and the air trigger is depressed, hold the torch so that the carbon is angled on the edge of the plate as close to 30 degrees as possible. Bevel across the entire length of the plate, figure 6-1.	3. On thick plate the entire bevel cannot be cut with one pass, figure 6-2. The following guide gives the approximate number of passes required for each thickness. Practice will establish closer application.

PROCEDURE	KEY POINTS
	3/16-inch plate . . . . 1 pass
	1/4-inch & 3/8-inch plate . . . 2 passes
	Over 3/8 inch . . . . . . multiple passes
	as required

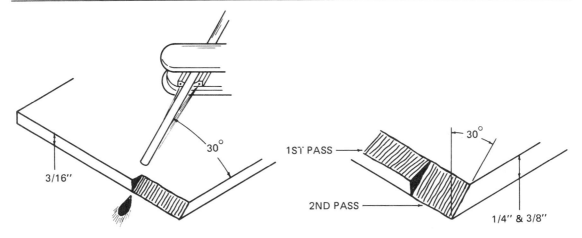

Fig. 6-1 Beveling plate with air carbon-arc process   Fig. 6-2 Cutting sequence for beveling plate

REVIEW QUESTIONS

Answer the following questions briefly.

1. Why is it necessary to bevel the edges of thick plate before welding?

2. Define an included angle.

3. How many cut passes are required on metal which is 3/8 inch thick?

Unit 7 Cutting Stainless Steel and Aluminum with the Air Carbon-Arc Torch

OBJECTIVES

Upon completion of this unit, the student will be able to:

- explain the properties which make these metals difficult to cut.
- cut stainless steel and aluminum plates.

Stainless Steel

Stainless steel is an alloy of iron, carbon, and chromium. In addition, some types of stainless steel also contain nickel and manganese. They resist most kinds of corrosion, resist oxidation in the presence of hot gases, and maintain their strength at highly elevated temperatures. The most frequently used classification is the alloy called 18-8 (18 percent chromium and 8 percent nickel). The 18-8 series is easily welded and nonmagnetic.

Cutting Stainless Steel

Stainless steels resist oxidation. The oxyacetylene cutting torch does a poor job of severing such material. When heated to near, or above, their melting point a refractory (heat reflecting) chromium oxide is formed. Since the oxide has a higher melting point than the steel itself, no cutting action takes place. Such steels can be cut with oxyacetylene, but a special cutting process called powder cutting is used. An iron powder is blown into the flame of the torch by compressed air and creates a cuttable influence on the steel.

The air carbon-arc process will do a satisfactory job of cutting stainless steel although the cut edges need cleaning when the cut is finished.

Aluminum

Aluminum is a nonferrous metal; that is, it contains no iron. It is the most weldable of all metals. But because of the oxides formed on it, it is difficult to cut with the oxyacetylene cutting torch. The oxides have a higher melting point than the metal and the oxyacetylene process will melt holes ahead of it, rather than making a straight, clean cut.

Air carbon-arc cutting does a satisfactory job of cutting on aluminum. The edges will need cleaning when the cut is finished.

JOB 7: CUTTING STAINLESS STEEL AND ALUMINUM WITH THE AIR CARBON-ARC PROCESS

Equipment and Materials:

Standard arc welding equipment with air carbon-arc attachment
Protective clothing
Stainless steel and aluminum scrap

PROCEDURE	KEY POINTS
1. Install the scrap on a cutting table or other installation.	1. Be sure the area around the cutting is clear of flammable material. The cutting area should be well ventilated.
2. For cutting stainless steel, the amperage setting should be lowered to 160 to 190 amperes, for a 1/4-inch electrode.	
3. Hold the carbon electrode about 1/8 inch *below* the metal thickness. Tip the torch in the direction of travel and maintain it at about a 60-degree angle to the workpiece, figure 7-1.	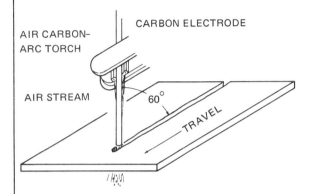
4. Make a cut across the workpiece, shut off the air and the welding machine and observe the cut. Much practice will be necessary in order to achieve proficiency with the process.	**Fig. 7-1 Application of the carbon-arc for cutting stainless steel and aluminum**
5. When the stainless steel cutting is satisfactory, proceed to the cutting of aluminum.	
6. Follow the same procedure as used for stainless steel and cut aluminum samples until satisfactory cuts have been made.	6. Aluminum which has been cut with the carbon arc process will have a black film caused by the carbon electrode left on it. This film must be ground off the aluminum before welding.

REVIEW QUESTIONS

Answer the following questions briefly.

1. What three substances are most stainless steels alloys of?

2. Is the 18-8 stainless steel series magnetic or nonmagnetic?

3. Why do stainless steels resist the cutting action of the oxyacetylene torch?

4. Why is aluminum called a nonferrous metal?

Unit 8 Scarfing
Fillet Welds

OBJECTIVES

Upon completion of this unit, the student will be able to:

- explain the air carbon-arc method of removing unwanted welds from fillet joints.
- successfully remove welds from a T weld and from a pipe welded to a plate so that the metal can be used for another welding project.

Repairing Welds

Welds which have flaws or need repairs are frequently removed from weldments by the use of the air carbon-arc process. In some cases, entire welds must be removed and re-welded. The air carbon-arc process can be used successfully to make repairs with no damage to the parent metal.

The air carbon-arc process is also used for gouging out flaws in castings which have come from the molds with imperfections.

When welds are removed with the carbon-arc process, care must be taken to remove only the weld so that the surrounding metal is not appreciably changed. Carbon-arc removal does not change the metal structure of the weldment and a new weld can be applied to the joint, thus saving the cost of a new part.

JOB 8: SCARFING FILLET WELDS

Equipment and Materials:

Standard arc welding equipment with the air carbon-arc attachment
Protective clothing
Scrap mild steel plate and pipe which have been welded with fillet welds

PROCEDURE	KEY POINTS
1. Obtain the specified scrap material.	1. If no scrap is available, the student must weld a T weld from 1/4-inch mild steel with a three or four pass weld on each side of the upright member of the T. The exercise also requires a fillet-welded pipe on 1/4-inch plate.
2. Beginning at the end of the T weld, gouge out the center of the fillet. Continue gouging, using the sequence shown in figure 8-1.	2. When the weld has been almost removed, be very careful not to gouge the pieces of usable steel. Make THIN gouges and watch carefully for the small line which indicates where the two pieces of metal were originally joined.

PROCEDURE	KEY POINTS
3. Remove the weld from the opposite side of the joint in the same manner. After the weld has been removed, a hammer will break the two pieces apart. 4. Perform the same procedure on the pipe weld, figure 8-2. 5. Continue practicing the cut. Have the work inspected by the instructor.	3. If properly done, both pieces of metal will be reusable. This is a common type of welding task. 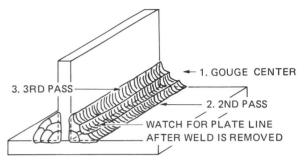 Fig. 8-1 Sequence for removing a fillet weld with the air carbon-arc process.

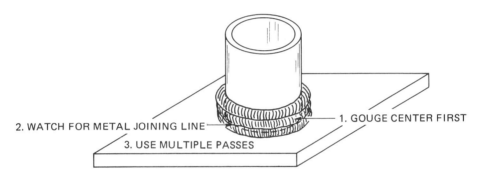

Fig. 8-2 Sequence for removing a fillet weld with the air carbon-arc process.

REVIEW QUESTIONS

Answer the following questions briefly.

1. Does the use of air carbon-arc for gouging welds change the structure of the metal surrounding the removed weld?

2. Should the metal in a three-pass weld be removed with one pass of the carbon-arc cut, or should more cuts be made?

3. Where should the first gouge be made on a wide weld?

Unit 9 Other Special Welding Processes

OBJECTIVES

Upon completion of this unit, the student will be able to:

- explain the special welding processes which are related to the electric arc welding process.

The Submerged-Arc Welding Process

Submerged-arc welding is done with a bare wire electrode. A blanket of grainy material (flux) which will melt is on, or around, or directly ahead of the arc, completely enclosing the molten metal to keep the outside atmosphere from being drawn into the puddle. After the molten puddle has become solid, the flux can be removed to expose the weld, figure 9-1.

The Bonded Submerged-Arc Welding Process. The bonded fluxes are finely ground chemicals which are mixed together with a bonding agent and then ground into a granular composition. Deoxidizers are added to the flux.

Fused Submerged-Arc Welding Process. Fused fluxes are formed by fusing the chemicals of the flux and then grinding the result into granular form. The flux is a form of glass.

Advantages of the Submerged-Arc Process

High currents can be used in the process. High heat is developed. Very high amperage can be used on small diameter electrodes. The high current allows for a fast melt-off rate. The flux that is used prevents fast escape of the heat and keeps it in the welding zone. The fusion of the process is deep into the base metal, using less filler metal and the weld is made rapidly so that heat distortion is reduced to a minimum. The submerged-arc process allows the welder to control the amount of penetration. Both ac and dc currents can be used. Direct current gives better bead shape, penetration, and welding speed and the arc starts more easily. The process can be used on all thicknesses of metal from 16-gauge sheet metal to thick plate.

The Flux-Cored Welding Process

The development of the flux-cored welding process came about through the desire to weld with a coil form of electrode which would be self-shielding and cut down the time lost in changing electrodes in the electrode holder, as well as eliminate the stub

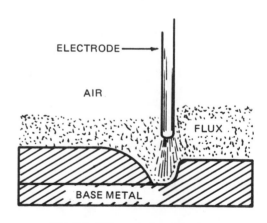

Fig. 9-1 Submerged-arc process

loss which occurs with stick-arc welding. It is a semiautomatic process in which an electrode which is many feet long and contains flux inside the wire is handled through a gun controlled by an operator, figure 9-2.

The electrode may be compared to the usual stick electrode by saying it is an inside-out electrode. That is, the flux is inside the wire, instead of on the outside. Substitution of the flux-cored process reduces welding costs wherever the stick-arc process is used for large amounts of welding.

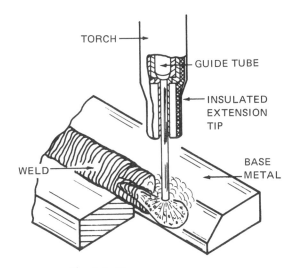

Fig. 9-2 Flux-core process

Advantages of the Flux-Cored Welding Process

Flux-cored welding makes possible a one-process weld in the building of structural steel. It reduces the amount of equipment delays in welding. Flux-cored welds can be made in a welding position. The process also makes it possible to perform strong, acceptable welds when fit-up is not exact.

The Electroslag Welding Process

The electroslag welding process is closely related to the submerged-arc process. The process resembles a casting operation. It is best suited for materials at least an inch in thickness. It can be used on steel up to 10 inches in thickness when multiple electrodes are used. A starting pad is used at the bottom of the joint to stop fall-through of the melted metal. Welding is done from the bottom to the top of the seam. Water-cooled dams are placed at each side of the joint. The dams move upward as the weld is made, helping to cool the deposit, figure 9-3.

At the beginning of the weld, a layer of flux is put in the bottom of the joint. When the arc is struck, it melts the slag and forms a molten layer. The slag shorts out the arc. The heat in the joint is made from the electrical resistance heating of the electrode sticking out of the contact tube and from the resistance heat within the slag layer which stays molten. As the electrode is melted away, the welding head and the water-cooled dams move up the joint.

Advantages of the Electroslag Welding Process

Weld quality is excellent with the use of the electroslag process because of the protective action of the heavy slag

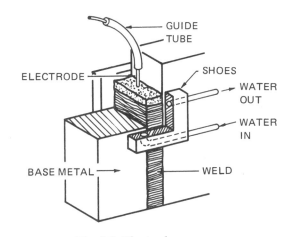

Fig. 9-3 Electroslag process

layer. There is no need for special preparation of the metal's edge and the weld is relatively free from porosity. However, special equipment is required for this process and most weld shops have very little need for the type of welding at which the process is successful.

The Stud-Arc Welding Process

Stud-arc welding is used for attaching studs, screws, and other fasteners to a workpiece. The stud (or small part) is put into a pistol-shaped tool called a stud gun and placed over the spot where it is to be welded onto the weldment. When the trigger of the gun is depressed, the electrical current flows through the stud. When the stud is lifted slightly, an arc occurs. The stud is then pushed down into the molten puddle and the gun is taken away. A ceramic ferrule is sometimes used around the stud where the contact to the plate is made. The gun is automatically timed to make the stud weld. The ceramic ferrule prevents atmosphere from getting into the molten puddle and also acts as a dam to hold it. Flux may also be used to cover the puddle. The stud-arc process uses the stud itself as the electrode for the weld, figure 9-4.

Advantages of the Stud-Arc Welding Process

Stud welding is widely used in the welding industry. It is a fast and efficient method of attaching fasteners to weldments and can be used for carbon and alloy steels, stainless steels, and aluminum. The equipment is portable and easily handled.

The Plasma-Arc Welding Process

Plasma-arc welding is a fairly new process. It is often used to replace gas tungsten-arc (TIG) welding. It develops temperatures as high as 60,000 degrees Fahrenheit. The heat in the plasma arc comes from the arc but is not spread out as much as other welds because it is forced through a small opening. Shielding gas may also be used for the process, figure 9-5.

Transferred-Arc Method. When the workpiece is part of the electrical circuit, the arc transfers from the electrode through the orifice (opening) to the work as in other arc welding processes, figure 9-6. This is called the transferred-arc method.

Nontransferred-Arc Method. When the nozzle around the electrode acts as the electrical terminal and the arc is struck between it and the electrode tip, the plasma gas carries the heat

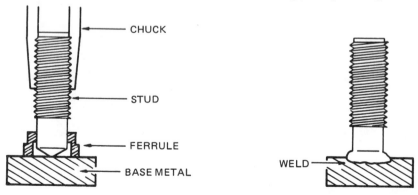

Fig. 9-4 Stud arc process

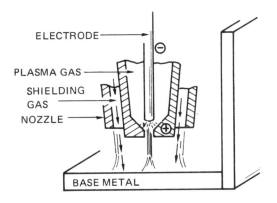

Fig. 9-5 Plasma-arc process

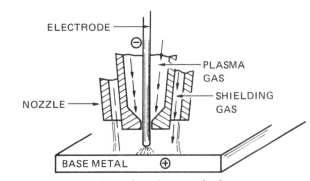

Fig. 9-6 Transferred-arc method

to the weldment, figure 9-7. The process is called the nontransferred-arc method.

Advantages of the Plasma-Arc Process

The plasma-arc process has greater concentration of the energy for welding, improved arc stability, it gives higher welding speeds, and controls the size of the puddle in the penetration.

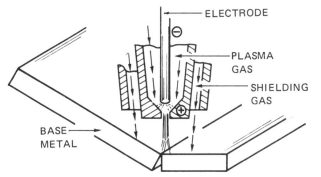

Fig. 9-7 Nontransferred-arc method

The Resistance Welding Process

Resistance welding, or spot welding, is performed by exerting pressure to clamp the metal pieces together while the electrodes pass a current through the metal to melt a small portion of the two pieces together. The machines are generally equipped with foot-operated levers. The time duration of the contact of the two electrodes is set so that welding only takes place for a given length of time while the arc spot is fused together, figure 9-8.

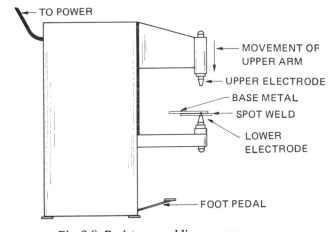

Fig. 9-8 Resistance welding process

Advantages of the Resistance Welding Process. Spot welding gives clean, strong welds. The operator-training time is short. For the welding of light-gauge metals it is an economical and fast process. When the machine is set correctly, warpage of the parent metal is reduced to almost zero. Some types of spot-welders are portable and may be used with ease on jobs which cannot be transported to the larger machines.

REVIEW QUESTIONS

1. What type of electrodes are used for submerged arc welding?
2. Does the submerged arc process use high or low heat to an advantage?

3. Which type of current gives more penetration on submerged arc — ac or dc?

4. Is submerged arc recommended for welding 16-gauge metal?

5. Is the flux inside or on the outside of the electrode when the flux-cored process is used?

6. Should flux be put into the bottom or top of the joint before electro-slag welding is done?

7. How high can plasma-arc welding develop temperatures (in degrees Fahrenheit)?

8. Is shielding gas always used for the plasma-arc process?

9. Give another name for resistance welding.

10. Is a resistance weld made with or without pressure?

Acknowledgments

The authors wish to express their gratitude to the following individuals and organizations for their assistance in the production of this text.

American Welding Society

Lincoln Electric Company

Society of Automotive Engineers

Union Carbide Corporation, Linde Division

Line drawings by Tom Bulmer

Reviewed by Gary Moldenhauer and Darrell Nelson

DELMAR STAFF

Source Editor — Mark W. Huth

Associate Editor — Frederick D. Musco

CLASSROOM TESTING

The material in this textbook was classroom tested by vocational welding students at the College of Southern Idaho.